Springer Praxis Books

Popular Astronomy

W0235085

This book series presents the whole spectrum of Earth Sciences, Astronautics and Space Exploration. Practitioners will find exact science and complex engineering solutions explained scientifically correct but easy to understand.

Various subseries help to differentiate between the scientific areas of Springer Praxis books and to make selected professional information accessible for you.

The Springer Praxis Popular Astronomy series welcomes anybody with a passion for the night sky. Requiring no formal background in the sciences and including very little to no mathematics, these enriching reads will appeal to general readers and seasoned astronomy enthusiasts alike.

Many of the books in this series are well illustrated, with lavish figures, photographs, and maps. They are written in a highly accessible and engaging style that readers of popular science magazines can easily grasp, breaking down denser aspects of astronomy and its related fields to a digestible level.

From ancient cosmology to the latest astronomical discoveries, these books will enlighten, educate, and expand your interests far beyond the telescope.

Michael Carroll

Planet Earth, Past and Present

Parallels Between Our World and its Celestial Neighbors

 Springer

Published in association with
Praxis Publishing
Chichester, UK

Michael Carroll
Parker, CO, USA

Springer Praxis Books
ISSN 2626-8760   ISSN 2626-8779   (electronic)
Popular Astronomy
ISBN 978-3-031-41359-9   ISBN 978-3-031-41360-5   (eBook)
https://doi.org/10.1007/978-3-031-41360-5

This Springer imprint is published by the registered company Springer Nature Switzerland AG
The registered company address is: Gewerbestrasse 11, 6330 Cham, Switzerland

Paper in this product is recyclable.

To the next generation: Alexandra and Harrison, Andrew and Stephanie, may you revel in this beautiful, unique world as you care for its inhabitants. And for Kirstin and Jeremiah, because good friends are like Venusian gold.

Other non-Fiction Springer Books by Michael Carroll

Alien Seas: Oceans in space (edited by Michael Carroll and Rosaly Lopes)
Antarctica: Earth's Own Ice World (with Rosaly Lopes)
Drifting on Alien Winds: exploring the skies and weather of other worlds
Earths of Distant Suns: how we find them, communicate with them, and maybe even travel there
Ice Worlds of the Solar system: their tortured landscapes and biological potential
Living Among Giants: exploring and settling the outer solar system
Picture This: grasping the dimensions of time and space
The Seventh Landing: going back to the Moon, this time to stay

Springer "science and fiction" novels by Michael Carroll

On the Shores of Titan's Farthest Sea
Europa's Lost Expedition
Lords of the Ice Moons
Plato's Labyrinth: dinosaurs, ancient Greeks and time travelers

Reviews of Planet Earths: Past and Present

Mike Carroll's combination of engaging writing and incredible artwork will help readers understand how our planet has become so hospitable for life and inspire them to take better care of our incredible world. Highly recommended.
— Jeffrey Bennett, astronomer and author

So far, the Earth is the only planet we know that fosters life. Liquid water, a breathable atmosphere, a steady magnetic field and a stable planetary tilt have all helped make Earth habitable. However, Earth hasn't always been a place conducive to life. At times in its history, it's been more like Mercury, Mars, Venus, and even Saturn's moon Titan.

Why is the Earth the only habitable body in the solar system? Why hasn't any other body in the solar system become habitable? What are the prospects for Earth's future? These are some of the questions Michael Carroll answers in his book, Planet Earths: Past and Present.

Carroll takes us on a grand tour of Earth's past and the solar system together, presenting us with vivid descriptions of the solar system's planets and how Earth resembled those planets in past epochs. His descriptions paint clear pictures in the reader's mind and he spices it up with just a touch of humor, making it a breezy, entertaining read. He also includes beautiful illustrations which complement his words.

Planet Earths: Past and Present is highly recommended for people who want to gain more insight into Earth's history, its place in the solar system, and its place in the galaxy.
—David Lee Summers, Senior Observing Associate at Kitt Peak National Observatory and author of *The Solar Sea*.

Michael Carroll is the consummate space artist, portraying things related to space both beautifully and correctly in terms of scientific reality. Now he has written a book about planets, both those in our solar system and beyond, in which his words depict these worlds in a stunningly visual way. Again, his adjectives never stray beyond a solid scientific foundation. He writes about and paints a diversity of rocky and icy bodies, both ranging spatially from our own planet Earth to those circling other stars and ranging temporally from billions of years ago to the present and even into the future. Neither the universe nor our local solar system are static so Carroll pictures times and places where other planetary environments might have resembled contemporary Earth as well as epochs in Earth's history when terrestrial places might have resembled, say, the surface of Saturn's moon Titan today. Highly recommended!

—Clark R. Chapman, Senior Scientist, Southwest Research Institute

You'd be forgiven for thinking Michael Carroll is a scientist who has a talent for writing; in fact, it's the other way around. That doesn't mean the reader will get shortchanged in the science – Mike conveys the absolute current ideas in the field in his science writing, as in his science art. It's in fact his ability to tell story that helps place the reader right in the thick of things, which in the case of this delightful book can be deep down in the center of a planet. His descriptions, movement and flow take you on a journey that starts at the very beginning – the big bang – and careens off into futuristic voyages. And he somehow manages to weave in compelling themes like sibling rivalries, escape artists and dog walking on an interstellar voyaging craft. With humor and immersion, Mike takes us on the ride of our lives in "Planet Earths".

—Jani Radebaugh

Introduction: An opening act you won't want to miss

Trying to see the Earth as it was long ago is a bit like trying to find a white Honda Civic in a blizzard: all kinds of things confuse the scene and get in the way of the destination. I know. I've been there. But while the search may sometimes be challenging, the payoff is well worth it. We will see—in our travels through space and time—our home world anew, as a planet different and more complex than we first thought. The journey will enrich our understanding of the cosmic history we are a small part of.

The Earth is a planet in flux. It is not the world it was, and it is not the world it will be. On Earth, change is the norm. Our planet is restless, in constant transition. Its crust is broken into interlocking pieces, a planetary Tetris game moving in the relentless motion called plate tectonics. At the Earth's rift zones, our planet generates new rocky material. Some of these puzzle-piece rock rafts jam into each other, lifting the highest mountains on Earth. (The Himalayas, home to Everest, result from the collision of the Eurasian and Indian plates). In other places, the plates grind against each other, triggering earthquakes. Some plates slip, sliding beneath one another and pulling other plates apart. The stony conveyor belt brings surface rock down into the Earth's hot interior, where it melts. As the rock melts, trapped gases—notably carbon dioxide and water vapor—escape. The heat and gas must come out somewhere, and they do so in the form of dramatic volcanic eruptions. Volcanoes on land and beneath the sea contribute new gases and water vapor to the atmosphere in a dynamic recycling of minerals and a rejuvenating of the air. From the throats of these volcanoes come gases that were imprisoned within rock. Most of the Earth's carbon dioxide resides in limestone, which acts as a chemical sponge. The process of plate tectonics, in concert with volcanism, frees the gases locked in the rock.

Our first atmosphere was rich in hydrogen and nitrogen. The hydrogen came largely from the solar nebula, but the nitrogen is a bit of a mystery. The nitrogen may have been delivered to our doorstep during the late heavy bombardment, a period of furious asteroid and comet impacts that tailed out roughly 3.8 billion years ago. Incoming rocks and cosmic icebergs may have supplied copious amounts of nitrogen into our infant atmosphere, or the nitrogen may have come from within, outgassed through volcanic activity. Like other gases, nitrogen has a propensity to drift off into space. But a dense atmosphere high in hydrogen or carbon dioxide may have protected the nitrogen from escape.

The Earth's atmosphere forms a blanket enshrouding the globe, offering shielding from solar wind. Its mélange of complex gases also has a subduing effect on incoming radiation, and burns up most of our cosmic intruders, the meteors and comets.

The Earth's core generates another natural shield. Our iron-rich center is partially molten. Currents within the iron "ocean" generate strong magnetic fields around the planet known as the magnetosphere. This magnetic bubble prevents the solar wind from directly impinging upon the upper atmosphere, which would otherwise strip away the lighter gases. It is important to note that the Earth today stands alone among the terrestrial planets in terms of its magnetosphere. While Mercury has a magnetosphere, it is comparatively weak. Both Mars and Venus lack an active one entirely, although Martian rocks may bear a record of ancient magnetism. Mars' small core probably cooled early, dooming its magnetosphere. Venus may never have had one. Although its core may be similar in size and makeup to that of Earth, any molten iron is relatively quiescent because of the planet's slow spin.

One remarkable event contributed to the Earth we experience today. It began 2.4 billion years ago, and as far as we can see, it happened nowhere else in our solar system. Known as the Great Oxygenation Event (GOE), something singular changed our atmosphere. Oxygen levels began to rise, and this rise was due to life. As Samuel Adams was the champion of the American Revolution, so cyanobacteria were the heroes of the GOE. Cyanobacteria were equipped to carry out photosynthesis, using sunlight and water as energy and producing oxygen as waste. The growing oxygen story is not as linear as it first seems: evidence points to mini-events of oxygenation long before (see Chapter Five). As oxygen began to increase, other things beat it down, geological as well as biological. Early Earth was largely covered in ocean water, with a few crater rims sticking out into the air. But as plate tectonics began to raise continents, minerals and various compounds washed into the oceans, becoming available to the bacteria there, and enabled life to thrive. In short, the Great Oxygenation Event was actually a team effort, a conspiracy between the Earth and its living seas, to transform our atmosphere into an oxygen-rich environment.

The protective qualities of air and magnetic fields are balanced by a relentless assault on the surface. Rain and wind wear away the peaks and valleys, carving spectacular pinnacles and deep canyons. Rain settles into cracks and freezes, fracturing even the toughest rock. Ocean waves eat away at coastlines. Alternating heat and cold cause expansion and contraction of surfaces, giving the wind and water a toehold to wear the surface down even more. Continents rise, volcanoes build, weather breaks down highlands into valleys, plains and desert sands. It is as if the planet rises and falls in the cadence of breath.

Change is the norm on other worlds, too. Some nearby planets suffer wild seasons due to their extreme axial tilts. Others spin consistently, but will not do so for long because they have no large moon to stabilize the precession (wobble) of their axes. The atmospheres of some are so dense as to create high pressure and temperatures too great to sustain liquid water. Others ache in the cold vacuum of space with no air to even out temperatures or support surface liquids amenable to life. For now, our world has a comparatively stable environment.

The Earth has not always been so lucky. In its infancy, long before life could arise, its landscapes resembled Dante's *Inferno*. Dark skies tented glowing landscapes of molten rock. Lava erupted into the sky, sending showers of embers into the inky firmament. No stars shone down to relieve that seared landscape; dense clouds of toxic gases blocked out any view of the sky beyond. Meteors streaked through the air, leaving incandescent trails, the only visible points of light in the blackened heavens. The world was a very alien Earth. What led to this severe environment is a story common to most rocky planets.

Over the course of its ~4.5 billion-year history, Earth has resembled other planets in our solar system, alien worlds that give us insight into the Earth's formation, its nature over time, and what our world may face in the future. The Earth has endured rains of asteroids and comets, and watched as its climate morphed from tropical conditions to wintry "snowball" phases. Its early surface conditions may have resembled Venus at one time, Mars at another. The Earth evolved early atmospheres quite different from the one we breathe today. Its biological processes—which may be unique in our solar system—likely began in a very alien environment which carried out chemical processes similar to those found on Saturn's planet-sized moon Titan. Planetary systems around other stars continue to teach us insights into our planet's development, and these studies will become richer as time goes on and our tools mature.

Our planet has many robust, interlaced systems, but it also has vulnerabilities. Our climate is in a delicate balance, driven by ocean currents and atmospheric flow, by the warmth of sunlight and the chill of night. The air above it dictates the climate on the surface. Greenhouse gases help to keep our temperatures warm at night, but they are rarified enough so as not to trap too much heat in the atmosphere. On Venus and Mars, processes that are familiar to Earth dwellers have resulted in quite alien environments. Understanding the worlds around us—their similarities as well as their differences—will inform our understanding of our own climate.

Finally, our world may eventually be headed toward a fate that faintly echoes current environs on Venus, becoming a desiccated planet heated by an angry, swollen red giant Sun, only to end as a cold, lifeless orb. But far from a doomsday scenario, this is the way a planet evolves, from energetic, blustery adolescence to aged, parched later years, from the heat of youth to the pall of twilight years. As we tease out details of our world's past, present and future, we unmask the natural path of a planet filled with the gift of living things. And that life may, in the natural course of things, move on to other worlds more like the past one from which it sprang.

Acknowledgements

Always to Caroline, whose 3x5 cards, file folders and creative organization saved this book (and Chapter 2 in particular). Thank you, my darling.

Jack Connerney at NASA's Goddard Space Flight Center for advice about Martian magnetic rocks. Special commendation goes to the "Mercury guys": ever-patient and creative Ron Vervack of the Johns Hopkins University's Applied Physics Laboratory for help on Mercury exospheric effects in sky and hollows, and cosmochemist Larry Nittler of Arizona State University for glimpses into Mercury's childhood—and the rest of the planets, too. Andrew Wilcoski lent wisdom about thin air. SwRI astrophysicist Robin Canup filled in the details of the Moon's creation at the hands of Theia. Bob Grimm gave me valuable guidance on cratering, sea floor spreading, and other exotic geological concepts. JPL's Mike Malaska and Rosaly Lopes offered vital help and expertise with my Titan labyrinths. Josh Colwell at University of Central Florida provided my story with rich background on Saturn rings and accretion disks. Lori Iliff, Senior Provenance Researcher at the Denver Art Museum, gave me insights into how scholars study ancient masterpieces, as did the Getty Research Institute's Sandra van Ginhoven. Marilynn Flynn and her new cat co-editor Rusty Mars were two of my trusty first readers, and Marilynn graciously supplied the photo of the "missing billion years" in layers of the Grand Canyon. Bill Gerrish rounded out my photo needs, thanks to a trip we took to Hoover Dam. If you have to be stranded in Las Vegas, Bill is the guy to be stranded with! Thanks to Fiona Yacopino for the usage of the wonderful and historic map of the ocean floors. Thanks to my good friend and talented artist Pat Rawlings for his spectacular Martian cave explorers. I am grateful to Dave Eicher and my friends at *Astronomy magazine* for letting me base some passages here on work I did on *Astronomy* articles.

Contents

1

Before Earth

You can't tell a book by its cover…or a planet by its crust.

When an art historian is tasked with evaluating the authenticity of an oil painting, she does not initially look at the brush strokes on the canvas, the color balance, or the subject and composition. These things all come into play, but appearances can be deceiving. What's beneath the surface tells much about the masterpiece's provenance. If the image is rendered on canvas,[1] is the canvas itself woven of the right material, and in the right pattern? If executed on a wood panel, is the wood a species from the right geographic area? What markings do the canvas supports show—areas where no one will be looking? Paint pigments can be tested for age and composition (some types of pigment came into use much later than others).

In the same way, clues within the Earth tell us about its origins. While its face provides part of the historical record, the structure, tectonics, minerals and gases within and beneath the landscape indicate initial conditions, what kind of temperature the planet issued from, and even something about its primordial travels and interactions with other worlds. Like a renaissance masterpiece, the makeup of a planet informs us of its provenance.

Just as the origin story of our solar system has a larger background, so does the Milky Way galaxy. It is one among many "star islands" scattered across the universe. Not all galaxy types are equally benevolent to living systems.

---

[1] Before canvas came into use, artists of the 13th to 16th century usually painted on wooden panels, paper, plate, metal, and even ceramics, all of which can be dated radiometrically or by other tests.

Figure 1.1 Awash in a torrent of lethal radiation, a comet strays near the galactic center of the Milky Way Galaxy. From this viewpoint, swirling tendrils of super-charged plasma spiral across the shrouded, powerful core of the galaxy, a black hole circled closely by careening stars in a region known as Sagittarius A*. (painting by the author)

Our Sun is a bigger-than-average star, a member of a club of 100 to 400 billion stars within the galaxy.[2] People have been wondering about our planetary system's origins—and the genesis of the universe at large—since the beginning. The second millennium BCE writings from the book of Job ask, "Who determined its measurements—surely you know! …who laid its cornerstone when all the morning stars sang together…?" These are good questions.

The first big event ever: putting us in our place

The beginning of all things is a strange story, one that seems counterintuitive and remarkable. Everything we see, all that exists or has ever existed, was once packed into a space smaller than an atom. This "cosmic egg" was infinitely dense. Something extraordinary visited this point-universe, triggering a violent, rapid expansion. In a sense, the universe is like a flower, its great petals unfurling from a center, expanding ever outward. But the universe didn't expand into nothingness; the universe was all there was. Before the event that started it, time and space were inert. Time as we know it did not yet exist. The evidence of the initial conditions has been wiped out in an information suicide, as the universe destroyed all trace of its earliest moments. What was the universe like in those first instants? Robert Jastrow, founder of NASA's Goddard Space Flight Center, commented that, "Now we would like to pursue that inquiry farther back in time, but the barrier to further progress seems insurmountable. It is not a matter of another year, another decade of work, another measurement, or another theory; at this moment it seems as though science will never be able to raise the curtain on the mystery of creation."[3] Princeton University astronomer James Peebles often said, "The equations refuse to tell us, and I refuse to speculate."

At the literal dawn of time, the universe was opaque to light. It remained so until some 380,000 years after its inception (its "Big Bang"). We can see the light from some of the most ancient galaxies and quasars, which brings us glimpses of a universe just a few hundred million years old. We estimate that the universe is roughly 13.7 billion years old, and the light from these objects

---

[2] Based on recent data from the European Space Agency's Gaia mission.
[3] *God and the Astronomers* by Robert Jastrow; 1978, Readers' Library.

has been traveling across the universe for more than 13 billion years, so it began when the universe was very young. Further observations from advanced observatories like the James Webb are pushing back our view, in both time and distance, even further. The glimpses offered by the ancient light of the universe give us a rough picture.

During its first second of expansion, the universe was incredibly hot, so hot that photons—"particles" of light—could transform themselves into matter and back into light. This still happens, but these days it is a rare occurrence. Natural forces like gravity and the strong and weak nuclear force were intertwined and could not yet operate until a millionth of a second had passed. As the universe spread out, it cooled. Just one second after the initial event, the universe had spread out so that its density was that of water. Its temperature had dropped by a billion degrees. From this raging soup, protons, neutrons, electrons, neutrinos and other subatomic particles condensed out.

Gravity and other forces began to diverge, becoming separate powers. The universe entered a period called the GUT era. Two forces now ruled the universe: gravity and the GUT force. The GUT force is a combined force of the strong, weak, and electromagnetic forces. The GUT force only remains unified at ultra-high temperatures. As the universe cooled, the GUT force split into its three constituents.

As temperatures reached 10 million degrees, at about the three-minute mark, subatomic particles began to combine into helium and hydrogen, ushering in the era of atoms. The universe was now in a runaway expansion, called inflation. Particles the size of an atom swelled to the size of our solar system in a fraction of a second.

Over the course of the first 380,000 years, a radiant glow infused everything. Atoms and smaller particles wandered through this broth of energy. The blinding glow of radiation began to clear, and the universe became transparent. Stars and galaxies formed, and the universe began to take on the appearance it has today.

Seeing red

How have we put such a picture together? The emergence of the universe around us has penned its saga upon the light that comes to us from every point across the sky. That light forms a spectrum, a parade of "colors" that include the hues we see in visible light, up into gamma and X-rays and down into infrared and microwave radiation. The waves in gamma radiation are tightly packed, while the waves of visible or infrared light are long and leisurely. Stars offer their spectrum of light—inscribed with dark and bright lines—as a glimpse of their composition. The shift in the lines of this spectrum, toward the red or the blue end of the spectral cosmic "rainbow", indicates that an object is moving toward or away from us. The light waves that make up the rainbow-like spectrum of light get stretched out as the object emitting it—a star or galaxy—moves away from us. Its light waves become stretched out toward the red end of light. In the same way, the light waves from an object moving toward us get bunched up together, appearing more blue. We see the same thing happen with sound waves. The tone of an approaching horn or siren sounds high-pitched as it approaches (its sound waves are bunched up as it advances toward us). As that noisy vehicle passes, it "pulls" the sound with it, stretching the waves into longer, lower-pitched ones. This shift in light or sound waves is called the Doppler effect. In the case of very distant objects, this redshift is dominated by the motion of the expanding universe itself, so we can get a hint of its age. The more red shifted a distant object is, the older it is, because it is closer to the visible "edge" of the ancient universe.

Early astronomers like Walter Adams (1908) and Vesto Slipher (1912) noted the "red shift" in their observations. A decade later, Edwin Hubble recognized the red shift as nature's clue that the universe was expanding. Everywhere he looked, light waves from the distant galaxies elongated toward the red, indicating that they were moving away from us. The farther the object, the more its light shifted toward the red. The entire universe, it appeared, was fleeing. Hubble concluded that the universe was expanding. This relationship between red shift and distance is known as Hubble's Law or the Hubble-Lemaitre Law.[4]

Hubble realized that—despite appearances to the contrary—we are not in the center of things. From any point in the universe, one's surroundings would appear to be doing the same thing: receding in all directions. If one ran the tape of the universe backwards, the logical result would be that at some point, everything began at a source, an infinitesimally dense seed that suddenly expanded outward from the center of the universe.[5] The event that began this expansion—the very birth of the universe—came to be called the Big Bang.

---

[4] Hubble's work confirmed that of Soviet physicist Alexander Friedman in 1922, and was confirmation of work by Georges Lamaitre, who published a paper in 1927 proposing that the universe is expanding.

[5] Because the universe is moving outward in all directions, we cannot actually locate a place in three-dimensional space where it all began. As Einstein would say, it's all relative.

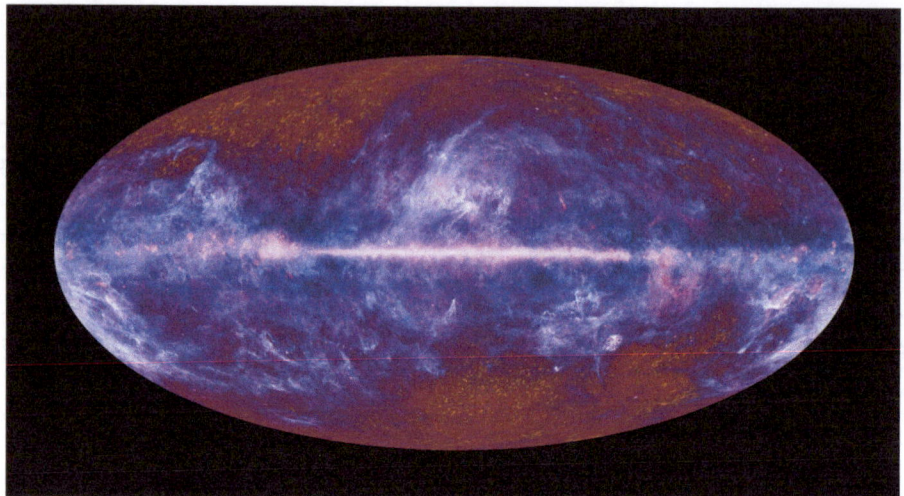

Figure 1.2 A map of the cosmic microwave background radiation across the entire universe, taken by the European Space Agency's Planck space telescope. This microwave image depicts the radiation remaining from the Big Bang. (Image credit: ESA/NASA/JPL-Caltech)

The red shift is one of many clues to the beginning of all things. Beyond the stars simmers a background buzz, an echo of the Big Bang that started it all (Figure 1.2). This cosmic background radiation—evidence of the moment of the Big Bang—smolders at 2.73°K, nearly absolute zero.

In addition to the cosmic background radiation, researchers have been able to measure the abundance of various primordial elements like helium, deuterium and lithium. Big Bang cosmology predicts certain ratios of these elements, and measurements come quite close to the predictions. The age of stars and the arrangement of galaxies are also in agreement with Big Bang predictions. It appears that from this singular event, our universe expanded in a tsunami of energy and matter.

The first stuff to condense out of this maelstrom was hydrogen, the building block of the universe.[6] Everything we see, the air we breathe, the rock we walk upon and the water we drink, began as hydrogen. But before the hydrogen, the entire universe consisted of subatomic particles drifting in a dense soup devoid of light.

---

[6] The Big Bang also generated trace amounts of lithium and helium.

The infant universe spawned something else, something quite mysterious. It is invisible to us, and yet it makes up the vast majority of all things. Astronomers call it dark matter. Observers use the motion and interaction of stars and galaxies to judge how much material the universe contains. But the "music of the spheres" misbehaved. Something wasn't adding up. A great deal of mass is influencing the drifting stars and spinning galaxies. This matter interacts with gravity, but not with light. If that weren't enough of a cosmological conundrum, another unseen force—which cosmologists refer to as dark energy—is driving objects away from each other. The universe is expanding even today, and its expansion rate is actually accelerating. Gravity should slow this rate. Instead, something is driving the universe apart, more and more quickly.

<u>The stars make their appearance</u>

Pondering the bejeweled night sky, Henry Wadsworth Longfellow penned these words: "Silently one by one, in the infinite meadows of heaven, Blossomed the lovely stars, the forget-me-nots of the angels."

Those fiery points of diamond light have a back-story. The universe arose as a dark, cold place, but those subatomic particles that led to hydrogen and helium eventually condensed into the first stars. And yet there was nothing to make an Earth from: elements across the universe were too light to form anything like a rocky globe.

Within that first generation of infant stars, nuclear fusion merged hydrogen atoms together to form more complex elements like helium. As they burned fuel, the pioneer stars became unstable. Eventually, they gave their lives to be new stars. Our Sun was born from the ashes of older stars. Its contents include heavy elements (like carbon and metals) not present in the first-generation hydrogen/helium stars. That is fortunate for those of us who live on a rocky planet; no terrestrial worlds existed in the first star systems.

The universe glistens with stars of many different sizes, colors and ages (Figure 1.3). The size of a star at its beginning determines much about its makeup, the length of its life, and even the nature of its death. The life cycle of any star (unless it's in Hollywood) begins as the center of a disk of gas coalesces into a sphere. The central sphere gains mass and gravity, and it collapses in upon itself, its gravity overwhelming the force that holds atoms together. As the atoms in its core fracture, they trigger nuclear fusion. Stars spend most of their time, roughly 90% of their life spans, burning hydrogen into helium. This is a very stable time of life, referred to as the "main sequence", where stars burn steadily. The energy of the hydrogen fusion supports the star's outer layers above its core.

This peaceful state cannot last. As a star ages, the helium begins to fuse, outpacing the hydrogen fusion, leading to dramatic changes. When the hydrogen runs out, what's left is a heavy shell of helium. The star's core begins to compress even more, abandoned by the bolstering force of hydrogen fusion. The helium shell on the outside, along with leftover hydrogen, inflates and heats up even further. The star grows larger and brighter, often transforming into a red giant. During its red giant phase, our Sun will expand to fill the orbits of Mercury and Venus, perhaps even making it out as far as the Earth's 1 astronomical unit (one AU, the distance from the Earth to the Sun). Stars the size of our Sun today also fuse heavier elements like carbon. Still larger stars produce a wide variety of elements.

Finally, when the star runs out of fuel, it departs from the "main sequence" and begins to die. When it comes to a star's fate, size does matter. Medium to low-mass stars like our Sun go fairly quietly into the night. They balloon into a giant red star, and eject their shell of burned hydrogen and helium into space, creating a spherical cloud expanding around them. This globe of radiant gas becomes one of the most beautiful phenomena in the universe, a planetary nebula.[7] Left behind is a cold, dim star called a white dwarf. The burst of luminosity concurrent with the star's ejection of its gas shell is called a nova.

Larger stars kick the bucket in much more dramatic ways. If a star weighs in at more than eight times the mass of the Sun, it will perish in a titanic explosion known as a supernova. A typical supernova may become as luminous as all the stars in the galaxy, putting out as much power in an instant as a typical star does during its entire lifetime. The resulting wave of gas may expand at up to 10% of the speed of light. The material disgorged from a supernova wafts into interstellar clouds adrift throughout the galaxy. Its suicidal detonations combine the hydrogen, helium, and other trace elements into new materials that enable rocky planets like

---

[7] The name comes from the fact that early observers thought the globe-shaped clouds were solid planets.

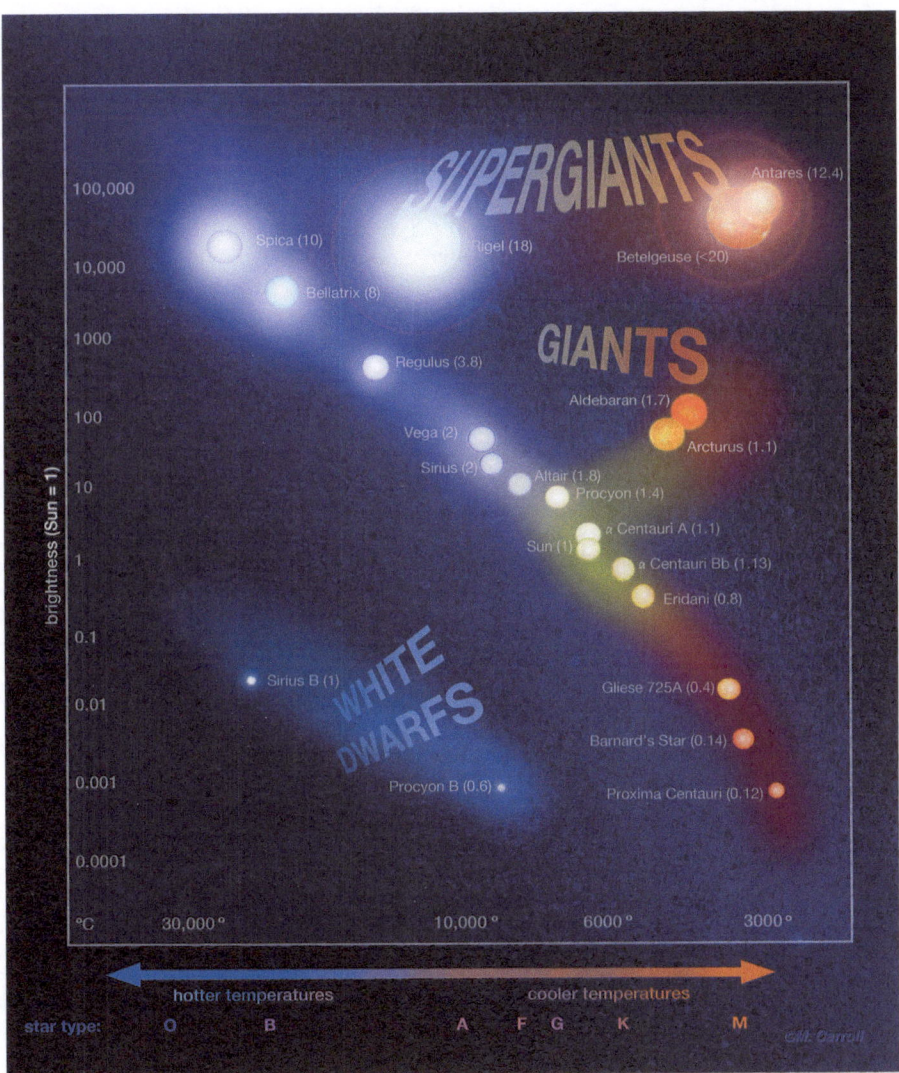

Figure 1.3 The Hertzspring-Russell diagram provides an overview of the life cycle of stars and their influence on the Earth's nature. The main sequence is the river of stars cascading from upper left to lower right, with several famous stars labeled. As a star ages, it leaves this pathway and becomes a giant (upper right) or a dwarf star (lower left). The numbers next to each star compare its mass to the Sun's (with the Sun equaling 1). The mass of the star affects its makeup; rocky Earthlike planets near metal-starved stars are unlikely. The Sun will become a red giant like Betelgeuse, though not as large. Still larger stars like Sirius will eventually become giants such as Aldebaran. Bellatrix will bloat into a blue supergiant like Rigel. Stars at the cool end of the main sequence (lower) live longest, while ones to the upper left burn hot and die young. Stars are *not* to scale. (diagram by author, originally from the book *Picture This: Grasping the Dimensions of Time and Space*, Springer 2015)

Earth to form around later generations of stars. This is the circle of life, stellar-style.

Not all stars are so dramatic. Smaller stars burn longer and cooler. Red dwarfs are the most common—and oldest—stars in the universe. Some may have been born over 10 billion years ago, making them nearly 2/3 the age of the universe.

Starry-eyed essentials

The Earth's nature is determined by the star it orbits and the solar nebula from which it came. Stars that differ from the Sun may produce fundamentally different worlds.

M dwarfs: Also known as M stars or red dwarfs, these are the coolest-burning stars, and among the most ancient. M dwarfs burn their fuel so slowly that some may be nearly as old as the universe itself. While the Sun will remain on the main sequence for ten billion years, M dwarf life spans may reach into the hundreds of billions. M dwarfs are by far the most common stars, making up over 95% of all stars in the galaxy. The largest weigh in at less than half the Sun's mass, and many are the size of a gas giant planet.

K type: Also known as orange dwarfs, these stars are slightly cooler and smaller than the Sun. Three to four times as common as Sun-class stars, orange dwarfs burn their hydrogen more slowly, staying in the main sequence lane for 15-30 billion years.

G type: Our Sun and its identical siblings fall into this category. Their main sequence lives last roughly 10 billion years. They offer planets like Earth a stable environment in which to develop. Sun-like stars make up 7 out of each 100 stars.

F type: Seven times as bright as the Sun, these stars make up only 2% of the stellar population at large. They may be as far across as 1.5 times the Sun's diameter. F stars burn so intensely that their stable period lasts only 2 billion years, not enough time to produce an Earth-like world for long.

A type: Twice as rare as F type stars, A stars glow twenty times as brightly as the Sun. They blaze furiously at about 8000°C (compared to the Sun's 5500°C), lasing only a billion years before burning out.

B type: Fierce solar winds gust from the blue giant stars known as B type. 3000 km/sec blasts of radiation sterilize any nearby planets. A thousand times as luminous as the Sun, they contain 2-16 times as much material as our own star, with surface temperatures reaching 15,000 °C.

O type: More deadly than the stars we've covered so far are the blue supergiants, classified as O types. With the light of a million Suns, these 50,000 °C behemoths contain 90 solar masses, but not for long. O stars die within half a million years of their birth. Earth-type planets are unlikely anywhere near these deadly stars.

For more on specific stars, see Appendix K: star types

Figure 1.4 Six types of galaxies top row, l to r: the vast Andromeda spiral galaxy was initially thought to be a twin to the Milky Way; our Milky Way galaxy more closely resembles the barred spiral NGC 1300; elliptical galaxy NGC 1316, tangled in cosmic dust, lacks a spiral or disk structure.Bottom row, l to r: NGC6861 has a lens shape typical of a lenticular galaxy; IC 4710 is an irregular galaxy lacking a central hub or spiral structure, and the peculiar galaxy M 82, also called "The Cigar Galaxy".Credits: top left, Wikipedia Creative Commons by David Dayag, public domain (https://commons.wikimedia.org/wiki/File:Andromeda_Galaxy_560mm_FL.jpg); top center, Hubble space telescope image NASA/ESA/STScI/Aura, public domain (https://esahubble.org/images/opo0501a/); top right, NASA, ESA, and The Hubble Heritage Team (STScI/AURA), public domain; bottom left: ESA/Hubble/NASA (http://www.spacetelescope.org/images/potw1502a/) public domain; bottom center: ESA/Hubble & NASA. Acknowledgements: Judy Schmidt, public domain; bottom right: NASA/ESA, AURA, Acknowledgment: J. Gallagher (University of Wisconsin), M. Mountain (STScI), and P. Puxley (National Science Foundation), public domain

A menagerie of stellar communities

Stars congregate in vast communities called galaxies. These star islands come in many forms (Figure 1.4), from tightly wound spirals to amorphous starry clouds. The universe at large is organized like an enormous sponge, with many hollows and empty cavities. Galaxies are arranged across the surfaces of the hollows. The galaxy "bubbles" surround empty voids and are connected via filaments and streams of stars and galaxies. Broadly, galaxies come in six major configurations (see Figure 1.4). Spiral galaxies resemble great, slowly rotating disks of stars, bulging slightly at the middle and thinning out toward the edge. Arms of stars spiral out from the central hub. This bright inner region, common to many galaxies, is called the galactic nucleus. For decades, astronomers believed that our Milky Way galaxy was a spiral, similar to our glorious neighbor, the Andromeda Galaxy. But as is true for all science, surprises were in store.

A variation on spiral galaxies is the barred spiral. Instead of a hub, the center of these galaxies is a bright bar. Arms spiral away from either end. Some type of bar is common to about two thirds of all spiral galaxies. The central bar is a veritable star factory, with gases and dust feeding into it from the outer spiral arms. Spiral galaxies may be the end product of barred spirals. The bars in the center of galaxies appear to be in different stages of decay, eventually dissipating into the more generalized arrangement of simple spirals.

Elliptical galaxies are the most common type in the universe. They contain an abundance of ancient, red stars. These star islands are egg-shaped in form, and lack any sort of disk. They are starved of cool gas and dust, so they carry on only rare star formation. But elliptical galaxies are charged with hot, ionized gases similar to those that emanate from exploding stars, or supernovae. One of the most massive galaxies in the universe, M87, is an elliptical galaxy some 132,000 light years[8] in diameter. Although not much larger than the Milky Way's longest dimension, it is not flat, but rather spherical, so it contains many more stars and perhaps ten times the mass.

Lenticular (lens-shaped or SO) galaxies contain mostly ancient, metal-poor stars. Like spiral galaxies, their core bulges, surrounded by a flat plain of stars. But their disk shows no sign of spiral structure, and lenticular galaxies are depleted of the kind of interstellar dust that leads to new stars.

Irregular galaxies tend to glow with the white light of young, massive stars. They contain copious amounts of star-generating dust. Most nearby irregular galaxies are smaller than the Milky Way, and less common than spiral and elliptical galaxies. But observations of more ancient regions in deep space show that distant galaxies are more likely to be irregular. Because the light from these far-off galaxies has taken longer to reach us, this indicates that irregular galaxies were more common in the young universe.

Peculiar galaxies—a type of irregular galaxies—round out the universe's collection of star islands. If spiral galaxies are omelets, peculiar galaxies are scrambled eggs, with their suns swirled into chaotic messes. Their form may arise from the collision or interaction of two galaxies. Star formation runs rampant within their cores, so they appear to be bluer than their galactic relatives (as many of their stars burn hot and blue).

---

[8] Interstellar distances are so great that miles and kilometers are inadequate. Instead, astronomers use the light year, the distance traveled by light in one year. This distance comes out to roughly 9.7 trillion kilometers.

In many cases, astronomers spy the aftermath of galactic crashes (Figure 1.5). When one galaxy passes directly through the center of a spiral galaxy, the result is a ring galaxy. The nucleus or hub of a ring galaxy is surrounded by nearly empty space, encircled by a rich, star-forming halo. The collision is similar to the interaction of a dart hitting the bulls-eye of the dartboard.

Not all collisions are so direct. In fact, most seem to be glancing blows, where galaxies zoom by each other in near-misses. The gravitational pull of the interacting star islands warps the shape of both. These encounters generate fascinating galactic shapes. A classic example is the Whirlpool Galaxy (M51), which was nearly struck by the passing of the smaller NGC 5195. Such close calls often pull streams of stars into tails or bars stretching from the galactic victims. If a marauding galaxy passes by in the opposite direction to a spiral galaxy's spin, the spiral will warp, often beyond recognition, morphing into an irregular galaxy.

Some galaxies resemble eyes, with tails from each member stretching outward like the lids of an eye, with dual nuclei as the irises. Some direct collisions meld the two galaxies together completely. Some observers suggest that elliptical galaxies are the consequence of the merger of two spiral galaxies.

It is clear that star formation is triggered by the collision of galaxies. Glittering streams of dust and gas flood into each other, while solid ices, stone and metal coalesce into new stars and planets. How many potential Earths lie within the glowing tendrils of interacting galaxies?

The Milky Way Galaxy: basics of our home

One of those Earths—our own planet—resides within the Milky Way galaxy, a great stellar disk some 120,000 light-years across. This is the dimension we can see in visible light, consisting of stars and nebulae. But an extended halo of dark, invisible gas and other matter stretches as far across as 1.9 million light years.[9] At the core of our galaxy lies a dazzling hub filled with stars and gas. A massive black hole nests at the center of this luminescent heart. We cannot see it with our own eyes; the entire region is swathed in dark cosmic dust, hiding it from sight. But if we could clear that dust away, the galaxy's hub would present a breathtaking sight, a huge mound of glistening stars in the night sky, rivaling the Moon in brightness.

---

[9] According to a team led by Alis Deason at England's Durham University. The team used computer models and speeds of nearby galaxies to lock down the outer edge of the Milky Way. Their work was reported in a February 21, 2022 paper posted on arXiv.org

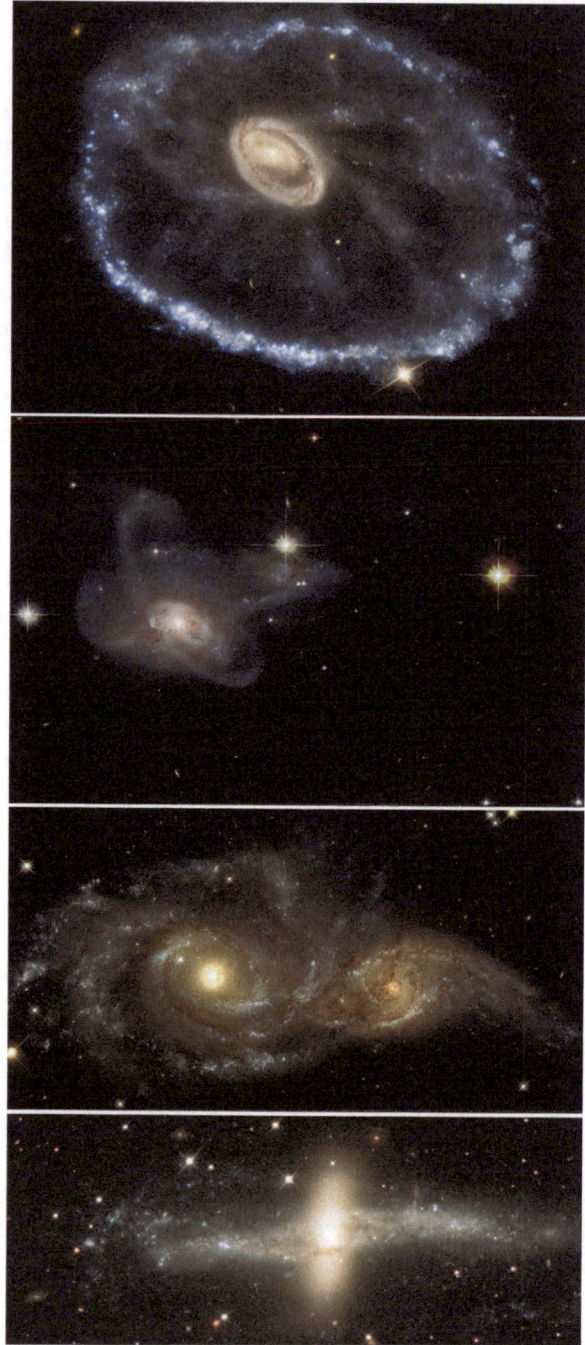

Figure 1.5 Bumper-car galaxies (top to bottom): The Cartwheel Galaxy is the remains of a spiral galaxy that suffered a direct collision with another galaxy about 200 million years ago. Note the rich star formation in the outer ring [Curt Struck and Philip Appleton (Iowa State University), Kirk Borne (Hughes STX Corporation), and Ray Lucas (Space Telescope Science Institute) and NASA/ESA]; galaxy CGCG 396-2, an unusual

Spreading out from that dazzling center, stars arrange themselves into a vast disk. The stars drift along in great tendrils, bracketed by the same dark dust that swaths the hub and hides it from our terrestrial view. A halo of dark matter surrounds the entire galaxy, populated by ancient stars.

At least 55 smaller galaxies accompany the Milky Way, and many orbit our own galaxy as planets orbit the Sun. The largest member of the club is the Large Magellanic Cloud, or the LMC. This disorganized star island spans 14,000 light-years across, and weighs in at 10 billion times the weight of our Sun. Although it lies 163,000 light-years distant, the LMC's stars are being disturbed by the gravity of the Milky Way. A smaller version of the LMC, the Small Magellanic cloud, is about 7000 light-years away. An amorphous stream of gases forms a bridge between the LMC and SMC. It is called the Magellanic Stream.

Dozens of dwarf galaxies tag along for the ride through the universe, including the Sagittarius Dwarf Spheroidal Galaxy. This conglomeration of at least nine globular clusters (globes of loosely-bound stars) sails above and below the poles of the Milky Way in a polar orbit lasting between 550 and 750 million years. Other exotic dwarf galaxies include the Sculptor, Fornax and Draco dwarfs.

Since we are inside of the Milky Way, looking out, it has been difficult to tell what the overall structure of our galaxy is. In the fall of 1609, Galileo Galilei focused his revamped telescope[10] on the Milky Way. He was the very first to see that the milky band cast across the night sky was not a glowing cloud, but rather a river of glistening stars.

Astronomers carefully charted the location and approximate distances of stars in an attempt to glean the structure of the Milky Way. Nearby, they

Figure 1.5 (continued) multi-armed galaxy merger which lies around 520 million light-years from Earth in the constellation Orion. [Hubble Space Telescope mosaic courtesy European Space Agency and NASA]; the galaxies NGC2207 and IC2163 form two "eyes" as they begin a merger which may last a billion years. [Hubble Space Telescope image courtesy European Space Agency and NASA]; a ring of material circles a galaxy passing through its center. The lens-shaped galaxy at center forms a bull's-eye piercing the other galaxy. [Hubble Heritage Team (AURA/STScI/NASA)]

---

[10] It is often reported that Galileo was the inventor of the telescope, but the honor probably falls to Hans Lippershey, a Dutch eyeglass maker, who applied for a patent in 1608. Traveling carnivals of Galileo's day sold them as toys, but Galileo—seeing the true promise of the instrument—improved upon the design and turned it skyward.

could watch the grand spiral of stars in the Andromeda Galaxy, and it made sense that ours was a similar affair. As the twenty-first century dawned, the Spitzer space telescope came online. Spitzer could observe in infrared light. Details of our galactic surroundings are obscured by dark clouds of dust, the same fodder that can lead to the creation of stars. But Spitzer's instruments could see through the dust, giving us a clearer picture of the cosmic neighborhood. Some observers suspected that the center of the galaxy was not a simple spiral, but rather a bar. Spitzer's keen eyesight confirmed the Milky Way's central "crossbeam". Spitzer and other advanced observatories are providing us with an accurate picture of the galaxy around us (Figure 1.6).

Our location in the galaxy is significant, as it appears that—like planetary systems—galaxies have habitable zones. Our world happens to inhabit a region of our solar system known as the "habitable zone", a specific distance that allows liquid water on the surface of planets. (Each star has a unique habitable zone; we will explore the concept further in Chapters 4 and 7). The Sun circles the hub of the galaxy just as a planet circles a star. It orbits roughly 25,000 light years out from the center, between two of the spiral arms. Here, the stars are not as tightly packed as they are at the galaxy's nucleus. This is

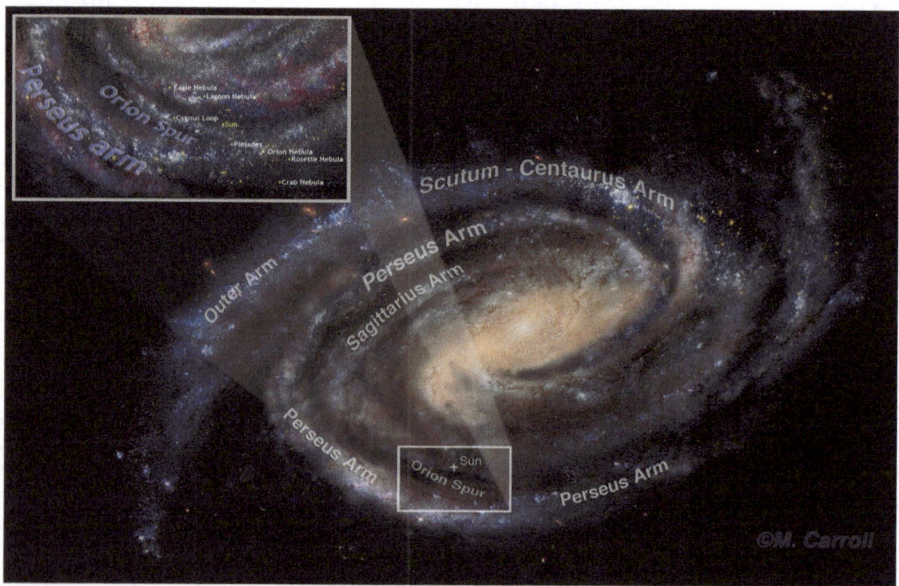

Figure 1.6 Where in the galaxy are we? Orbiting the galactic hub at a distance of 25,000 light-years, our planetary system shelters in the relative calm of space between two busy streams of stars, the Perseus Arm and the Orion Spur. In this image, many details on the opposite side of Milky Way can only be guessed at, but new observatories are lifting the veil of obscuring dust. (art by the author)

fortunate for planet Earth. An astonishing 95% of the Milky Way's suns may not be able to sustain habitable planets, because many orbit the galaxy in paths that carry them through the deadly spiral arms. Any star that passes through one of these starry swarms is subject to deadly radiation from the congested stars. Our own solar system orbits far enough from the center to keep it in sync with the rotation of the rest of the galaxy, so that it remains in the quieter space between the spiral arms. The Earth and its planetary siblings are well placed in a quiet, resource-rich niche of a vast and complex galaxy. The interior of the galaxy is a dangerous place to be, as radiation levels are high from all that stellar overcrowding. Larger stars have a bad habit of blowing themselves up in supernova events, but we are a distance away from the dangerous, unstable ones in our somewhat lonely neighborhood. At the very center of the galaxy lies a mysterious and deadly energy source known to astronomers as Sagittarius A*, or Sgr A* (pronounced Sagittarius "A"-star). In that high-energy core, strong magnetic fields trace lines and webs through incandescent gases (Figure 1.1). Intense gravity imprisons stars, sending them lurching around a central point that was, until recently, invisible to us.[11] That source, along with millions of stars floating cheek-to-jowl, pours into its surroundings a flood of X-rays, gamma rays, and other lethal radiation. That kind of energy is not conducive to life; the presence of life on Earth owes a debt of gratitude to the paucity of stars inhabiting our region.

But if the Earth's distance were much farther out in the galaxy than it is, our planet would have different problems. The Earth and other terrestrial planets are built of heavy elements, and those elements come from the same exploding stars that would endanger us if we were closer to the galactic hub. But in the outer galaxy, stars are so few and far between that these critical elements become more rare. As distance increases from the galactic hub, the relative sum and substance of elements denser than hydrogen and helium declines. In the outskirts of the Milky Way, the heavy elements that contribute to metal and rock may be too low to create Earth-like rocky worlds. In the outskirts of the Milky Way and other galaxies, terrestrial planets are likely smaller than the Earth, and may be quite rare as well.

Life on Earth also depends on our liquid metallic core, which generates a protective magnetic "bubble". This magnetosphere shields planet Earth from incoming radiation.[12] But the core is kept hot by radioactive elements like

---

[11] The black hole and accretion disk at the core of Sagittarius A* was first imaged in 2017 by the Event Horizon Telescope (EHT), an array of eight linked radio observatories across the Earth which form a single "Earth-sized" virtual telescope.

[12] Some evidence suggests that radioactive elements and the heat they shed contribute to plate tectonics, which appears to be critical to life on Earth.

[13] Chris Kirkland, et al, *Geology*, August 23, 2022.

uranium and thorium. These elements become more rare with distance from the galactic hub. The Earth is clearly in a region that keeps the two extremes— too many deadly stars, and too few heavy elements—in balance.

The galaxy's link to the Earth's crust?

The Earth—along with the solar system around it—circles the Milky Way galaxy once every 200 million years or so. Scientists have recently linked this time interval to a recurring natural phenomenon: the generation of new crust on Earth. In 2022[13] an Australian team reported that vestiges of the Earth's primordial crust known as cratons reveal a fascinating confluence: new crust was produced along the same time periods as the Earth's passage around the Milky Way. Each 200 million years, the Earth overtakes the spiral arms of the galaxy, as it travels more quickly than the galaxy spins. These travels bring the planet through the dense spiral arms. Some scientists have long thought that comets and asteroids may have triggered or husbanded the process of new continental crust formation, but they had no solid explanation for why the periodic bombardment happened. Now, researchers have been able to tie these timelines together, suggesting that as the Earth passes through a spiral arm, the dense material there would have gravitationally tossed comets inward from the outer solar system. The incoming comets would have split the oceanic crust, melting regions of the Earth's mantle. Lighter material would have risen, forming new continents. Researchers are studying Moon rocks to see if a similar pattern arose there.

The Earth's "bobbing" through spiral arms may also result in periodic mass extinctions. There have been at least fifteen that appear in the fossil record. The five major ones occurred at 450 million years ago, 365 million years ago, 252 million years ago, 201 million years ago, and 66 million years ago. The last event came at the hands of an asteroid that destroyed 95% of all species, including the non-avian dinosaurs (see Chapter 2). But a link between these mass extinctions and impact events has only been confirmed with the last event, called the Cretaceous/Paleogene extinction.

A closer look at the center

Jane Austin said, "There is no charm equal to tenderness of heart." Austin had never seen the heart of the Milky Way. There is no "tenderness" to be found in that maelstrom of swirling, treacherous radiation and matter. The glimmering stars of the Milky Way Galaxy spiral majestically around a central province of terrifying extremes (Figure 1.7). Even from the earliest days of radio

Figure 1.7 (continued) The image covers as much of the sky as a full Moon. The heart of our galaxy glows with incandescent tendrils, clouds and knots of gases. Some are in the process of forming stars, while others are the remnants of exploded supernovae. The bright swirl at lower center is the black hole in the center of Sagittarius A*. Compare this image to the painting at the opening of the chapter. (NASA, ESA, SSC, CXC, STScI) (image courtesy NASA/JPL-Caltech/ESA/CXC/STScI, public domain)

Figure 1.7 In honor of the International Year of Astronomy in 2009, NASA combined images from three of its "Great observatories", the Hubble Space Telescope, the Spitzer Observatory, and the Chandra X-ray observatory. All three space telescopes were trained on the galactic center. Their infrared and X-ray collectors could see through the opaque cosmic dust for a clear view toward the central hub of the galaxy.

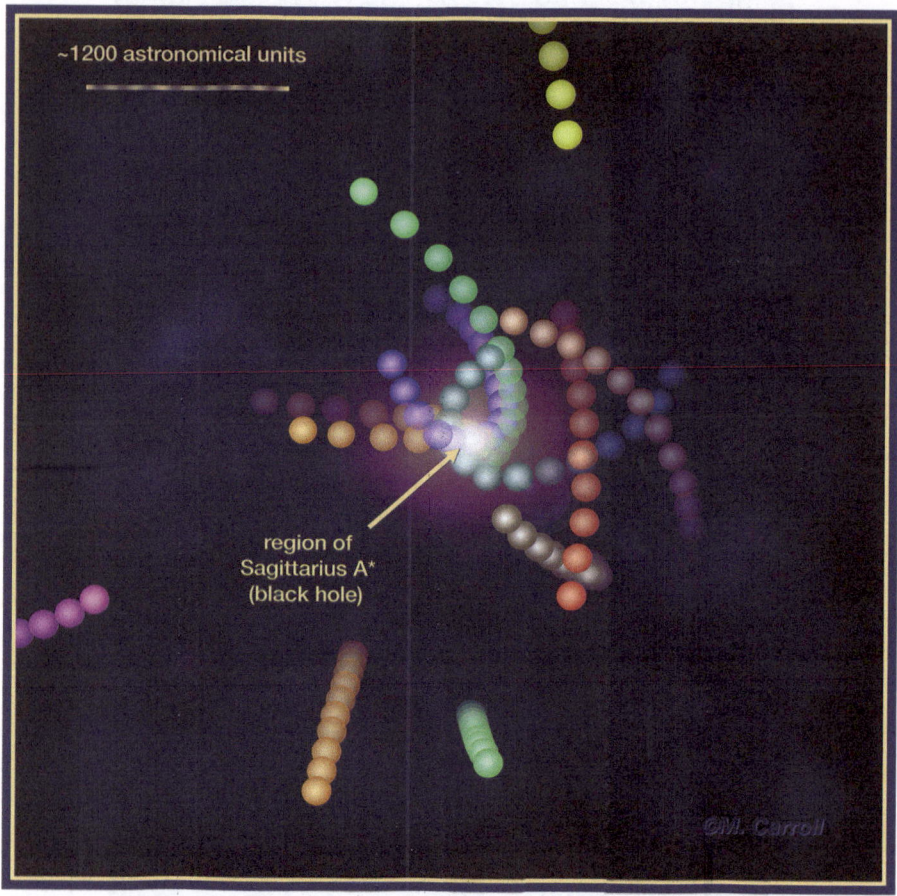

~1200 astronomical units

region of
Sagittarius A*
(black hole)

©M. Carroll

Figure 1.8 The dance of stars around the **Sagittarius A*** galactic center over an eight year period (1996-2004). A few hundred stars are actually packed in around the center, but we have chosen only a few to demonstrate the orbital disturbances there. Extreme gravity from the black hole at the Milky Way's core causes nearby stars to travel at breakneck speeds—some at 8% the speed of light—whipping around it in tight orbits. For scale, it would take a beam of light 28 days to cross the frame of the image. Orbits are to scale; stars are not. (art by the author, based on data from European Southern Observatory and the Keck Telescope)

astronomy, observers saw the glare from a region in the constellation Sagittarius, in the direction of the center of the galaxy. But clouds of cosmic dust obscure our view of whatever lies there.

Our first clear views of galaxy central came through X-ray and infrared telescopic eyes. As our imaging capabilities progressed, observers charted stars dragging trails of gas as they careened through interstellar dust (Figure 1.8). Their tight hurtling circuits, and the gusts within the surrounding gases, betrayed the gravity of the Milky Way's beating heart. It became clear that the central force, whatever it was, was as massive as millions of stars, packed into a very small region. There were not nearly enough stars in the area to explain their movements, and the stars moved at unimaginable speeds. Years of observation brought the realization that the mysterious object weighs in at 4.2 million times the mass of our Sun. It is now known to be a highly energetic black hole. Don't look for an Earth twin there: any worlds that may have formed in the region are long gone, flung away by gravitational slingshots within the remarkable Sgr A*.

Black holes are among the most exotic of all objects in the cosmos. The universe is littered with the corpses of stars. Some have compressed into tiny white dwarfs or diamond-like neutron stars as they blow off their outer layers, collapsing into themselves. But when the core of a dying star is large enough, its gravity becomes so strong that it continues to crumple under its own "weight", essentially squeezing itself out of existence. In reality, a black hole has some dimension, but we measure the size of a black hole by a glowing halo around it. Surrounding the black hole is a dark region where its gravity is so strong that no light can escape. The edge of this invisible realm is called the event horizon, and it is a bizarre universe unto itself, where no light enters and time stands still. Usually, the strong gravity of the black hole will pull surrounding material into a disk, and this whirlpool of matter—the accretion disk—is the part of the black hole that is visible.[14] Accretion disks are like a circle on a diagram telling the reader to "look here."

For some time, astronomers have suspected that the star motions at the Milky Way's core reveal a black hole. The visual proof of the existence of one came in a paper published on Halloween of 2018. A European Southern Observatory team of observers combined an interferometer[15] called GRAVITY with four telescopes of the Very Large Telescope to study Sgr A*. They clocked masses of gas traveling at 30% of the speed of light. Super-charged electrons close to the black hole showed as three bright flares, in the same configuration as computer models predicted for "hot spots" in the accretion disk of a black hole weighing in at four million solar masses (Figure 1.9). The gravity of the black hole heats the surrounding gas and dust to 18 million degrees F (10

---

[14] The gravitational force of a black hole actually warps time and space. Scientists can use this phenomenon as a tool, searching for invisible objects by charting distant stars or galaxies whose images are warped by an unseen, heavy object between the distant object and us.

[15] Interferometers merge several light sources to set up interference pattern. The technique is very sensitive, and can measure more subtle changes than other techniques.

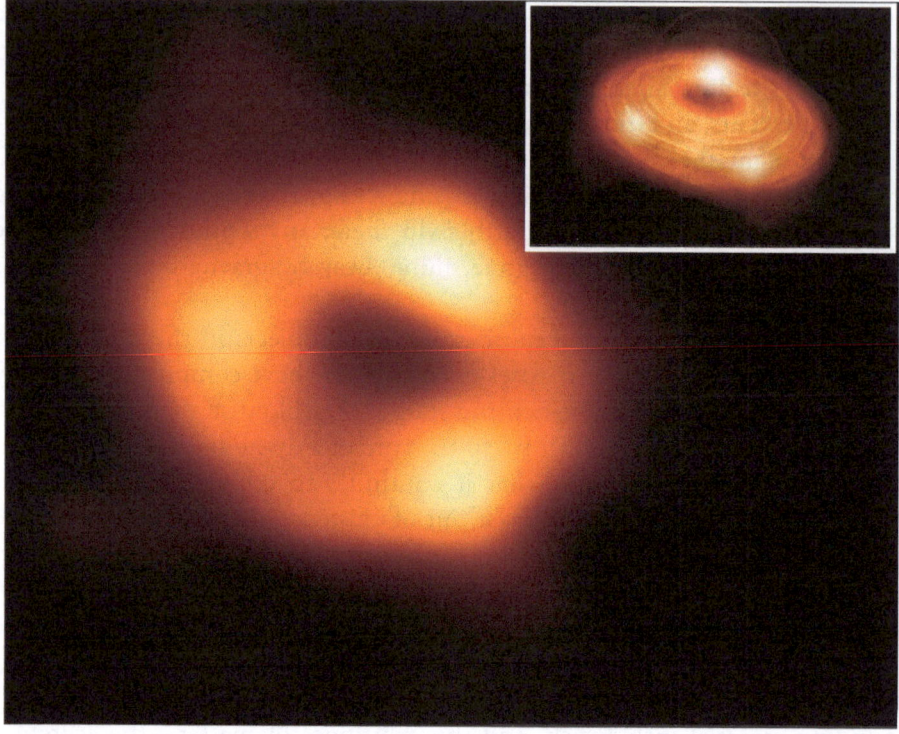

Figure 1.9 The first image of the black hole at the heart of the Milky Way. The invisible black hole at center is surrounded by a glowing ring of gas and dust falling into the black hole's event horizon. Note the bright flashes within the surrounding ring of debris. (Courtesy the European Southern Observatory/Event Horizon Telescope collaboration, public domain. https://www.eso.org/public/images/eso2208-eht-mwa) Inset: artist's conception of the details (art by the author)

million C) as it pulls the material into itself. The heat shows as flashes of infrared and radio energy.

The black hole's diameter is estimated at 23.5 million km. The bizarre object was captured in a series of images taken by the Event Horizon Telescope array, a network of eight major observatories. Over 300 scientists collaborated on the final image. The source of all this excitement is invisible, as light cannot escape from a black hole, but material swirling around Sgr A* tells the tale. A radiant donut-like structure surrounds a dark central region containing the black hole itself.

Building better worlds

Among the 100-400 billion suns in the Milky Way, planets circle the vast majority. With the advent of telescopes, the scientific community came to a

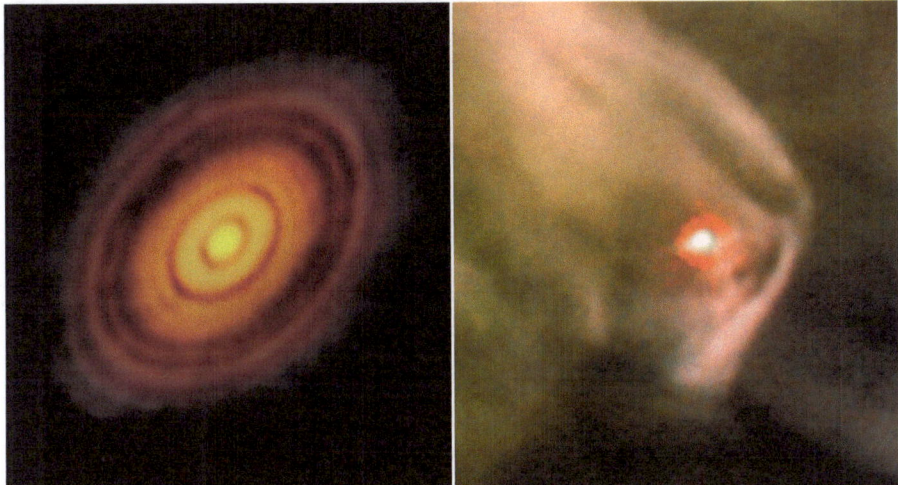

Figure 1.10 Left: Chile's Atacama Large Millimeter Array imaged this protoplanetary disc surrounding the infant star HL Tauri. The image shows details in the disc's structure such as cleared regions (dark) and possible locations of planets forming within the cloud. (courtesy ALMA [ESO/NAOJ/NRAO]); right: a proplyd in the Orion Nebula. Shock waves of gas bend around the forming star and surrounding dark disk, blown from a nearby star. (courtesy NASA/ESA and L. Ricci [ESO])

fairly early consensus that stars form within the elegant, glowing clouds of space dust and gas called nebulae. But the next step, that of planetary formation, stirred more controversy. Beginning in the18th century, theorists[16] put forth a series of hypotheses to explain the formation of planets (see Chapter 2). Thanks to advanced orbiting and ground-based observatories, we now have first-hand observations of solar system formation in action. The Hubble Space Telescope has imaged proplyds—protoplanetary disks or still-forming planetary systems—in the Orion Nebula. Chile's Atacama Large Millimeter Array and other facilities have successfully snapped gaseous disks of material where objects within have cleared orbits (Figure 1.10).

The picture we have today of our early solar system is one of an elegant dance of dust and gas, moving to the music of inertia and centrifugal and other forces. Before the Sun and planets, there were only clouds of cold gases and dust particles, the solar nebula. Physics dictates that as a cloud condenses, it contracts. This contraction results in a gradual spin.[17] Eventually, that

---

[16] Emmanuel Swedenborg in 1734, Immanuel Kant in 1755, and Pierre Simon Laplace in 1796.

[17] This phenomenon is beautifully exhibited by an ice skater. If the skater begins a spin with her arms out, and then pulls her arms inward to her body, her spin becomes more rapid.

Figure 1.11 Two possible analogs to the Sun's early star cluster include the high mass grouping NGC2244 in the Rosette Nebula (left) and the lower-mass M44 star cluster (right). (Rosette nebula by Stephen Rahn via Wikipedia Commons: https://commons.wikimedia.org/wiki/File:Rosette_Nebula_in_Monoceros.jpg. M44 by Fried Lauterbach via Wikipedia Commons: https://commons.wikimedia.org/wiki/File:M44_47x300s-10%C2%B0C_O30_G0_PM_RGB_03032022.jpg)

Figure 1.12 Left: The Stardust spacecraft returned precious fragments of the comet Wild 2. The comet's nucleus is seen here in a spacecraft view. (NASA/JPL) Center: Genesis with panels—resembling tennis rackets—deployed for capturing interstellar dust. (NASA/JPL art) Right: Despite the sample capsule's failed parachute and crash-landing in the Utah desert, Genesis scientists were able to collect and study material from the interstellar medium, teasing out secrets of the primordial universe. (388th Range Squadron, public domain, https://commons.wikimedia.org/wiki/File:Genesis_crash_site_scenery.jpg)

Figure 1.13   A protostar is forming within the dark cloud L1527, shown in this infrared image from NASA's James Webb Space Telescope Near-Infrared Camera (NIRCam). The infant star, which is not yet on the main sequence, is still feeding from its surrounding solar nebula. The disk of the solar nebula appears as a dark line across the narrow "neck" of the hourglass, with light from the star shining out above and below it. Ejections from the star have cleared out voids above and below it. The star's age is estimated to be a scant 100,000 years old. Courtesy NASA/GSFC, ESA, CSA, STScI. Image processing: Joseph DePasquale, STScI, Alyssa Pagan, STScI, Anton M. Koekemoer, STScI)

spinning motion causes the cloud to flatten. Heavier material sinks toward the center, while lighter gases swing to the outer regions. The central hub of this spinning cloud grows fastest, becomes more massive and compact, and begins to glow with its own heat. Eventually, its atoms cannot hold themselves together as individual components. The collapse of the atoms triggers nuclear fusion, and a star is born (Figure 1.13).

The gaseous disk surrounding our Sun—known as a protoplanetary disk—began to change as the star warmed its center. The spiraling disk was inconstant, with clumps of denser materials, eddies of swirling gases and variable density and consistency. Its uneven character eventually led to the planets, as those clumps of dust and gas fused into larger and larger bodies.

The Sun and its emerging planetary system were not alone. Recent work indicates that the Sun had company, birthed in a cluster of stars drifting through space. Astronomers are studying the chemical makeup of various stars in an attempt to locate some of the Sun's early companions (see A Stellar Solar Family: companion stars of the Sun?)

Our solar nebula had only been in existence for 10 million years by the time most of its material had collapsed into embryonic planets or was thrown from the system. Over 98% of the material in the Sun's nebula was gas. The other 2% or less consisted of solid material, either condensed from the Sun's cloud or wandering in as grains of interstellar dust (born in the hearts of more ancient stars).

A Stellar Solar Family: companion stars of the Sun?

For dynamicists who study workings of the cosmos, our solar system presents a puzzle: why is the outer edge so sharp? The planetary system builds in mass as it moves out from the Sun, with Jupiter making up more mass than all the other planets and moons combined. But after Neptune, there is no gradual tail-off. Instead, the mass distribution of our system shows a precipitous drop beyond the major planets. The cause of this strange distribution of bulk presents an intriguing possibility: our Sun may have been born in a cluster of stars (Figure 1.11). That cluster may have been as large as 10,000 loosely associated stars, lazily drifting through space, or the stellar clump may have been much smaller, perhaps only 1,000 members, and of a much more violent nature. A close drive-by on the part of one of those stars could be responsible for our solar system's abrupt drop-off of mass on its outside edge.

The Sun's formative years do matter. If massive stars surrounded the young Sun—the kind that explode and manufacture heavy elements—their proximity would have infused the Sun with planet-making elements. But young groupings of stars spread out very quickly, scattering within a matter of a few million years, so the process must have begun early.

A recent study finds that with such stellar flybys, the Sun's cradle could have been one of only two types of star groupings, either "high-mass, extended associations", or intermediate mass, compact clusters. In both cases, close encounters of stars with the infant Sun take place often enough to sculpt a star's environment, but not so often as to be destructive to new planetary systems.[18]

The Sun was born 4.6 billion years ago. In that time, the Milky Way Galaxy has rotated twenty-some times, giving the Sun's native cluster plenty of time to disperse. Nevertheless, astronomers are on the trail of the Sun's siblings, searching for stars about the right age with the right balance of elements that resemble our Sun's constitution. Two candidates have been found so far: HD 162826 (slightly warmer than the Sun with more heavy elements) and HD 186302 (similar in composition and size to our Sun).[19] There may be thousands of others, researchers contend.

The material from our ancient solar nebula seems inaccessible. The cosmic leftovers float among the planets in amorphous, rarified clouds. And finding samples of the material that led to our infant Sun is even more difficult. How can we access matter that came before the Sun existed as a star? Astronomers have come up with several strategies. Getting our hands on actual samples seems a daunting, nearly impossible task. But there are locations in our solar system where fine material actually streams in from interstellar space, the kind of dust that presumably leads to star formation. It would take an interplanetary expedition to grab some. So, said the scientists, let's go get some.

Our picture of the solar nebula's initial makeup came into finer focus with the advent of two spacecraft missions: Stardust and Genesis (Figure 1.12). Engineers crafted both missions to gather interplanetary material to return to Earth. Stardust's mission included a fly-through of Comet Wild-2, thought to contain material from the early solar system—followed by a sampling period in deep space at a region where interstellar material streams into the solar system from beyond.

Both spacefarers carried tennis-racket-sized paddles to collect minute particles in interplanetary space. The paddles were covered in aerogel, an ultralight substance that could snag fast-moving particles, trapping them within the glassy material. After the harvesting phase of the mission, the paddles would be stowed in a capsule that would return to Earth, descending on a parachute to land in the deserts of Utah. That was the plan.

For its part, Genesis engineers tasked their spacecraft with garnering material issuing directly from the Sun. Research and models intimate that the outer layers of the Sun preserve the ingredients of the early solar nebula. Knowing the elemental and isotopic composition of the outer layer of the Sun is, essentially, the same as appraising the composition of the solar nebula.

Flight controllers sent NASA's Genesis into a halo orbit around the Earth-Sun L-1 point[20], a sort of cosmic parking garage where objects remain "parked", balanced between the gravity of the Earth and the Sun. There, Genesis collected solar wind particles over 887 days, from 2001–2004. Designers crafted different arrays that could collect varied types of particles. After its multi-year cruise, Genesis left the L-1 point and sent its capsule on a return journey to Earth with its precious cargo.

---

[18] *Cradle(s) of the Sun* by Pfalzner and Vincke, Astrophysical Journal, volume 897, number 1; July 2, 2020

[19] A 2019 paper argues against the star being a possible sibling to the Sun, because its path through the galaxy is quite different from the Sun's.

[20] Short for "Lagrangian Point 1". The Earth/Sun system has five Lagrangian points.

The sample return shell survived its flaming reentry, but all was not well. A switch that detects acceleration was installed backwards, so the capsule never deployed its parachute. The recovery teams watched in horror as the craft plummeted through the sky, impacting the ground at nearly 200 mph. The protective outer shell cracked open like an egg, shattering many of the collection plates inside. Desert soil encrusted much of the interior of the craft. But because the solar wind particles hit the aerogel at such high speeds, they were buried deep inside, protected from terrestrial contamination.

The microscopic treasures carried by Genesis held many surprises. After a long period of frustrating cleanup and sorting of materials, researchers subjected the stellar samples to spectrometry and other tests. Their analysis revealed that oxygen and nitrogen in the solar wind are lighter, atomically, than similar oxygen and nitrogen atoms found within the terrestrial planets. Members of the solar system, from the sun to the planets, possess differing amounts of nitrogen, oxygen, and other elements. But each of those elements come in lighter or heavier varieties, called isotopes (depending on how many neutrons they have). The proportions of the Sun's lighter to heavier nitrogen, oxygen, and other elements were thought to be roughly equal to those of the terrestrial planets, because they shared a heritage of birth in the same nebula of gas and dust.

But Genesis samples reveal significant differences. Earth is richer than the Sun in two important oxygen isotopes, oxygen-17 and oxygen-18. The difference in nitrogen is even more marked: particles from the Sun held 40 percent less of an isotope called nitrogen-15 than the Earth, Mars, the Moon and the majority of meteorites. The implication is that something starved the solar nebula of certain elements before it had the chance to form the dust grains that would eventually coalesce into the inner planets and asteroids.

At the same time, measurements of neon and argon match lunar rock samples that contain remnants of the young solar wind. The results show that the makeup of the Sun's outflow of gases has not changed significantly in the last 100 million years.

Somehow, the Sun ended up with differing relative amounts of elements than the planets did. How and why this is so is under debate, but it is clear that we still have an incomplete understanding of the early solar nebula.

In the case of Stardust, one face of its paddles would be exposed to the primordial dust of Comet Wild 2. If asteroids are mountains of rock and metal, comets are the solar system's icebergs. Residing primarily in the cryogenic gloom of the outer planets,[21] comets are thought to preserve elements

---

[21] And beyond. The Oort cloud is a shell of comets that may extend over a quarter of the way to the nearest star, Proxima Centauri.

from the earliest days of planetary and stellar formation. They are ice-rich, with volatiles (gas and water vapor) that evaporate as the comet nucleus heats up during its approach to the inner solar system. A great cloud called a coma forms around the center (nucleus). The pressure of the solar wind pushes that coma into a long, drifting tail. The tail has a bluish gas component extending directly away from the Sun,[22] as well as a yellower dust component that drags behind as the comet orbits the Sun. While a nucleus may be only a few kilometers across, its tail may spread across the planetary system for one hundred million km.

During several stages in its cruise, the opposite faces of Stardust's panels would collect interstellar dust, much finer than what was expected from the cloud of debris surrounding the comet. Interstellar dust drifts in throughout the solar system, but the Sun's solar wind repels it from the inner system. Because of this, Stardust could only collect the rare dust outside of the orbit of Mars, en route to the comet. After the comet encounter, when the aerogel panels were flipped around to capture comet particles, plans called for returning the panels back to their original orientation to collect more interstellar dust. But after the successful encounter of Wild 2, flight engineers and the science team decided to play the mission conservatively, so the craft was commanded to stow the panels into their protective capsule early.

The team had reason to be nervous: Stardust's return capsule was equipped with the same parachute system as Genesis had been. Would it work correctly, or be a rerun of the crash landing? As it turned out, Stardust's systems worked beautifully, deploying a parachute for a soft landing as they were designed to. Science team member Benton Clark witnessed the opening of the capsule weeks later at the Johnson Space Center. The interstellar dust side was revealed first. "It was absolutely pristine," says Clark, "which one would expect since the particles are thought to be on the order of one micron." For a brief, terrifying moment, Clark wondered if all had worked as it was supposed to. But the scientist was delighted to see the comet collection side, unveiled fifteen minutes later. "You could see big splotches, almost like bird droppings. It looked like the thing had flown through a flock of geese."

The prized interstellar particles—seven have been located so far—were so small as to be invisible, but under scrutiny they are giving up their secrets. The microscopic grains can tell us what substances float between the stars. We are

---

[22] Because the ion tail is pushed directly away from the Sun, there are times when it actually extends in front of the comet.

learning about the deaths of stars, as this material comes from the violent end of distant suns. The particles also fill in some of the details in our narrative about planetary formation. In short, these seven samples are the actual building blocks of our primordial solar system, relics of dead stars whose lives preceded our Sun's birth. Each particle is composed of minerals containing silica, oxygen and metal—called silicates—but they vary in composition. This likely signifies that they have distinct histories. Some may have been born in the hearts of cold interstellar clouds, while others issued from ancient stars. All of them drifted for millions or billions of years in the interstellar medium, perhaps mixing together from multiple sources.

Of course, the solar system itself has plenty of its own dust, so scientists had to first establish that each grain actually came from outside of our Sun's neighborhood. To do this, Stardust team members came up with a set of prerequisites. First, the dust mote must have been moving at high speed when it impacted the aerogel. Particles orbiting within our system would be traveling at lower velocities.

A second of the team's criteria was that the particle must be composed of material consistent with a formation in space. Its material must not resemble any of the substances that make up the spacecraft itself.

Perhaps the most convincing tell for a particle's extrasolar birthplace would be the presence of an oxygen isotope not normally occurring within our solar system. But if a particle's oxygen isotopes match those within our system, the test would not definitively rule out an interstellar origin.

Investigators got to work studying three fragments of probable cosmic dust embedded in the aerogel. They estimated the entry speed of two particles—dubbed Hylabrook and Orion—at about 10 km/s. A third particle, labeled Sorok, came in at over 15 km/s. All three of these qualify as interstellar particles under the velocity guidelines.

Likewise, the three passed the second test; their contents did not resemble artificial materials like spacecraft metals or insulating blankets.

Frustratingly, the oxygen isotope tests were less conclusive. Measurements either matched isotopes of our solar system, or there was not enough data for a definitive answer. Nevertheless, considering their compositions, impact speeds and angles within the aerogel, and comparisons to models that predicted entry angles and speeds of particles originating from the solar system, researchers are confident that these tiny dust grains were born in the deep cosmos well beyond our own planetary frontiers. They put the possibility of the particles arising within our own solar system at less than 0.03 percent.

Investigators are also studying four particles that impacted on Stardust's aluminum foil frames that hold the cells of aerogel in place. The size of these grains more closely aligns with predictions of interstellar particle sizes.[23] Researchers continue to scrutinize Stardust's cosmic stardust. Their work will provide many insights into the formation of planetary systems in general, and our own Earth's planetary nursery in specific.

After its collection of interstellar particles, Stardust's mission of discovery entered a new phase. After years of collecting interstellar dust, the spacecraft reoriented its tennis-racket-sized collector panels in preparation for its appointment with Comet Wild 2. Before Stardust's return of comet material, the common wisdom within the scientific community assumed that comets represent original raw material of the solar nebula. The floating icebergs formed in the outer solar system, so the story went, incorporating materials from the coldest regions of the early solar nebula. Before Stardust's expedition, meteorites were thought to be the most primitive material we had, but they were "processed" by heating (for example, some contain iron). The rock/metal meteors were also altered by the interaction with ices. The asteroids and meteors formed inside the "snow line", the boundary between frozen and vaporous water in the early solar nebula, so meteorites were thought to represent the inner solar system. Comets were thought to form outside the snow line, taking in samples of the original materials from the frigid outer fringes of the infant solar system. But in addition to volatile-rich matter from the outer system, particles from Wild 2 include materials that formed at higher temperatures within the inner region of the solar nebula. Scott Sandford, a research astrophysicist at NASA's Ames Research Center, has been analyzing the particles returned by Stardust. Wild 2contains materials formed throughout a wide spectrum of temperatures and conditions. "It looks like these particles formed in different places, stuck together in the nucleus, and then nothing happened to the components," Sandford says. "In other words, the comet sampled material from all over the solar nebula and did a good job of preserving them."

If the solar nebula had different compositions in the inner and outer regions, how could Wild 2 contain material from both regions? Researchers have put forward several possibilities. There may have been a high temperature spike in the outer solar nebula. Alternately, there was abundant mixing throughout the nebula early on. Scientists once believed that as matter coalesced from the solar nebular cloud, denser material migrated toward the center, while light volatiles drifted to the outer regions. But research shows that the genesis of our planetary family was not so simple. One model

---

[23] Citizen scientists across the world continue to pour over a million microscopic images of the aerogel in search of more particles. The NASA program is called Stardust@home.

suggests that cells of turbulence may have moved inner materials toward the outer system in a cosmic relay. Another model suggests a sort of conveyor belt from inner to outer system, powered by jets of material pouring from the polar regions of the early Sun. The stellar jets would cast matter into the outer solar system.

The major component of Wild 2 is rock, and that rock formed in red-hot conditions above 1000°C (1832°F). But some of the ices that later mixed in with the rock actually formed at temperatures near absolute zero. The implication is that a broad mix of materials was transported into the outer solar system, past Pluto, where Wild 2 formed. Stardust revealed that the inner solar system was not isolated from the outer solar system, as was once thought. Instead, primordial dust and gases were mixing across provinces from near the Sun to the dark deep-freeze of the outer system.

The amount of mixing that took place is still a mystery; experts don't know the abundances and ratios of inner and outer materials. Stardust samples have been found to include minerals that form at high temperatures (refractory minerals) like olivine, pyroxene, and anorthite. But the majority of the comet's constitution—frozen water and gas—is rare in the samples. Most of Wild 2's volatiles were lost to space as they warmed after being ejected from the comet, but before Stardust had a chance to capture them. Additional volatiles were probably lost when particles collided at high velocity (6.12 km/sec) with the aerogel collectors. However, traces of helium and neon gases have been detected, along with a variety of organic compounds.

Wild 2's variety of complex organics proved to be one of the most provocative discoveries from Stardust. Astrobiologists see organic material as the building blocks of life, so it is critical to our understanding of the genesis of life on Earth. Researchers have discovered non-aromatic (volatile) organics. Where do they come from? In laboratory studies, when organic molecules embedded within ice are bombarded by radiation, they break apart. As the ice warms, they reassemble into more complex chains. Membranes form. The new chemical combinations forced by these cosmic deep-freeze mischiefs have led to new organics never seen before. While research is ongoing, investigators have identified a new class of organic material more primitive than the organics previously found in meteorites. Within Stardust's trove, they have also discovered CAIs, calcium/aluminum-rich inclusions, particles numbering among the most ancient in the solar system.

One proposed scenario for the appearance of life on the early Earth (see also Chapter Five) is that carbon-based materials rained down on the infant landscape from comets in an organic mix. In this scenario, as the organics settled to the Earth's cooling surface, they met clays and other veneers that preferentially arranged the organics into complex chains, which in turn led to biological organics. But to even assess the likelihood of this scenario, we need to know what organics are hiding in the comets.

Stardust and Genesis have been followed by other spectacular missions, including the European Space Agency's Rosetta mission to the rubber duckie-shaped Comet 67P/Churyumov-Gerasimenko, and asteroid sample return missions such as Japan's Hayabusa and NASA's OSIRIS-Rex (see Chapter 2). With advanced technologies used by these and other robotic explorers, our picture of the infant solar system is becoming more and more clear.

The star islands cartwheel across the cosmos. Their ancient stars infuse the galaxies with heavy elements from their dying explosions, and in one of those galaxies, our Sun is born. Within its solar nebula, the stage is set for the birth of the Earth. Our world's naissance issues from the maelstrom of rock and metal and ice whose remnants reside, even today, within the asteroids and comets.

2

Earth=Asteroid belt: the impact of uneasy relationships during the great solar system cleanup

It all started out as an unassuming boulder tumbling through space. Who could have guessed that such a simple rock could result in seas and mountains, trees and giraffes, cars and buildings, literature and music and masterful paintings? After many false starts, that little boulder pulled enough material to itself to become the third planet from the Sun, our own Earth. But the journey wasn't easy, and it wasn't calm.

<u>Planetary birthdays: hot or cold?</u>

For centuries, the origins of our solar system seemed out of reach. In his 1698 treatise *Celestial Worlds Discover'd*, Dutch astronomer Christiaan Huygens lamented, "I have often wonder'd how an ingenious man could spend all that pains in making such fancies hang together. For my part, I shall be very well contented…if I can but come to any knowledge of the nature of things, as they now are, never troubling my head about their beginning, or how they were made, knowing that to be out of the reach of human Knowledge, or even Conjecture." But not all were content in knowing only how things stand now. Many yearned to see into the great dark past, into those mysterious moments of cosmic creation.

The lustrous clouds of gas called nebulae provided clues to early theorists about how our Solar System came together (Figure 2.2). Among them, Swedish scientist Emmanuel Swedenborg envisioned a hot globe of material

Figure 2.1 The dust and gases of the solar nebula begin to sort themselves into rings of matter. Within that flattened accretion disk of material, eddies and knots fuse into planetoids and bona fide planets. Here, planet "embryos" the size of our Moon collide, fragmenting into clouds of debris, only to recombine later. (painting by the author)

around the infant Sun as the birthplace of the Solar System. German philosopher Immanuel Kant (1755) and French astronomer Pierre-Simon Laplace (1796) added more detail, suggesting that the primordial cloud surrounding the Sun somehow flattened into a disk, eventually leading to the genesis of the planets.

Soviet theorist Otto Schmidt submitted that the primeval Sun passed through a cosmic cloud of gas, dragging it along in a great tail which ultimately consolidated into the planets we see today.

In 1778, the French mathematician and naturalist Comte de Buffon described the formation of the solar system as a hot Sun ravaged by a series of comets.[1] He theorized that passing comets pulled mass from the Sun, leaving blobs of hot material to cool into planets. Buffon was not alone in

---

[1] *Les époques de la nature,* by Georges-Louis Leclerc, the Comte de Buffon, 1778

Figure 2.2   Views of our origins. (top to bottom): Kant and LaPlace's vision of a cloud-like protoplanetary disk; James Jean's close encounter with a passing star, with modern planets superimposed; Schmidt's wandering Sun passing through a cosmic cloud of gas; Chamberlin and Moulton's fly-by star pulling the Sun's material into mingling planetes-imals a few tens to hundreds of kilometers across. (art by the author)

thinking that the Earth's formation took place in a very hot environment. Astronomers soon abandoned his idea of planets torn from solar material, instead proposing that the planets condensed from heated clouds of dust and gas in the solar nebula, a vast disk surrounding the early Sun. Buffon's hypothesis was seen as unlikely because comets have very low gravity, and so many near-misses seemed dubious. Still, planets holding on to the warmth of their creation made sense. After all, where did all that internal heat come from?[2]

Playing off of this idea, William Thompson, the honorable Lord Kelvin, attempted to calculate the Earth's age by looking at its size and temperature. He assumed that the early Earth began as a globe of molten rock. He then estimated how long it would take to cool the planet to its current state. In 1897, he proposed the age of the Earth at 20 to 40 million years.

In 1917, James Jeans added another idea to the mix: that a star passed close to the Sun, pulling material from it. This theory seemed to fit the outline of the planets, as if a donut-shaped cloud surrounding the Sun, thickening toward its outside edge, led to the small terrestrials on one side, and the large gas and ice giant planets on the far rim. But as the twentieth century dawned, the scientific community began to realize that a more likely scenario had the terrestrial planets forming in cold clouds of material rather than a searing cloud of sun-like material. The planets' internal warmth came not from leftover heat in the solar nebula—although there was some—but primarily from radioactive material within their iron cores. Thus, they said, the heat of planetary interiors grew after the planets themselves formed.

A geologist and an astronomer at the University of Chicago teamed up to formalize the idea. Geologist Thomas Chamberlin and astronomer Forest Moulton suggested that a passing star pulled material from the Sun, and that this material condensed into small solid bodies a few tens to hundreds of kilometers across. They called these floating cosmic mountains *planetesimals*. As these bodies collided and stuck together, the larger ones grew quickly, their increasing gravity pulling more material to themselves. Eventually, the theory went, they grew up to be planets.

---

[2] Buffon estimated the age of the Earth at 75,000 years, based on his laboratory experiments of the cooling of iron.

Some fifty years later, astronomers sharpened the ideas under the banner of the nebular hypothesis, which described a disk of dust flattening as it spun. The central hub—where the material was densest—collapsed into a large sphere and became the Sun, while eddies in the cloud condensed, pulled material (including planetesimals) into themselves, and became the planets. The modern view was coming into focus (Figure 2.1).

More recent work indicates that ingredients within the solar nebula differed with distance from the Sun, leading to some of the variety we see in the makeup of the planets today. Ices were abundant in the outer part of the cloud, far enough away to be less affected by solar wind and heating. Closer in, the nebula was ruled by heavier materials, while water and lighter elements like hydrogen drifted away on the Sun's outflowing gales.

While some moons in the outer solar system are clearly captured vagabonds, others appear to have formed within the same cloud that led to their parent planets. These natural satellites give us a sort of check for our nebular theory. For example, the four major moons of Jupiter, called the Galilean satellites, form a miniature version of the solar system. The innermost two closest to Jupiter, Io and Europa, have far more rock than volatiles (a bit like the terrestrial planets) while the outer two, Ganymede and Callisto, are largest and least dense, made mostly of water. They find their analogs in the gas and ice giants of the solar system.

Carbonaceous chondrites chronometers

The planets and moons have been through a lot since their inception over 4.5 billion years ago. Solar radiation and natural chemical processes have changed their mineral makeup. Their geological faces have been sculpted by meteors, weather and tectonics. Much of the record of the infant solar system, once preserved within them, is now gone. But that record has been preserved in the form of the most primitive meteorites, the carbonaceous chondrites. They are leftovers from the age of planetary accretion, and within them can be found calcium-aluminum-rich inclusions—CAIs—which arose from the earliest epochs of the solar nebula.

Carbonaceous chondrite meteorites formed in the cooler regions of the Sun's early solar nebula, and were never subjected to chemical-changing high temperatures. These precious meteorites contain the elements present near the beginning of the worlds: carbon, graphite, and even some amino acids. They also contain silicon carbide, a mineral from the era before the Sun's formation. Included in the meteorite mineral parade are microscopic diamonds that formed within other stars before the advent of our own solar nebula.

CAIs have been used to age the solar system. They date to 4.56 billion years ago, at a time when the Sun was new and the planets were just beginning to accrete. Unusual isotopes of magnesium, beryllium-boron and oxygen imply that CAIs may have formed close to the Sun, and then spread out to meteor-forming regions farther from the hot central hub of the solar nebula.

The solar nebula: true beginnings

The narrative we are left with portrays a Solar System that began as a great cloud of interstellar gas just over 4.5 billion years ago. As the cloud collapsed, physics dictated that its material began to spin, and the great pirouetting cloud flattened into a disk.[3] The thickest and densest portion nested at the center. Once it gained enough material, this central hub collapsed under its own weight. Gravity became strong enough to tear the hydrogen atoms apart, and nuclear fusion began. A star was born.

Within ten thousand years, the infant sun had separated from the nebular cloud. Farther out, eddies within the cloud continued the same process of collapse. One hundred thousand years along, planetesimals the size of large asteroids were common, and the gas giants began to form in the outer cloud. It was time for Earth and the other terrestrials to arrive at center stage.

Astronomers recognize three stages in planetary formation. In Stage One, that 2% of nebular material that is solid drifts to the midplane of the solar nebula disk. There, it accretes (condenses) into planetesimals, rocks a few kilometers across. Stage Two ushers in run-away growth as the planetesimals grow, pulling more and more material to themselves. During this period, Moon- or Mars-sized planetary embryos arise. During Stage Three, the solar system is a terrifying place. Planetary embryos circling the Sun are ripped from their orbits as they interact with the gravity from each other. Some careen into the outer reaches of the system, while the majority plow into each other. These random, ferocious collisions destroy many, but some suffer only glancing blows. Embryos combine to form larger and larger planets. Planets with the most circular orbits suffer fewer collisions than those in orbits that cross other planetary paths. During this time, the terrestrial planets arise in the warm central region of the Sun's disk. The outer planets also form, but the outer disk is cold and their formation is as different from the terrestrials as their nature is (Figure 2.3).

A critical component of planetary formation is called accretion. The process involves solid particles and larger fragments growing into a mass as they collide and stick together. A form of accretion led to everything from small rocks to asteroids and comets to full-blown planets and even stars. The

---

[3] Telescopes like those in the Atacama desert (Chile) or Mauna Kea, Hawaii, along with space telescopes like the James Webb and Hubble, have imaged such clouds.

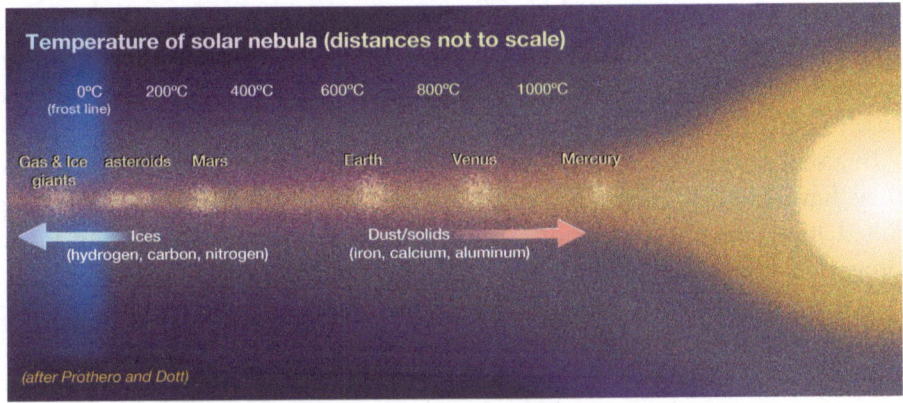

Figure 2.3 Conditions within the great disk surrounding the Sun changed with temperature and location. While rock and metal ruled the inner cloud, volatiles like water condensed in the outer realms of the nebula. (diagram by the author, after Prothero and Dott)

microscopic particles of solid material condensing from the gases of the solar nebula continued to orbit the Sun in the same orderly fashion as the gas in the disk. Since all the particles circled at about the same velocity, collisions were gentle bumps. Particles of such size had essentially no gravity to attract each other, but electrostatic forces caused them to congregate like motor heads to a used car parking lot. The gathering particles increased in mass until their gravity became a force to be reckoned with. Particles became boulders, boulders became mountains, mountains became planetesimals (planet embryos).

Each planet grew within its own "feeding zone", a ring of material in the solar nebula that extended roughly halfway to the next forming planet. Seen from far above the Sun's pole, planetary feeding zones would appear as the concentric circles of a dartboard. Since the composition of the solar nebula changed with distance from the center, each band consisted of slightly different material, and led to planets with varying constitutions. The Earth would not be the planet it is today if it had come from a different feeding zone. If our world had formed farther out, in the neighborhood of the asteroid belt, it would have obtained higher amounts of carbon, and a global ocean hundreds of kilometers deep would have inundated the planet. These factors would have combined to give Earth a dense, hot carbon dioxide atmosphere filled with water vapor, probably at pressures similar to those on Venus. At a different part of the solar nebula, the Earth might have ended up with far more carbon than it has today. Carbon within the graphite-rich crust and mantle would have led to diamonds and silicon carbide, preventing the volcanism that replenishes our atmosphere. The world as it is today owes its nature, in part, to where it grew up in the solar nebula.

In a sense, today's asteroids and comets are the solar system's escape artists. They have eluded impacting upon larger bodies. They've dodged getting pulled into planetary accretion. They've skipped out before burning up in someone's atmosphere. They are the ones that got away.

Within a million years of the beginning, the terrestrial worlds had accreted and begun to differentiate, with heavy material sinking to form cores. The heat from the planet's initial creation along with the heat from impacts, combined forces with the energy of radioactive elements like uranium to melt metals, freeing them to sink, while molten rock rose. Interior heating is the trigger—and a requirement—for differentiation, a process that sorts interior material within large asteroids, moons and planets. All the terrestrial planets are arranged like a three-layer onion. The highest density material makes up the core, consisting mostly of metal like iron and nickel. The mantle overlays and surrounds the core. This intermediate layer contains moderately dense rock, including silicon, oxygen and other elements. The lowest density rock—like the volcanic basalts and the continent-building granite—rises to the top to form the crust. The character of a planetary crust depends on differentiation, as does the geological activity within a world.

How planets heat up and chill out

The geologic activity of a planet (the ongoing changes to the surface of a planet) is an expression of how a world tries to get rid of heat. It may take on the guise of mountain building, volcanism, earthquakes, erosion, and other forms. The Earth is geologically active: its mountains, continents, oceans and canyons are dynamic and fluctuating. The surfaces of the Moon and Mercury—both worlds that are geologically quiet—have remained nearly unchanged for billions of years.

A planet's internal heat drives geology. Erupting volcanoes appear only when a planet's interior temperature is high enough to melt rock into magma. Plumes of hot magma cause uplifted landforms, and movement of plates is another expression of heat energy moving out from the interior. The fire within the hearts of planets comes from several sources. The heat of accretion builds up from collisions of planetesimals. The energy of impacts converts to heat, and that heat lasts throughout a major part of a planet's childhood. Differentiation, the sorting and movement of materials, also contributes heat. As dense material sinks and less-dense material rises to the crust, the concentration of the planet's mass moves inward. Materials separate, and friction builds up more heat. Finally, radioactive elements decay, sending out subatomic particles at high speeds. In an echo of accretion on a much smaller scale, the collision of these tiny particles with surrounding atoms heats up the interior of the planet.

Nature abhors imbalance. The air currents in an atmosphere move in an attempt to even out temperatures from night to day, from frigid poles to warm tropics. Helium in a birthday balloon escapes with the intention of equalizing the gases inside the balloon with the ones outside. As a planet heats up, it also continually seeks to cool off, always trying to find equilibrium (in this case, equilibrium in temperature). So, just as heat from accretion, radiogenic material and sinking mass raise temperatures, planetary interiors cool by three general methods.

The first is convection, a process that mixes materials within. Hot matter expands and rises. At the same time, cooler material contracts and descends. This migration of planetary ingredients pulls heat up. Convection serves as a great carousel cycling hot and cold mater, but as heated material rises, it loses energy through the crust. Any cook can tell you about convection: it is evident in a pot of boiling water. That heat is lost through the movement and transformation of the crust.

Another planetary method for heat loss is called conduction. Conduction happens when heat from warm regions to cooler ones transfers through direct contact. Conduction comes from the collisions of molecules: the faster-moving (hotter) molecules convey some of their energy to the more sluggish (cooler) molecules.

Finally, radiation rounds out our triad of a planet's heat-reduction techniques. Planets radiate infrared energy and light into space, and that loss cools the planet.

In the Earth's case, the most important of the three cooling processes is convection. Hot rock from deep down in the mantle rises, cooling as it ascends (in convection, the solid rock of the mantel is involved, rather than liquid magma). By the time the heated rock arrives at the top of the mantle (where the crust begins), it has shunted its heat to its surrounding mantle, so it cools off. The cooled rock then begins to sink again. The circular movements within the mantle are called convection cells. Mantle convection is a long-term process. Convection causes the mantle to move at about a centimeter each year, so it takes 100 million years for rock to move from the bottom of the mantle to its outermost layers. From its high perch atop the mantle, conduction takes over, bringing the heat to the surface.

Planetary formation? Take your time!

Astronomers thought that planet-forming disks dissipated after 1 to 3 million years. Once the accretion disk surrounding a star wanes, planets have no building material from which to form. Less than three million years is a short geological time to create a solar system, a fact that has baffled the experts. But new discoveries have changed that timeline.

A survey of young star clusters finds that low-mass stars like red dwarfs hold onto their protoplanetary disks longer than the easier-to-see large stars. Earlier studies skewed toward larger stars, but new studies include smaller stars, which constitute the majority of stars in the galaxy. Within 650 light-years of Earth, observers estimate that disks surrounding low-mass stars may have lifetimes up to 10 million years. It is possible that brighter, high-mass stars disperse their disks more quickly (perhaps because their solar wind and more powerful light push the dust and gas away in short order).

If further study confirms that low-mass stars keep their planet-birthing clouds longer, the revelation could explain a difference between our own planetary system and those we've charted around red dwarf stars. Smaller stars seem to lack Jupiter-sized gas giants, instead forming Neptune-class worlds at the top of their scales (see Chapter 8). Planets of this size may be plentiful because they take longer to form, and evolve more readily in older disks. Further studies with advanced instruments like the James Webb Space Telescope will probably solve the mystery.

Models of solar system formation imply that the relative proportions of metal and rock should have been consistent throughout the inner solar system during the Sun's planet-forming epoch. In general, the terrestrial planets do follow the pattern, but not exactly. The Moon has a smaller core than it should (see *Theia and the Birth of the Moon*, below), while Mercury has a much larger one (Chapter 3). These discrepancies point to the differences in the development of each.

A recent microscopic discovery sheds light on one source of heat in the process of differentiation, at least here on Earth. In laboratory experiments, scientists studying conditions in Earth's lower mantle synthesized a mineral called davemaoite. Their studies indicated that 5-7 percent of the lower mantle should be made of the stuff. But none had been found in nature until recently, when miners unearthed a diamond in Botswana. Diamonds migrate from the Earth's depths, carrying along minerals trapped within their matrices until they reach the Earth's crust. The Botswana diamond formed more than 660 km below the Earth's surface, at the upper edge of the lower mantle. Within it, tiny fragments of davemaoite, a silicate mineral, waited to be revealed. The davemaoite holds miniscule quantities of uranium, thorium and potassium, all thought to contribute to the heat that enabled the planet to differentiate.

Botswana's story doesn't end there. The gem also contains a high-pressure form of water ice, as well another mineral associated with high pressures, called wustite. The two minerals help to locate the depth at which the davemaoite formed. It's a nightmarishly dense environment with a pressure between 24 and 35 billion pascals (sea level pressure is about 101,000 pascals). The davemaoite studies confirm the role of radioactive elements deep within the terrestrial planetary mantles.

Planet-building and hot times

At 4.45 billion years ago, Mars had settled to half the size of the Earth or Venus, with Mercury trailing behind. The more mass a planet has, the more planetesimals it will call to itself, and those planetesimals will hit harder in the stronger gravity. Impacts impart heat, and some of that heat is sequestered within, warming the core. More mass brings that second heating element: more radiogenic material. Added to these factors, larger planets conserve heat more readily. Stronger gravity holds denser atmosphere, which acts as a blanket against the cold of space. On Earth, hot, less dense rocky material from deep within migrated upward through the mantle in the process of convection. Through its surface crust, the planet could transport and discharge heat from the interior. All that heat drove many tectonic processes on the crust, including faulting, volcanism, and Earth's unique plate tectonics. The Earth has the strongest convection of all the terrestrials, reaching nearly to the surface at the mid-Atlantic and other ocean ridges where new crust constantly spreads. The mantles of tiny Mercury and the Moon probably ceased convection altogether. Mars may still have some convection today, and Venus may have nearly as much as the Earth. The thicker Venusian crust prevents the forms of heat loss used by planet Earth; it exhibits heat loss in other ways (see Chapter 4).

The violent epoch of intense meteor bombardment influenced the interiors of all the terrestrial worlds. Heat from impacts moved down into the mantle as well as across the surface, leaving slurries of molten rock within and about gigantic impact basins. In the Earth's case, one impact outshone all others, as we will soon see.

At the end of a billion years, the planets had swept up the major planetesimals, although it would take some time to clear up all those pesky asteroids (a process still taking place). We come to the solar system of today, neatly arranged with terrestrial (rocky) planets near the warm Sun and the gas and ice giants in the cool outer system, kept company by a cloud of icy comets and dwarf planets.

Every fairy tale has its problems

The planetary arrangement of warm terrestrials forming close to the Sun and giant worlds arising toward the outer system edge is a nice story, but it has some major issues, as do other tales. Why is Cinderella the only person in the entire county who wears that particular shoe size? Why would the prince open Snow White's glass coffin and proceed to kiss a presumed corpse? Why are the planets arranged the way they are? The problems in our understanding of the evolution of the solar system come into focus in light of new dynamic computer models. These simulations tell us that the planets may not have formed in the places they inhabit today. It appears that the early planetary system carried out a game of "cosmic bumper cars", with Jupiter, Saturn, Uranus and Neptune migrating toward and away from the inner solar system. As they did so, their gravity acted as snowplows, moving material from the inner to the outer solar system.

We get a hint of the madcap development of our planetary system from the Kuiper Belt, a band of icy comets and planetoids (small versions of planets) beyond the orbit of Neptune (most of the comets we see come into the inner solar system from the Kuiper Belt). Many Kuiper Belt objects (KBOs) follow paths that are "in resonance" with the planet Neptune. This means that for each three times that Neptune orbits the Sun, a more distant object "in resonance" will circle the Sun twice. Pluto is one of the objects in resonance with Neptune. Computer modelers and planetary dynamicists realized early on that the only way to trap Kuiper Belt objects in resonances with Neptune would be to slowly move Neptune outward from a location closer to the Sun than it is today. The Kuiper Belt architecture gave rise to an important new concept: planetary migrations. Gone was the clockwork solar system with its planets following their well-organized orbits, says astronomer David Jewitt. "The high population of resonant KBOs could only reasonably be explained by moving Neptune outwards slowly into the region now occupied by the Kuiper Belt." Jewitt compares the planetary wanderings to a giant snowplow moving outwards and picking up snow on the front of the plow. The KBOs all pile up on the resonance as it moves outward with Neptune.

The notion that the planets were not always where they are now may be the most radical concept to arise in Solar System studies in the last forty years. The revelations gave birth to a theory called the *Nice model* (named after the French city), in which the planets' orbits not only migrate (i.e. the sizes of the orbits change) but the planets actually pass into and out of resonance with each other, so that Jupiter and Saturn might get into a mean motion resonance where the ratio of their orbits is 2:1 or 5:2. If that happened, the Nice model argues that the shifting arrangement would dramatically shake up the Solar System and give rise to many of the phenomena that dynamicists see today.

Another model, called the Grand Tack, shows Jupiter and Saturn marching in toward the Sun, followed by Saturn pulling Jupiter outward at just the right moment. The Grand Tack results indicate that Jupiter robbed Mars and its surroundings of rocky, planet-building material, sending it toward the inner system. Thanks to Jupiter's gravity and its migration in the early Solar System, about one out of every one hundred water-rich C type asteroids[4] dispersed into the outer fringes of the asteroid belt, the torus of boulders circling the Sun between the orbits of Mars and Jupiter. But for each one of those, at least ten spiraled sunward, delivering water and minerals to the terrestrials.

The Grand Tack version of our planetary scrapbook has the advantage of explaining the petite size of Mars, the architecture of the asteroid belt and the theorized birth of terrestrial seas at the hands of incoming space rocks.[5] The Grand Tack and the Nice model are both incremental advances in our mapping of the Solar System's shifting past. And where once we had worlds dutifully following their eternally charted orbits, we are coming to view the system as a dynamic, changing arrangement of shifting worlds.

---

[4] Some C type meteorites and asteroids are up to 20% water, mostly locked into minerals.
[5] But for problems with the seawater/asteroid connection, see Chapter 4.

Planetary migration: making the habitable zone habitable

Our Earth orbits the Sun at a distance that enables surface water to remain in a liquid state. This is rare in our solar system. Mars is on the outer fringe of this zone; if the planet had a bit more atmosphere, its surface might just support liquid water. Venus may be at the inner edge of the Sun's habitable zone; if its atmosphere was less dense and its day shorter, it might support lakes and oceans on its torrid surface (see Chapter Four).

But it takes more than the right temperature to make a region habitable. If the Earth was going to be a life-filled world, it needed the right biogenic (biology-related) ingredients. Throughout the solar system, the building blocks of life abound. Organic matter, carbon, methane, nitrogen, precursors of amino acids, all have been discovered floating in the interplanetary medium or hidden away in comets and meteorites. As in real estate, location is everything. Ironically, when a planet forms close enough to its star to orbit within the habitable zone, it is usually starved of water and the elements that make up important life-related matter. If a planet is in the right location for temperature, it will typically be in the wrong location for life-building construction materials. At the Earth's distance from the Sun, temperatures were too high to bind carbon, nitrogen and water into the fragments that would accrete into planets. Ices that are abundant in the outer solar system (water, methane, nitrogen) remained in a vaporous state in the inner system. The composition of carbonaceous meteorites, thought to be representative of the majority of asteroids, gives us an idea of what the Earth was missing. These meteorites hold up to 20% water, locked away in minerals. They also contain 4% carbon, from which they get their name. Compare these numbers to bulk levels on Earth: our planet hosts a tenth of a percent water and a scant .05% carbon.

How did a planet that formed in a mineral-desolate region get critical elements for life? We have seen that some planets may have been vagabonds in the early epochs of our solar system. The wandering worlds likely shepherded precious compounds from the cold outer system. These included nitrogen, water and carbon compounds, all elements crucial to living systems. They were delivered aboard asteroids and comets, which have left their cratered fingerprints throughout the entire planetary system.

Traces and signs

The development of our solar system has left fingerprints among the planets today. Our first insight comes from the spin of the worlds and their pathways around the Sun. From a viewpoint above the solar system (above the Earth's north pole), all of the major planets circle their star in a counter-clockwise direction. Likewise, the vast majority of the moons orbit their host planets in the same counter-clockwise direction. Additionally, all the planets orbit essentially in the same plane, known as the ecliptic, which is parallel to the Sun's equator. These characteristics are what we would expect if all the members of our solar family formed in the same disc-shaped cloud, moving in the same direction.[6]

As we have seen, where a planet forms in the solar nebula has a great deal to do with its composition. The Sun warms the inner solar system, so many volatiles such as water and reduced gases (hydrogen, helium, methane) have boiled away or pushed outward. The outer system, much colder, is ruled by hydrogen, helium, methane and ammonia. Cold temperatures pave the way for solid bodies to have mountains of water ice rather than rock, and seas of cryogenic methane instead of liquid water. The larger worlds have no surface at all, being titanic spheres of gas.[7] The difference between the inner rocky planets (the terrestrials) and the outer gaseous worlds (gas and ice giants) offers yet another clue that the family of planets formed in an organized cloud of dust and gas.

While many planetesimals came together in violent collisions, some may have merged more gently. We have learned this not from the inner solar system, but rather from the cold, dark outskirts of our solar family, the realm of comets called the Kuiper Belt. After its spectacular encounter with Pluto, the New Horizons spacecraft went on to explore the tiny, peanut-shaped ice chunk Arrokoth.[8] Farther from the Sun than Pluto, this strange Kuiper Belt Object may help to narrow the many concepts we have of planetary formation (Figure 2.4).

Arrokoth is a cosmic time machine, revealing the dawn of our solar system. Investigators conclude that the KBO is largely unchanged since its birth some 4.5 billion years ago. Its surface is uniformly red, bearing the chemical signature of methanol ice. Methanol ice is significant, because it commonly resides in protoplanetary disks. Arrokoth's surface is surprisingly smooth, with few obvious craters and soft contours. Multiple clues point to a gentle meeting of Arrokoth's two lobes. Computer models that faithfully reproduce Arrokoth's shape simulate the two objects initially orbiting each other, slowly meeting in the middle at less than 9 mph (4m/s). It is likely that the two impacted at a slight angle.

---

[6] There are, of course, exceptions to every rule. While nearly all planets spin in a prograde direction, Venus turns slowly in the opposite direction, and Uranus is tipped over on its side; its axis lies at 98°. And while all the largest moons orbit their planets in a prograde direction, Neptune's large moon Triton orbits the other way (i.e. retrograde, or clockwise as seen from the north). Triton is likely an interloper from the outer solar system, and was captured by Neptune's gravity. Smaller bodies like asteroids, dwarf planets like Pluto, and comets and asteroids, orbit in all sorts of directions and angles.

[7] Gas and ice giants do have dense, solid cores, but there is no clear demarcation between their atmospheres, liquid-gas seas and solids.

[8] Originally labeled $2014MU_{69}$.

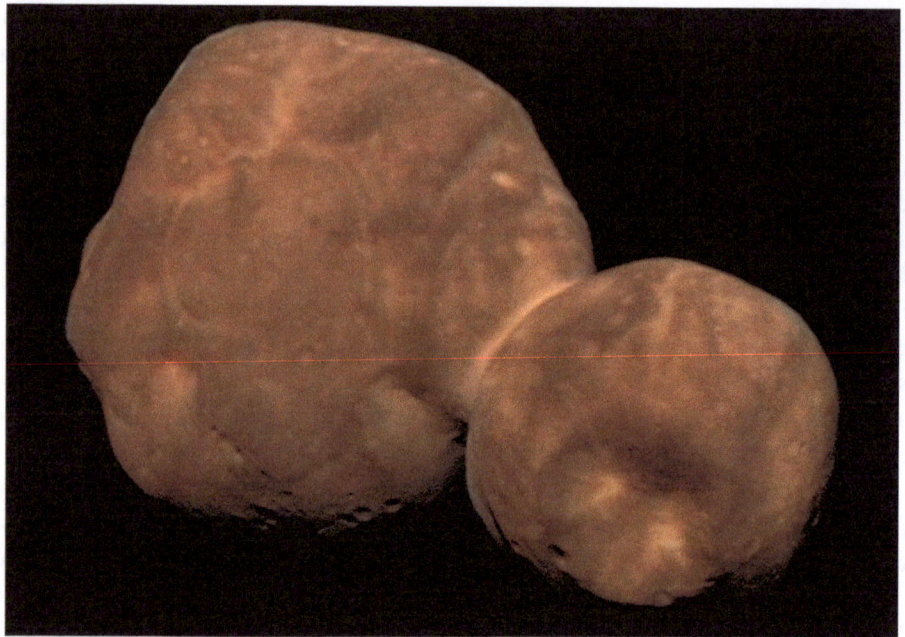

Figure 2.4   New Horizon's image of Arrokoth (NASA/Johns Hopkins University Applied Physics Laboratory/Southwest Research Institute/Roman Tkachenko, public domain)

Getting from dust bunnies to planets

As a solar nebula forms, tiny particles condense out of the cooling gases. The disk of gas and dust particles continues to rotate around the star, and the gas "herds" the dust particles into concentric circular orbits. Astronomer William K. Hartmann has been studying low velocity impacts to gain insights into this ancient environment from which the planets arose. He likens the primordial particles to "cars around a race track. As long as the objects remain on circular paths, neighboring objects can't collide with them." But with any turbulence, they bash into each other (Figure 2.5). The business of making a new planet is a messy affair. Some of the collisions occur at very low velocities, while others are violent.

The planets of our solar system accreted in a slurry of material from microscopic dust to gravel to sarsens, eventually forming thousand-mile-wide planetesimals. Those miniature planets bumped or crashed together, resulting in planets. As the planetary embryos swelled, so did the strength of their gravity. Later stages of planetary accretion became chaotic and ferocious. Glancing blows peeled crusts away or donated core material into crash victims. Not all

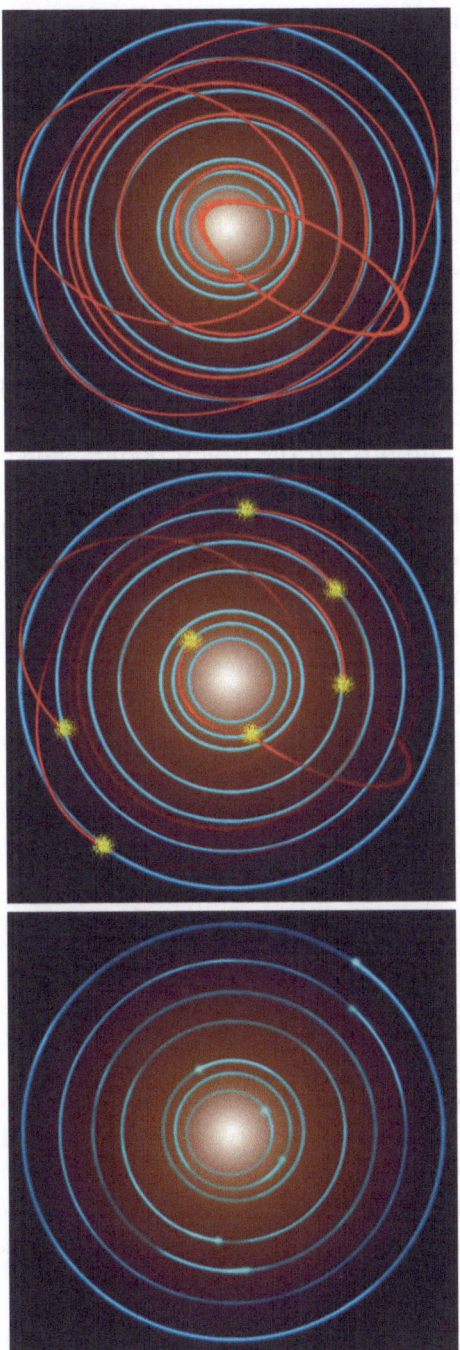

Figure 2.5   By 3.8 billion years ago, most of the large planetesimals (red) that did not follow the concentric orbits of the major planets (blue) had been "cleaned out" by collisions. (art by the author)

planetesimals collided; there were frequent near-misses, and these hurled the embryonic planets into disheveled orbits, sending them on paths to new collisions or voyages toward the outer fringes of the system.

Initially, scientists thought particles grew progressively larger, gaining steadily in mass. Particles wandering close to a forming planetesimal would be pulled into the Hill radius, the area influenced by the gravity surrounding an object. As their mass increased, so did their gravity, completing an ongoing—and ever-growing—cycle. This was the picture painted by Kant and other early theoreticians. The dust of solid minerals was thought to concentrate in the midplain of the solar nebula, where it stuck together into multi-mile/km sized planetesimals. The larger of these continued to expand—in a process called oligarchic growth—becoming planetary embryos the size of the Moon or Mars. Random collisions left the largest bodies, which became the terrestrial worlds. The infant planets continued to pull material into themselves from their surrounding feeding zones, clearing paths through the solar nebula.

Planetesimals circling the Sun four and a half billion years ago had one of two likely fates: either to bump into the growing core of a planet and become part of it, or to be scattered by that core and ejected from the Solar System completely (or tossed into the outer nether regions of the system). Of all the objects that were kicked outward, at least 90% left the Solar System. Today, they are off drifting among the stars as interstellar asteroids or comets (the likes of which we have seen only twice[9]). A small fraction, less than 10% of those ejected bodies, were redirected a second time by nearby stars or vagabond worlds in the outskirts and captured into the Oort cloud. As we saw with members of the Kuiper Belt, the Oort Cloud inhabitants began closer to the Sun, and then were scattered out by the planetary snowplow events. But first, there had to be planets.

The rise of the planets

The rise of the planets seems fairly orderly, but the devil is in the details. The transition from dust to planetary embryo is not so straightforward. One of the stickiest problems with the scenario is called the meter problem. As primordial flotsam increases to the meter range, objects tend to bounce off each other rather than sticking together. The problem is exacerbated with larger

---

[9] The bizarre interstellar object Oumuamua passed through the solar system in 2017. The asteroid or comet has a remarkably elongated shape, with estimates for its long axis ranging from 100-1000 meters (although it may be flattened like a pancake). It is reddish like many Kuiper Belt objects. In 2019, a Crimean astronomer spied a comet on such a fast approach to the inner solar system that it could not possibly belong to this system. The comet is called 21/Borisov.

planetesimals possessing higher gravity, making impact speeds faster. The question becomes, *Why did anything grow larger than a meter?* Or, more simply, *why are there planets at all?* The pebble accretion model calls for something to slow down those interactions, but what force could do that?

A host of astronomers have been looking into the problem. William K. Hartmann was one of the earliest. Hartmann carried out experiments at the Planetary Science Institute and at the vertical gun facility at NASA's Ames research center. "Everybody else was using that facility to try to study higher and higher velocity collisions, but, being me, I came in asking for the lowest possible velocities!" The important result from the series of laboratory trials, says Hartmann, is that "if two bare, solid rocks collide at low velocity they tend to bounce off each other. If they are irregular, some energy is lost in rotation and and they fall back together." So irregular shape may account for a fraction of the planetesimals combining. But it can't account for the majority, Hartmann finds. "However, through statistical or electrostatic attraction, small particles can form dust layers of 'regolith' on the surface of larger particles. And the big result of my simple experiments was that if you have a surface regolith layer of particulate material on the surface of a "target" rock, even only 1/10 the depth of the slow impactor's diameter, the impactor loses energy moving the granular stuff around, so the impactor (even at speeds larger than escape velocity of the target) tends to bury itself in the regolith of the target rather than bounce off." Hartmann suggests making a comparison of dropping a rock onto a clean concrete surface and then a concrete surface with a thin layer of dirt. A layer of regolith—finely crushed rock dirt—can account for some of the congregating particles as well.

Asteroid expert Dan Durda has been observing high-speed impacts using the Southwest Research Center's gun facility in San Antonio, Texas. His experiments were designed to investigate how regolith might form from impacting bodies. Durda's team shot a one meter diameter sphere of granite with a two centimeter diameter aluminum ball at a speed of 2 km/sec. "We witnessed the process firsthand," Durda says. The impact test resulted in a "Star Wars-style Death Star crater; half the hemisphere was blown away. But looking at the fracture mechanics around the side of the crater, you got these beautiful radial fractures and concentric fractures and little spall sheets that wanted to come off. A lot of them were barely hanging on."

Between some of the fractures, a tiny chip of rock had become trapped, confined by the other bits of rock that didn't quite peel away. The fragment was, in effect, trapped within the cage of the other fragments. "That would have provided next impact with the opportunity to make loose stuff. If it had been 1 kilometer sphere, that's the beginning of making regolith on a small object. You almost can't *not* make regolith if the object has at least some gravity to hold it." Durda's results imply that "you don't need to start with a loose rubble pile like Bennu or Ryugu (recently visited asteroids). Even with a solid, monolithic boulder, you will build regolith over time. You almost can't avoid it." The gun impact results demonstrate that regolith—with its cushioning effects on collisions—began early in the process of planetary formation.

Saturn's moons may provide us with a window into the process of a regolith cushion and its effect upon collecting material to itself. Two of the small Saturnian moons orbiting in the same plane as the rings wear bands of material around their equators (Figure 2.6). Images taken in 2005 by the Cassini Saturn orbiter reveal Atlas as a roughly spherical rock pile at the center of a smooth equatorial ridge. The little moon is sweeping up material from the rings, and the fine dust accumulates around the equator. The "sweeping up" echoes the accretion that commonly occurred in the primordial solar system as planetoids grew. The moon Pan, just 35 x 23 kilometers, shows similarities, and in addition leaves a gap in the rings as it arcs through them, called the Enke Gap. We have seen similar pathways cleared in the disks of distant stars, where planets or planetesimals may be tidying material from the solar nebula as they grow. In the center of the Enke gap, Pan trails a stream of debris which either contributes to— or issues from— the equatorial material around Pan's midriff (Figure 2.7).

Formation of a new moon may be taking place as we speak, at the outer edge of Saturn's A ring. Cassini images reveal a bright clump of material 1200 km long, moving around the ring. Nicknamed "Peggy", the clump likely is the sign of a moon forming within the ring particles (Figure 2.8). If a moon does exist at the brightened location, its diameter is estimated to be only half a mile/1 km across. But it's enough to disturb particles around it, making it visible from a distance. The forming moon may be in the process of migrating completely out of the ring system, as other moons may have done in the past. Peggy may be a case of accretion in action.

In addition to irregular shapes and a cushion of regolith, a third factor influencing accretion may be found in the particles and gases surrounding planetesimals. Researchers recently put forward a response to the meter problem,[10] a sort of slow-motion version of the classic model. It is elegantly called the *streaming instability model*. In this version, gases within the spinning solar nebula dampen the momentum of pebble-sized particles. Some continue to

---

[10] *Science*, February 28, 2022

Figure 2.6   Cassini images of Atlas (above) and Pan, seen from above and in profile, betray a process of accretion reminiscent of the early solar system. (Cassini images courtesy NASA/JPL-Caltech)

Figure 2.7 The central cores of Saturn's small moons Pan and Atlas are likely rubble piles or more solid remnant asteroid material, either captured wanderers or formed on site as Saturn condensed into planet and moons. The satellites have accreted fine dust around them, forming a belt of fine material around their equators, as seen in this cut-away view (top left). As Pan circles Saturn, it clears a gap in the rings known as the Enke Gap (lower left). Similar gaps have been seen in the accretion disks surrounding young suns. At right, a thin, bright trail of debris follows Pan through the dark Enke Gap. In this view, Pan is off of frame. The gas in protoplanetary disks would prevent such trails associated with forming planets. (top art by author; bottom left and right photos, Cassini images courtesy NASA/JPL-Caltech)

speed along, while others drag behind. The resulting streams of material would be unstable, rapidly collapsing into hordes of larger objects. Many of these objects would circle each other closely, spiraling into each other and coming into contact at low speeds. The result would be a double lobed object, just as we see in Arrokoth. Computer models have simulated embryo growth by pebble accretion,[11] and when gases are present to slow the incoming

---

[11] See Levison, et al: *Growing the terrestrial planets from the gradual accumulation of submeter-sized objects,* Proc. Natl. Acad. Of Sciences, volume 112, 2015

Figure 2.8 A disturbance in the rings may mark a moon's formation. The bright trail of material may follow a small, forming moon, which researchers have nicknamed Peggy. (Cassini image courtesy NASA/JPL/Space science Institute)

particles or fragments, the models result in planetary systems that resemble the solar system's outer and terrestrial planets.

A familiar ring to it

Saturn's glorious ring system—the most extensive in our solar system—affords us a hint of the early days of planet Earth's formation. "Saturn's ring system is not just an object of beauty," says Larry Esposito of the Laboratory for Atmospheric Space Physics in Boulder, Colorado, "but a complicated physical system that provides a local laboratory and analogy for other flat cosmic systems like spiral galaxies and planet-forming disks."

Josh Colwell[12] agrees. Colwell is an expert in the fields of the origin and evolution of the solar system and its planets. His studies have also made him an expert on planetary rings, and he was a Co-Investigator on the Cassini Saturn mission. Colwell suggests that many of the same physical processes occurring within the rings will be seen in a protoplanetary disk. The first

[12] Colwell is professor of physics at the University of Central Florida.

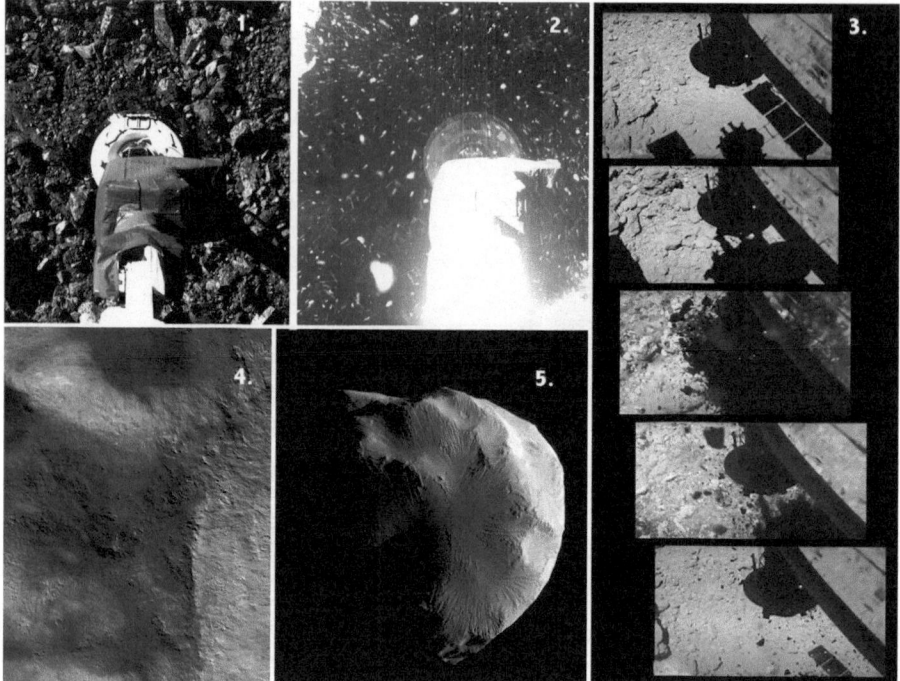

Figure 2.9 (1) The regolith on the surface of the small rubble-pile asteroid Bennu before and (2) after the touchdown of the Osiris-Rex sample return spacecraft. The loose nature of the regolith is apparent as the craft departs. (NASA/Goddard/University of Arizona) (3) February 22, 2019, the Japanese Space Agency's Hayabusa II gently touched down on the asteroid Ryugu. In the top view, the spacecraft shadow is visible at the bottom of the frame on final approach. In the third frame from top, rocks spray out from the sample retrieval "foot". Rocks continue to float around the craft as it backs away (bottom frames). The sample capsule was recovered on Earth in December of 2020. (courtesy JAXA) (4): regolith on larger asteroids, where electrostatic forces influence the fine material, can pool in topographically low areas, as is evident in this ponded material on the flank of Marcia Crater on the asteroid Vesta. (NASA/JPL-Caltech/UCLA/MPS/DLR/IDA) (5): Saturn's moon Helene exhibits multiple types of regolith. Note the flows of material, despite the low gravity. (NASA/JPL/Space Science Institute)

major shared trait is gaps. Gaps in the rings of Saturn are caused by either the cleanup of small moons or by the gravity of larger moons farther out. But in the case of a planetary accretion disk, Colwell explains, "because it's a planet orbiting a star, there are two players in the game, and the dynamics are such that the planet opens a gap. When we see moons opening gaps in the Saturn rings, we're seeing a similar dynamic."

Incomplete gaps, known as "propellers", provide another analog. Smaller bodies within the ring material cause propellers within Saturn's rings. "We see these little moonlets too small to open a complete gap, but big enough to open an incomplete gap. We see them drifting due to their interactions with neighboring ring material. That's a type of migration that occurs in the proto-planetary disks as well, with larger objects interacting with the gas and dust in the disk and migrating."

A third parallel involves aggregation or clumps of material, Colwell says. "We do see ephemeral clumps. Things can't accrete into very big objects because they're too close to Saturn. The tidal forces are just too large." Instead, observers see aggregations due to gravitational instabilities that cause clumps. "That's a big part of physics that explains the formation of planetesimals, the first building blocks of planets. On a larger scale, these would lead to planets, but at Saturn, the tidal forces of the planet prevent moon formation."

This phenomenon—Saturn's gravity preventing the formation of large moons within its rings—underlines the fact that there are also major differences between protoplanetary disks and the Saturnian rings. One obvious difference has to do with distance to the center of the cloud. In the case of Saturn's rings, the ring particles are close to the planet, within Saturn's Roche limit (a distance too close to the planet for moons to form). But protoplanetary disks remain well outside of the Roche zone of their star, allowing for a stable gravity environment of planetary formation.

A second important distinction relates to gas mixed in with dust particles. Saturn's ring system consists of ice particles ranging from the consistency of cigarette smoke to chunks over a meter across. Within protoplanetary disks, similar particles are suspended within gas, which has a profound effect on their movement and ability to stick to each other, and in their ability to accrete to form larger particles. The gas is the dominant part of the mass of the pro-toplanetary disk. "In Saturn's rings there's no gas," Colwell says. "A lot of the physics of the protoplanetary disk involves gas pressure pushing particles around, and we don't have that at Saturn at all."

Colwell says that the current thinking is that our planetary nebula pro-gressed from pebble-sized particles to planetesimals quite rapidly, on the order of thousands or a few million years. It was thought that turbulence from the all-pervasive gases would prevent the self-gravity of particles from being able to coalesce into a planetesimal, because "the gas is blowing everything around so it can't settle. Then the (theory of) streaming instability came along. The analogy is like the Tour de France peloton. Everybody's in a front group, but behind that front row of riders, the people behind aren't feeling the drag force because they get a sort of windscreen. You sort of get a windscreen in the disk that enables particles behind to glob together to form a planetesimal. This is probably how the meter barrier is overcome. You go from tiny particles to a planetesimal pretty quickly. But there is no streaming instability in Saturn's rings.

Despite the disruption of Saturn's tidal forces upon its close-in ring system, observers do see evidence of some limited accretion in Saturn's rings. This suggests a sort of particle sticking, that there is more than gravity holding the particles together. This same phenomenon would occur in planetary disks.

One of the older ring theories suggests that the rings formed 4 ½ billion years ago, and spread naturally. As the outer edge moved out, ring particles formed objects like Peggy. Eventually the moon a ring's edge would become large enough that it detached from the disk. The model proposes that this is the process that led to so many little moons like Pandora and Prometheus and Atlas. These moons, then, would issue from Saturn's most ancient rings, which were very massive. As the ancient ring system evolved, so the theory goes, it was spitting new moons out from the edge like a conveyor belt. "It's a very elegant theory," says Colwell, "but it's also not super-detailed. It has appealing elements, but there are all sorts of difficulties." The ancient ring hypothesis and others like it continue to attempt to explain all the nuances of planetary accretion.

Fleeting beauty

Saturn's rings may have formed during the Jurassic or Cretaceous periods, when stegosaurs and tyrannosaurs ruled the Earth. "There is a lot of circumstantial evidence that the rings are very young, perhaps less than 100 million years old," says Cassini's Josh Colwell. The ring particles are very reflective, very bright. Our understanding of micrometeorite bombardment on the rings is that the process would darken the ice particles from pure, highly reflective water ice to their current level in less than 100 million years. This is the strongest piece of evidence for ring youth. And new research[13] suggests that they will not be around for a geologically long time. A team of observers charted the migration of Saturn's ring particles using the Keck Telescope at Hawaii's Mauna Kea observatory. The observers, led by James O'Donoghue of NASA's Goddard Spaceflight Center, combined their observations with data from the Cassini mission. The team concludes that ring systems are "temporary features."

O'Donoghue's colleagues discovered a sort of "ring rain", a hail of icy particles dropping from the inner edge of the ring system onto the planet itself. There's plenty of material to continue this cosmic rainstorm; Saturn's rings

---

[13] *Icarus*, Volume 322, April 2019; "Observations of the chemical and thermal response of 'ring rain' on Saturn's ionosphere" by O'Donoghue, et al.

Figure 2.10 Less than 100 million years ago, an event led to the spectacular ring system surrounding Saturn. The birth of the rings may have been triggered by the breakup of a moon that spiraled in toward Saturn reaching the planet's "Roche limit" where gravitational forces demolished it. Judging by the mass of the rings today, the moon may have been as large as Saturn's third-largest moon, Iapetus (1469 km/913 mi). If so, the moon may have had a complex structure, perhaps holding a subsurface ocean, as many of Saturn's other moons do. (art by the author)

span thousands of kilometers from inner to outer edge, and may contain half as much ice as the entire Antarctic ice shelf. The mass is equal to half that of Saturn's moon Mimas. The Goddard team estimates that the rings lose 2 tons of water ice every second (enough to fill an Olympic-sized swimming pool every half hour). Ultraviolet sunlight causes the orbital downpour. UV light imparts an electrical charge to the ring particles. The ice particles are trapped within Saturn's powerful magnetic field, which carries them along the magnetic field lines, down toward the planet. At the present rate, the team estimates that the rings will disappear within the next 100-300 million years. As the book of Proverbs reminds us, "beauty is fleeting."

Although Saturn's rings may look like an accretion disk, they are actually an example of the opposite process: rather than planetoids coming together, the rings are the result of something coming apart. An ancient moon of Saturn likely spiraled in toward the planet, eventually reaching a region known as the Roche limit. Here, gravitational forces between Saturn (and possibly some of Saturn's larger moons) tore it apart, spreading its fragments into the vast ring system we see today (Figure 2.10). A 2022 MIT study asserts that the destruction of the moon—unofficially dubbed Chrysalis—took place in the geologically recent past, perhaps just 160 million years ago.[14] This timeline would explain the pristine, "fresh" appearance of Saturn's rings. The other giant planets, Jupiter, Uranus and Neptune, all have ring systems as well, but all are in various stages of decay. The study submits that Chrysalis had as much mass as Saturn's third largest moon Iapetus, and that some 99% of the moon ended up incinerated in Saturn's atmosphere. The rest remained aloft in the form of the planet's golden rings.

There may be even more to the Chrysalis story. Saturn's "moment of inertia", the amount of energy it takes to tip over a spinning planet, is about what it would be if Saturn were in resonance with far-away Neptune.[15] The gravity of Chrysalis could have teamed up with the gravitational forces of Saturn's other large moons—most notably Titan, Iapetus, and Rhea—forcing Saturn into its tilt and forcing it out of its resonance with Neptune, some studies show. As Titan spiraled its way out from Saturn, something moons tend to do naturally, the planet-sized moon eventually synced up with Chrysalis, disrupting its orbit and sending it into Saturn's Roche limit, where Saturn's gravity tore it apart. Skeptics of the scenario point out that of 390 computer simulations, only 17 resulted in Chrysalis becoming a ring. But despite their different genesis, the rings of Saturn still stand as a glorious portrait of planetary formation, a complex dance of gravity, swirling clouds of debris and gas, and convoluted interactions of physics and chance.

Accretion interrupted

Because of planetary migration, Mars was unable to grow as large as Venus and Earth. The raw materials that would have built it into a larger world are scattered between Mars and Jupiter in the asteroid belt. But it wasn't for lack of trying. In fact, it appears that many asteroids grew to the size of small planets. They were large enough to differentiate, forming a light crust and heavier core. Laboratory research using chemical and isotope investigations of nickel/iron meteorites has revealed that at least 50 distinct parent bodies were the sources of the meteorites. In other words, there were once at least 50 asteroids

---

[14] Analysts arrive at the number by calculating how quickly Titan is moving away from Saturn, a process that presumably was influenced by Chrysalis' demise.

[15] The long-term wobble of a planet's axis is called precession. Saturn's precession is nearly identical to the precession of Neptune's entire orbit, which also wobbles like a bobbing hoop.

Asteroid Belt architecture

The band of rock and metal circling the Sun between Mars and Jupiter is not a uniform crowd. Some asteroids are great mountains of stone; others glisten as massive bulwarks of iron and nickel metals. Still others may contain abundant water in the form of ice and minerals, and smaller ones take on the form of loosely confined rubble heaps. The largest asteroids orbit in distinct clusters, but the smaller asteroids have been mixed by planetary migration and the influence of gravity from surrounding planets—most notably Jupiter. Jupiter's mighty gravity may well be the force that prevented the asteroids from combining with Mars to form a much larger planet.

The main section of the asteroid belt circles the Sun between the orbits of Mars and Jupiter, extending 225 million km from the inside edge to the outside boundary. Astronomers divide the asteroids into eight groups, or regions, each named after a dominant asteroid or asteroid type within their group: the Hungarias, Floras, Phocaea, Koronis, Eos, Themis, Cybeles and Hildas. Although the asteroids do not orbit in a flat disk, but rather in a donut-shaped torus, their numbers are separated by gaps, just as the rings of Saturn are separated by the gravity of moons. The gaps within the asteroid belt—known as the Kirkwood gaps—are pulled apart by the gravity of Jupiter. Jupiter itself is trailed in its orbit by the Jovian Trojans, and is led by another group of asteroids called the Greeks. Each grouping orbits within a stable point in Jupiter's orbit called a Lagrangian point. The Trojans and Greeks measuring over 1 km diameter are nearly two million strong.[16] If Jupiter's Trojans and Greeks make up the top two forks of a "Y" and Jupiter marks the center, on the opposite side of Jupiter's orbit (at the "base" of the Y) scatters another field of asteroids called the Hildas triangle. The Hildas, Trojan, and Greek groups move around Jupiter's orbit at constant gaps, remaining in a stable configuration as Jupiter orbits the Sun. Still farther out, between the orbits of Jupiter and Neptune, the Centaurs spread around the Sun in a ring ruled by the gravity of the giant planets. Centaurs are comet-like, being rich in volatiles (elements that turn readily to vapor). The planet Neptune itself has its own Trojans, though not nearly as densely packed as those at Jupiter.

Sunward of the main belt, the planet Mars is followed by the Mars Trojans, a cloud of asteroids that follow behind and lead Mars at 60° intervals along the red planet's orbit. Another group of scattered asteroids, called Near-Earth Asteroids (or NEAs), swoop inward of Mars and cross the orbit of the Earth. Some 29,000 Near-Earth Asteroids are currently being tracked. Ones with diameters greater than 460 feet/140m are considered potentially hazardous to Earth. Objects smaller than an asteroid or comet are referred to as meteoroids. As a meteoroid enters the atmosphere and leaves a burning trail, it becomes a meteor. If it makes it to the surface, it's called a meteorite.

---

[16] This number may be too high; several recent studies put the total Trojans over 1 km diameter at closer to 10,000, depending on their distribution and brightness, both of which are unknown quantities.

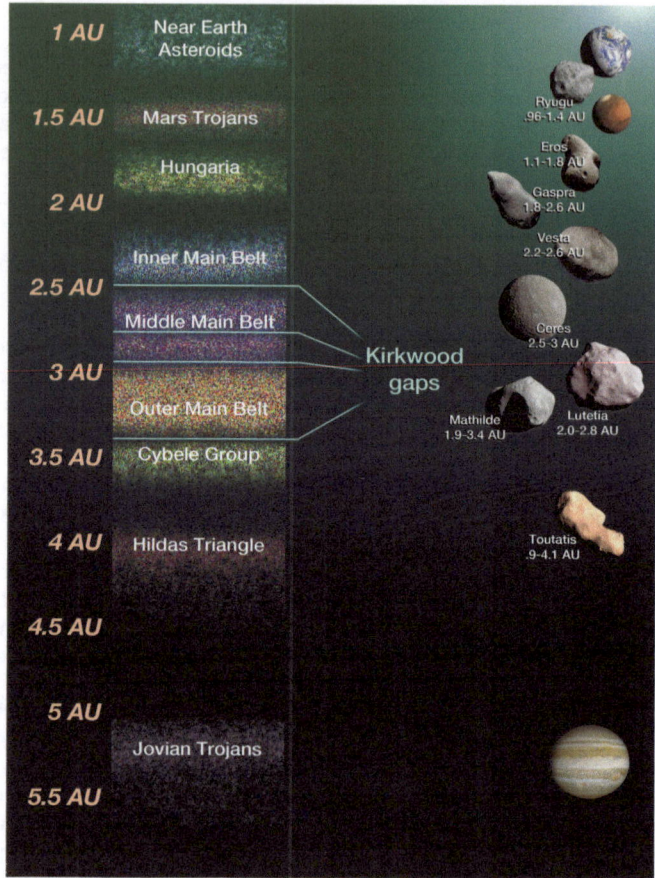

Figure 2.11 The various asteroid groups shown by distance from the Sun in Astronomical Units (scale at left). The Kirkwood gaps are caused by Jupiter's gravity. An assortment of asteroids is seen at right (not to scale), along with their distances from the Sun in Astronomical Units. (art by the author)

We would love to know more about this unique moment in solar system history, a singular epoch when planets emerged from rubble. To that end, space-faring nations have dispatched spacecraft to scrutinize the asteroids and comets (Figure 2.9). Among them have been missions to actually return fragments of these primordial planet building blocks for study here on terra firma. Our search for ancient solar system samples began when the Japanese Space Agency (JAXA) dispatched its Hayabusa craft to the asteroid Itokawa, a pile of dust and rock just 513 meters long. Despite some engineering challenges, Hayabusa successfully returned some micro-particles from Itokawa.

Hayabusa was followed by the island nation's second asteroid sample return mission, Hayabusa II, an emissary to the surface of the asteroid Ryugu. After deploying a series of miniature surface rovers, the main spacecraft managed to grab 5.4 grams of pristine rock and dust from the diamond-shaped rubble pile, returning the samples to Earth in 2020. Analysis reveals that the bits of Ryugu are perhaps the most primitive residue from the solar system ever studied. Some of the particles actually predate the formation of the Sun. Chemically, the Ryugu debris closely matches carbonaceous chondrite meteorites,[17] the most ancient known meteorites. These meteorites are so rare that less than ten have been found on Earth so far, although the total number is under debate. But the chondrites found on Earth have been modified by weather and chemical reactions. Hayabusa II's asteroid samples are untainted by atmosphere or human contact, offering us a pristine record of the early planetary nebula where Earth and its siblings arose. Isotopic studies of the asteroid samples suggest that Ryugu formed in the cold, dark gases of the presolar nebula's outskirts. Some grains of material contain carbon-rich organic matter, called Aliphatic organics. Researchers have charted over 20 amino acids within the asteroid specimens. Amino acids, the building blocks of life-related enzymes and proteins, can arise under a variety of conditions.

The OSIRIS-REx (Origins, Spectral Interpretation, Resource Identification, Security, Regolith Explorer) Mission carried out its assignation with the Near Earth Asteroid Bennu on October 20, 2020 after a reconnaissance lasting 22 months. The craft touched down and successfully retrieved samples, which are currently en route back to Earth for a scheduled landfall September of 2023. The Bennu samples are likely as old as those returned from Ryugu.

Images returned by both Hayabusa II and OSIRIS-Rex look like rugged, alpine landscapes. But in the strange microgravity environs of the asteroids, looks can be deceiving. Some of the close-up vistas viewed by recent asteroid missions are evocative of the sites explored by the Apollo astronauts, says Southwest Research Institute scientist Dan Durda. "It looks like a pile of rocks that you could hike through, but that's not what is going on at all. It would be more akin to those funhouse tents full of plastic balls. If you went to one of these regoliths, you could probably just sink yourself into it; spread your arms and just crawl in." Durda's experiments in microgravity demonstrated to him that, "just the barest of contact sends things flying."

We have other asteroid samples, and these are meteorites. Multiple amino acids have been detected in the famous carbonaceous chondrite Murchison meteorite. These include nine hydroxyl amino acids and seven diamino acids. Many are racemic, meaning that their structure has a mirror-image symmetry that is both "right-handed" and "'left-handed". This "handedness" is called

---

[17] also known as C1 meteorites

chirality. Although amino acids exist in both left- and right-handed arrangements, terrestrial life is made up almost entirely of the left-handed flavor (for more details on chirality, see Chapter 5). Biogenically related terrestrial amino acids are left-handed, so these are clearly extraterrestrial in origin. Some are complex, with from two to ten carbon atoms in length. Experts estimate that Murchison contains thousands of individual amino acid types. Tests show that these compounds were enriched in certain isotopes, proving that they are of extraterrestrial nature rather than contamination from Earth. Searches for organic material in other meteorites have found many types of organic matter. Outstanding examples come not only from Murchison, but also Allende (the meteorite that fell over Chihuahua, Mexico), and the Murray meteorite from Lake Murray in Oklahoma, US. Amino acid studies have gotten a huge boost with the recent meteorite searches in Antarctica. There, meteorites are rafted to the icy plains from higher ground, and then eroded out of the ice by Antarctica's furious winds, exposing them as obvious solitary stones on the frozen surface. The organics found in carbonaceous chondrite meteorites reflect a diverse history of meteor formation, starting during the formation of the solar nebula cloud (where conditions were cool) and moving into much hotter epochs of planetesimal formation, as found in the Genesis, Hayabusa and OSIRIS REx samples (Figure 2.12).

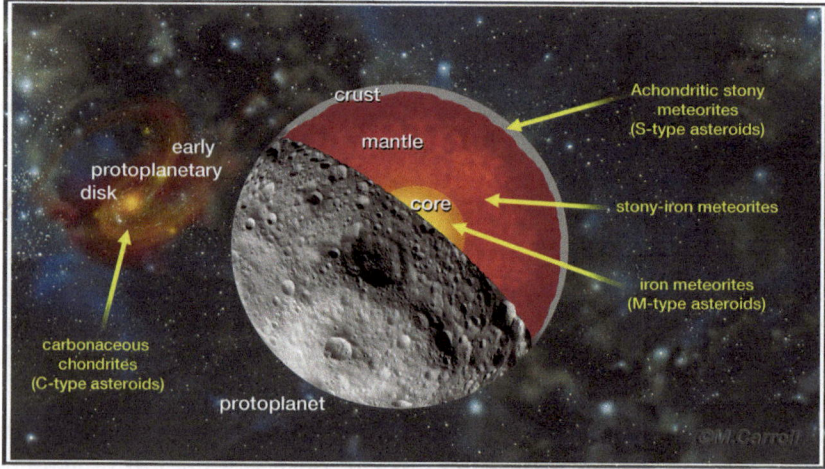

Figure 2.12 As asteroids gain mass and become protoplanets, they begin to differentiate. Internal heat melts rock and metal, and the heavy material sinks to form a core. If the planetary embryo is then destroyed by an impact, its remnants become different types of meteor fragments. Scientists have identified at least 50 distinct past embryos as sources of today's asteroids. Chondrites form independently, outside of any developing planet. They are probably older than other types. (art by the author)

Exchange Program

It was the only pond as far as the eye could see. Most of the surface water had frozen solid and would remain solid with the coming ice age and thinning of the Martian air. Along the margins of some of the ice fields, or by the rims of the dying geothermal vents, trickles of water drew borders between rock and glacier. But the rivers and lakes and seas were long gone.

The meteor announced itself as a scorching flash across the gunmetal sky. It plowed into the rusted plains, vaporizing rock and hurling stones, pebbles and water vapor high above the desert landscape. One of the ejected rocks—the size of a basketball—made it up beyond the atmosphere and eventually out of Mars' gravity field. But its long journey had just begun. After tens of millions of years of drifting sunward, the rock entered Earth's sky, trailing sparks, plasma and molten rock as it slammed into the ground in a lonely arctic desert. It was not the first meteor to make the trip from Mars to Earth. Others had, and in fact quite a few meteorites had come from other celestial objects. Even the Earth, with its more selfish gravity, still gave up a few of its own rocks, sending them as meteorites all the way to Mars.

Today, some 100 tons of meteoritic material rains down from Earth's skies daily. Some of it comes from the nearby Moon, the asteroid Vesta, Mars and an assortment of other asteroids, comets and even more distant moons. In some cases, isotopes or spectra can identify their sources. The solar system, it seems, has delivered to us samples of distant worlds, free of charge.

Frigid frontiers

Once the planets and asteroids of the inner system had settled down from their wandering ways, things became more organized in the outer system as well. Beyond the arena of the nomadic asteroids and marching ice giants lies the reservoir of the short period comets like Comet Halley and 67P/Churyumov-Gerasimenko. These comets visit the inner solar system on the order of years or centuries, sailing through the planets on oblong paths. Even the more distant of the Kuiper Belt objects orbit in elongated ellipses rather than the fairly circular orbits typical of the planets (Figure 2.13). Pluto travels around the Sun two times for every three circuits made by Neptune in one of those famous resonances. Pluto's is a 3:2 resonance, and nearly ¼ of all known KBOs are in this specific orbital relationship. They are referred to as Plutinos.

Researchers estimate that some 1400 Plutinos have diameters greater than 95 kilometers. This compares to just a few percent of the total material in the Kuiper Belt. Though they are small in size, they are mighty in number.

Even farther out, beyond the Kuiper belt, a distant sphere of comets encircles our entire solar system, extending perhaps as far as a quarter of the way to the nearest star. The Oort Cloud is left over detritus from the formation of the

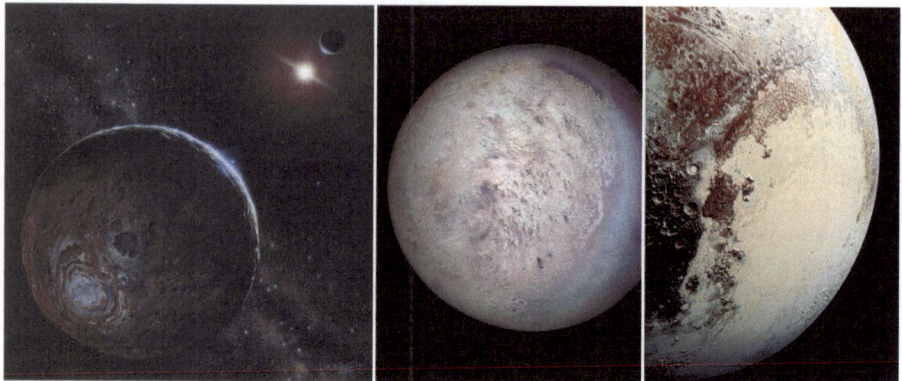

Figure 2.13   (Left) Thanks to planetary migration, material from the Kuiper Belt probably scattered into the inner solar system, where it fed the Earth and the other terrestrials. Today, over a thousand Kuiper Belt Objects span greater than 100 km, large enough to become spherical. Many may have one or more moons. Neptune's moon Triton (center) is likely a captured KBO. Pluto (right) is an example of the larger KBOs, with its planet-like geology, dynamic atmosphere and multiple moons. (Painting by the author; Triton Voyager image NASA/JPL; New Horizons Pluto image NASA/Johns Hopkins University Applied Physics Laboratory/Southwest Research Institute)

solar system, too far away to have flattened out like the solar nebula. Like members of the Kuiper Belt, the Oort cloud is populated by great floating icebergs, frozen mountains lurking in the darkness. Occasionally, one of them is disturbed by the gravity of a close encounter with another, or perhaps by a passing rogue planet or star. These encounters may toss the object toward the inner solar system, where—for the first time in its existence—it feels the heat of the Sun and a transformation occurs. The object warms, begins to vaporize, and develops a coma and tail, becoming a long period comet. Comets that come in from the Oort cloud tend to be rich in volatiles, as they have spent their lives frozen in the cosmic wilderness. They often outshine periodic, or "short period", comets, whose volatiles are warmed with each pass through the inner solar system, typically repeating in periods of less than 200 years. The long period Oort Cloud comets have orbital periods lasting from centuries to millions of years.

Comets are on the list of important vacation spots for spacecraft as well. Stardust's successful mission to comet Wild 2 (Chapter 1) was not alone. In 1999, an international group of scientists and engineers combined forces to convey a regatta of spacecraft to Comet Halley. Other comet encounters took place in the form of extended missions whose primary flights had been accomplished (see Appendix D for the full list). In 2004, the year of Stardust's comet encounter, the European Space Agency dispatched its Rosetta comet orbiter,

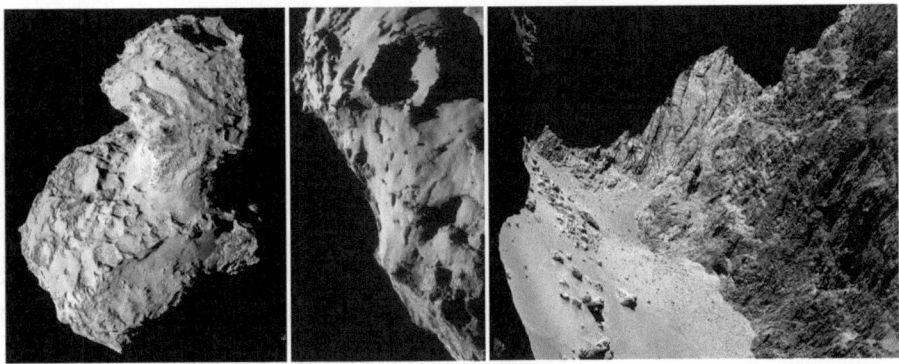

Figure 2.14 Left: Comet 67P Churyumov-Gerasimenko seen through the eyes of Rosetta from 285 km. The rubber duck shape is evident from this angle. Center: Rosetta charted alien features like these sunken pits, perhaps formations left over from the comet's geyser-like jets of gas and dust. Right: Looking like an alpine vista, Comet 67P displays peaks, valleys, canyons and crags. (Images courtesy ESA/Rosetta/MPS for OSIRIS Team MPS/UPD/LAM/IAA/SSO/INTA/UPM/DASP/IDA)

which entered orbit around the comet 67P[18] in August 2014. After observing the comet, studying its geology and activity, and searching for landing sites, Rosetta deployed a lander called Philae to carry out in situ measurements of the comet's surface.

Remarkable Rosetta: science from above a comet

As we consider the many asteroids and comets that threaten the Earth (see **DARTs in Space**, below), it is important to not only chart their courses, but also to understand their composition, their densities and other physical traits. To that end, the European Space Agency mounted a daring expedition to follow a comet through much of its orbit, observe it as its activity increased and ebbed, and sample the comet's material from orbit and on the surface (Figure 2.14).

Rosetta found that 67P has no intrinsic magnetic field. Solar wind flows around the nucleus, but does not interact with any magnetic "bubble" generated by the comet itself. The discovery reveals that at the time Comet Churyumov-Gerasimenko formed within the solar nebula, the surrounding magnetic fields coming from the infant Sun were too weak to magnetize the dust particles that would eventually make up the comet. This helps

---

[18] Comet Churyumov-Gerasimenko

astronomers to draw boundaries around when and where the comet formed, and confirms that its constituents came from the outer nebula.

Rosetta's sampling of the comet's surrounding coma demonstrated that the elements within the coma match the abundances of elements of the solar system in general, intimating that 67P is very ancient, dating back to the birth of the Sun and formation of our planetary system. The ingredients pouring from the comet's surface varied by location and spacecraft distance. Investigators suggest that either the gases undergo differing chemical reaction at different distances from the nucleus, or that the nucleus itself is outgassing varying mixes from different locations. If the latter is true, then the comet is made of assorted compounds unevenly mixed throughout. This may mean that comets like 67P accrete from dissimilar materials existing at different locales within the solar nebula.

Nearly half the mass of the comet's dust consisted of carbon-based, complex molecules resembling those found in the most primitive meteorites, the carbonaceous chondrites. These types of molecules may have played a critical role in the biogenesis of Earth. Other organic compounds included ammonia, methylamine, and ethylamine. Hydrocarbons were also plentiful, and would have made a successful backyard grill. They included methane, ethane, acetone, propane, butane and others. In all, instruments charted 16 organic compounds, along with water vapor, carbon dioxide and carbon monoxide. Among Rosetta's findings, astrobiologists were intrigued to discover the amino acid glycine, commonly found in proteins, and the presence of phosphorous, a critical component of DNA and cell membranes.

University of Bern researcher Katherine Altwegg, a principal investigator on the Rosetta team, held an early press conference in which she compared the comet's compounds to cologne. "The perfume of 67P is quite strong, with the odor of rotten eggs (hydrogen sulfide), horse stable (ammonia), and the pungent suffocating odor of formaldehyde. This is mixed with the faint, bitter, almond-like aroma of hydrogen cyanide. Add some whiff of alcohol (methanol) paired with the vinegar-like aroma of sulfur dioxide and a hint of the sweet aromatic scent of carbon disulfide, and you arrive at the 'prefume' of our comet."[19] Her imaginative description emphasizes the toxic environment of comets, despite the fact that their chemistry includes life's construction materials.

---

[19] As quoted by Peter Bond in his book *Rosetta: the remarkable story of Europe's comet explorer* (Springer Praxis, 2020)

<u>Philae; landfall on a new kind of world</u>

The Philae comet lander was dispatched from its Rosetta mother ship (which had been orbiting the comet since August of 2014) in November 12 of the same year (Figure 2.15). Upon landing, the craft was designed to fire a small

Figure 2.15 Top left: Philae lander image of the Agilkia site—its intended landing site—from an altitude of 51 meters. Part of Philae's landing leg is seen at upper right. Top right: view during the lander's first bounce, which sent the craft on a two hour, kilometer-long detour to its ultimate resting site. Bottom: Philae settled in a rugged hollow at Abydos. Its panorama revealed a craggy terrain of pebbles, stony blocks, cliff faces and linear cracks across the surface. Philae's perch against a cliff left one of three landing legs dangling, one of seven CIVA cameras pointing at the sky, and its solar panels in almost complete shadow, but the craft prevailed. (Images courtesy ESA/Rosetta/Philae/CIVA via NASA Planetary Photojournal.)

rocket mounted to its top; at the same time it would fire harpoons into the surface to secure itself in the low gravity. At the end of its 7½ hour descent, after making a pinpoint touchdown within its targeted landing ellipse, the lander's harpoons and thruster malfunctioned, and Philae bounced off the surface. Despite its slow descent, it nearly reached escape velocity.[20] It began its science investigations as soon as it detected the surface, despite the fact that it had immediately begun pirouetting away from the surface. For over 90 minutes, the craft sailed for a kilometer over the rugged comet nucleus. It skimmed the surface a second time, and then lodged itself against a rock face after another seven-minute-bounce. Remarkably, the Philae lander and its entire suite of instruments survived its descent, at least two bounces, and final landing almost fully intact, with only minor damage to one footpad's support strut.[21] Throughout its bouncing flight from the Agilka region to its final landing site at Abydos, and for its 56 hour 28 minute long surface reconnaissance at Abydos, the intrepid lander performed almost exclusively on battery power. Sheltered from sunlight against a cliff wall, rendering its solar panels almost useless, Philae's science sequence had to be revamped on the fly. Mission designers sent commands to curtail some experiments, skip others, and adapt the mission operation into a prepared emergency sequence. Every instrument (even a sampling drill that ended up out of contact with the surface) was able to carry out at least one science cycle. Philae's precious science return included data gathered during descent, unscheduled flight, and surface operations at both Agilka (briefly) and Abydos.

Philae radioed to Earth—through the Rosetta orbiter—an extraordinary scientific bonanza.

The ESA lander characterized the physical composition of both Agilkia, the area of the first touchdown, and of Abydos, the final touchdown location. The upper surface layer at Agilkia appeared to consist of "air-fall product," particles erupted by the comet's geyser-like jets, which fell back to the surface in a fluffy coating. But the powder-like fallout did not cover the surface uniformly. Particles raining down upon Agilkia experienced some type of sorting; larger grains (ranging from millimeters to greater than a centimeter) made it back to the surface, while most of the smaller dust grains issuing from the surface escaped into space.

---

[20] The craft contacted the surface at a speed of 38 cm/sec. If it had been traveling just 6 cm/sec faster, it would have escaped the comet's weak gravity completely.

[21] The entrance port of one sensor may have been blocked by soil from its rough landings.

In concert with Rosetta,[22] Philae was able to beam radio waves through the comet nucleus, sensing a rigid crust 10-50 cm thick. Below is an interior of ices blended with very low density, "fluffy" dust. Oddly, the comet's density *decreases* toward the center.

Philae's remarkable panorama of its final landing site on Abydos shows jagged cliff walls, boulders and gravel, forming a stark contrast to Philae's first landing encounter of the relatively smooth Agilkia. Fractures of varying lengths score the face of the rocks. The cracks are likely caused by the thermal stress of daily temperature swings, along with changing light and heat levels throughout the comet's periodic trek around the Sun. Another difference between Abydos and Agilkia is that Abydos is as dark as asphalt (Agilkia is considerably lighter). Some areas of Abydos' rocky landscape seems encrusted by bright material, which could be ice or salt deposits. Dark, reddish material appears to blanket the entire comet nucleus, and may be carbon-rich organic compounds (as confirmed by Rosetta's chemistry findings).

Like Philae, the Rosetta orbiter itself was finally commanded to a gentle landing to close out its remarkable mission. The orbiter continued to transmit images and data on gas and dust in the comet's environment until moments before impact. Rosetta has given us valuable, comprehensive insights into the nature of comets. Bodies like Comet Halley, Wild 2, and 67P/Churyumov-Gerasimenko come from beyond the frost line, and represent those earliest, planet-forming epochs of our solar system's history. Thanks to advanced remote sensing and the recent spate of on-site sampling, these cosmic vagabonds—so critical to the building of planets and moons—are gifting us with a deeper understanding of the Earth's genesis and evolution over time.

Distant disks of suns and planets

The planets, asteroids and comets inform us about the nature and extent of the early nebular disk surrounding our own Sun. While observers have spotted planet-forming disks around other stars such as HL Tauri, researchers studying the young star AS 209[23] have discovered the next step in planetary formation: a disk around a forming planet (Figure 2.16). While circumstellar disks lead to planets, circumplanetary disks lead to families of moons, and are an important stage of planetary evolution that—up to now—we had not actually witnessed. AS 209's solar nebula disk is very well organized, with seven concentric rings

---

[22] Philae actually beamed radar through the nucleus to Rosetta.
[23] Also known as IRAS 16464-1416 and PDS 92

of gas and dust surrounding the star. Within the gaps between the rings, infant planets appear as small irregularities. The gaps themselves are tracks through the solar nebula that have been cleared by the developing planets (just as Saturn's moon Pan clears the Enke gap within its ring system). In this young star system some 395 light-years from Earth, scientists point to a location that appears to harbor a Jupiter-sized world only 1.5 million years old. The planet is so young that it is likely still accreting, pulling material into itself as it continues to grow. Within this material, moons are undoubtedly forming, or will soon. AS 209 has a mass just 1.2 times that of our Sun, so gazing into its nascent planet-birthing system is like looking at our solar system's past when the Earth and other terrestrials were embryos of worlds to come.

Pure lunacy: the Moon's coming-of-age story

Back here at home, the surfaces of all the terrestrials continued to suffer bombardment from the flotsam and jetsam of the solar disk. Large planetesimals on collision courses with their siblings carried out glancing blows, fused together or obliterated each other outright. The survivors followed concentric orbits around the Sun, while the orbit crossers removed each other from the playing field (see Figure 2.6).

Figure 2.16 Artist impression of the circumstellar disk surrounding the star AS 209 in the constellation Ophiuchus. A circumplanetary disk surrounds the youngest planet yet found, a Jupiter-class world just 1.5 million years old. A new moon forms within the disk surrounding the giant planet. (art by the author) Inset: The feeding zones of planets are evident in the disk surrounding AS 209, here seen by the Atacama Large Millimeter/submillimeter Array (ESO/NAOJ/NRAO), A. Sierra (U. Chile)

The Earth's story, had it continued on this quiet path, would have been quite different than it turned out. But one sunny day, a Mars-sized planet rose above the horizon to change the course of our world. The impactor is informally known as Theia. Theia was responsible for the existence of our Moon, and all it has given us.

The Earth owes a lot to its Moon. Aside from Pluto, no other world has yet been found with a natural satellite so close in size to its host. The Earth and Moon are, essentially, a double planet, with the Moon spanning about a third the diameter of its host planet. The Moon has influenced poetry, romance, horror stories, and migration patterns of various creatures. Some aquatic species, like Grunion, spawn or lay eggs at high tide on the full or new Moon. Sea turtles and several types of crabs show similar preferences.

But more critically (unless you are a sea turtle), the Moon also has great influence on the Earth's spin axis, called the obliquity. While the Earth's slant remains at a steady 23°, all of the other terrestrial planets have suffered wobbly tilts. Venus is tipped over so far that it technically spins backwards. The Martian axial tilt has ranged from 0° to 60° over the past 100,000 years, triggering ice ages and alternately pumping up and depleting the entire atmosphere. (see Chapter 9). The slant of our axis is important, as it defines the extent and severity of our seasons, which in turn determine the nature of our long-term planetary climate. Without the Moon, the Earth's spin would tip wildly from 0° (vertical) to 85°. Even with the Moon's dampening effect, subtle changes in the planet's tilt have led to ice ages here.

Jacques Laskar, the French astronomer whose pioneering calculations showed the Moon's effect on Earth's obliquity, said, "...the Earth is very peculiar. The common status for all terrestrial planets is to have experienced very large-scale chaotic...obliquity, which, in the case of the Earth and the absence of the Moon, may have prevented the appearance of...life. We owe our present climate stability to an exceptional event: the presence of the Moon."[24]

Some research indicates that the Moon may have played a role in winnowing out some greenhouse gases from Earth, distancing our nature from that our sister planet, Venus. And the ocean tides caused by the Moon may have played a pivotal role in the rise of life here. But the "gentle goddess of the night" had a tempestuous arrival, interrupting the Earth's development and resetting its unfolding timeline completely. As we have seen, some zircon crystals in Earth's rocks date to before the impact that created the Moon (~4.4

---

[24] As quoted in *Rare Earth* by Ward and Brownlee (Copernicus books, 2000), page 224.

billion years old), and they show evidence of contact with liquid water. Even on the infant Earth, seas or oceans washed the ancient shorelines. But those oceans would vaporize in an instant.

Our Moon is exceptional in another way: it is an outlier as natural satellites go. Conditions among the outer planets were ripe for moons. Among the four giants, Jupiter, Saturn, Uranus and Neptune, some 216 moons inhabit the region.[25] Some are the size of planets. But among the four terrestrial planets, there are only three. The Martian moons may be captured asteroids, and they are pint-sized rocks less than 20 km across. But our Moon is the outlier. Seventh-largest of all solar system satellites, it spans a third the Earth's diameter, making the Earth-Moon system something like a double planet. Pluto and Charon are the only other substantial worlds that come close to this relationship. In their case, the two really are a dual planet: they rotate around a common point between them.

Theia and the birth of the Moon

The Mars-sized planetoid Theia had been living out its own planetary life story for millions of years, growing, differentiating, building an atmosphere, and perhaps even sporting seas or polar caps. Its surface was undoubtedly cratered. Did it have deserts and sand dunes? Soaring snow-capped crags? Arctic wastes? Explosive volcanoes and deep canyons? We will never know.

Theia came barreling in out of nowhere, bulldozing into the infant Earth at a glancing blow (Figure 2.17). The Earth was recently differentiated, and any seas were freshly minted from the clearing skies—our planet had only been a planet for a few million years. The collision peeled off much of the Earth's outer layers of crust and mantle. The core material from the incoming planetoid combined with that of the Earth, while lighter material fanned out into a ring system vast enough to make Saturn envious. That ring would become the Moon of today.

The ravages of the great impact have left their subtle mark in many clues, says Robin Canup, astrophysicist at the Southwest Research Institute. "The Earth-Moon pair has an unusually large amount of rotational energy, which is combined in both the Moon's orbit around the Earth and Earth's spin." Theia's impact added rotational energy to the Earth-Moon system. The Moon's orbital path and Earth's spin are both consistent with the Giant Impact hypothesis, she says.

---

[25] 92 moons for Jupiter, 83 for Saturn, 27 for Uranus and 14 at Neptune.

Figure 2.17 Hours before impact, the Mars-sized planet Theia looms over the horizon of the early Earth. Its collision will strip Earth of much of its crust, leaving the planet a molten globe of loose material. Much of Theia's heavy core material will bury itself into the Earth's fractured heart. As the Earth reforms into a planet, the remnants of Theia—mixed with the Earth's own crust material—will leave a ring that eventually coalesces into our Moon. (painting by the author)

Canup also points to the Moon's odd-sized core. "The Moon has a much smaller core than our Earth. This, too, is consistent with the model; the impact stripped part of the outer layers of the impacting object and Earth to form the Moon. The core of the colliding object combined with the Earth's own dense core. The Moon was formed with much less iron and other heavy elements to form its core." The earlier theories of lunar birth posited that the Earth had not yet fully accreted at the time of the great impact, but recent work counters that version. "The impact during Earth's accretion is by now a pretty old (and discarded) idea," she says. Instead, the latest models involve either a Mars-sized Theia, or a collision between 2 half-Earth-sized bodies (Figure 2.18).

The entire process from impact to debris ring to a bona fide Moon may have been rapid. A 2022 computer simulation with higher resolution than earlier work[26] shows a moon forming just days after the impact. The simulated

[26] Immediate Origin of the Moon as a Post-Impact Satellite, Kegerreis, et al, *Astrophysical Journal Letters*, Volume 937, number 2

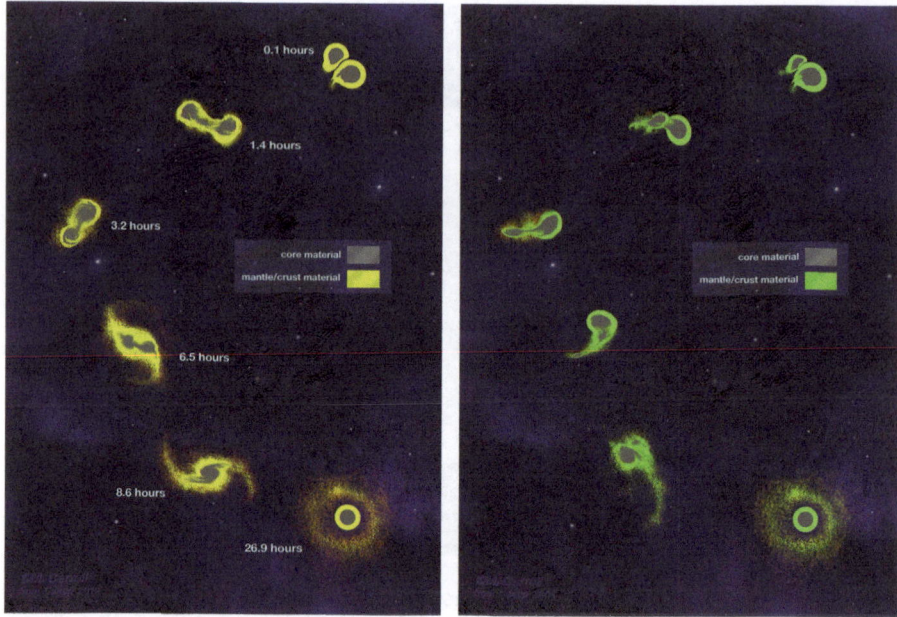

Figure 2.18 Updated versions of the Giant Impact scenario, offered by a team led by Robin Canup and Al Cameron. Right: An impact occurring between two half-Earth-sized planetoids. After a glancing, low-velocity blow, the two planet cores separate and then circle around to connect again. The metallic cores of the two planets (gray) form Earth's large core. Portions of lighter material from both mantles (yellow) form a ring around the Earth that eventually condenses into the Moon. Left: a similar time scale shows a fully accreted Earth hit by a Mars-sized planetoid, resulting in the Moon/Earth system of today. (art by the author after computer simulations by Canup, 2004 and 2012)

satellite fit the isotopes found in the Earth/Moon system, and showed that the new Moon would consist of the correct iron content and mass. The same simulation demonstrated that even protomoons ending up in close orbits would rapidly settle into stable orbits outside the dangerous Roche limit, where Earth's gravity would tear them apart.

Shortly after the Moon's formation, the Earth's natural satellite loomed overhead just 32,000 kilometers away. The ocean tides on Earth—today undulating by a few meters—may have risen by hundreds of meters. Even the solid ground on our world flexed by many meters, heating some solid rock to molten magma. To get an idea of the power of these tides, one need only look to Jupiter's moon Io. The gravity of Jupiter and nearby moons Europa and Ganymede tug at the very core of the little moon,[27] setting up tidal friction.

---

[27] Io is about the size of Earth's Moon.

The rise and fall of the land, over a meter each Ionian day, triggers the pervasive volcanism across the tiny globe.

The colossal tidal extremes would have settled down quickly, dampened by the Earth's prodigious oceans and the moving out of the Moon as it gradually spiraled away from Earth. The rise and fall of water and stone slowed to roughly modern levels within a million years or so.

The Giant Impact concept remained unconfirmed until after the Apollo era. Apollo astronauts and Soviet lunar sample return probes returned samples that could be studied in the lab, first-hand. Lunar rock samples and lunar meteorites contain the chemistry of the Moon and support the Giant Impact theory as well. The searing heat of the impact drove away most of the gases and liquids, leaving a relatively dry world. Moon rocks have very little of the water and gases found in Earth's rocks.

Other studies reinforce the scenario, including chlorine levels in Apollo Moon rocks. Chlorine exists in two stable forms, one heavy and one light. While terrestrial rocks are rich in light chlorine, lunar material has an abundance of the heavy type. The Giant Impact model suggests that both forms were present in the molten blobs that would become the new Earth and the Moon, but as the planets re-coalesced, the Earth's stronger gravity pulled in more of the light, volatile chlorine while the Moon became starved of the volatiles like chlorine and water. Judging from the lunar samples, this is apparently precisely what happened.

As with any theory, questions remain to be answered. These have inspired a host of variations on the great impact, all attempting to address the apparent discrepancies found in the dynamical models and in the samples of meteorites and lunar rock.

Some 4.4 billion years ago, the Earth was settling down nicely to become a stable, round planet. An atmosphere—probably rich in hydrogen, methane and water vapor—now blanketed the world. Earth's molten surface began to cool into a broadly basaltic skin. This shell would become oceanic crust of the type that makes up the modern sea floor. As it cooled, it began to melt from the heat of radiogenic material in the deep interior. Silica-rich rock similar to granite rose up through the layers, forming the first continental crust.

The earliest landforms peeking from the primordial seas differed from the world's modern islands and continents. The first land to appear after the water cycle inundated the planet was generated by volcanism or impacts. Volcanic islands would have been mostly black basalt, rocky islands eroding into coarse gravel and sand. The larger impacts would have raised crater rims, circular islands of stone and sand. Dunes may have piled up in the toxic winds, and banked along shorelines to make the first beaches, but the entire young planet would have been devoid of the familiar mountain chains, tectonic canyons, and "fruited plains" we are so used to today.

We find a record of this ancient crust-forming activity, a process similar to one that takes place on all differentiated planets, in tiny grains of minerals called zircons (technically zirconium silicate). As those silica-based rocks formed within Earth's cooling crust, zircons crystallized within them. Zircons are cherished treasures for geologists because of their remarkable nature. As they crystallize out of magma, they fold uranium into their structure. Their crystals are too fine to allow lead in, and this limitation makes them into a sort of geological clock. Radioactive forms of uranium (like uranium-235 and uranium 238) turn into lead over time. This transition from uranium to lead is very consistent and predictable. The rate of decay—or transition from uranium into lead—is well known. In the case of uranium-238, half of the uranium turns to lead in 4.47 billion years. In similar fashion, half of uranium-235 will decay into lead in 710 million years (it has a "half-life" of 710 million years). The two types of uranium can act as a sort of cross-check as geologists measure the age of zircons. Since lead cannot enter into zircons from outside, any lead inside must have come from the decaying uranium, creating a straightforward timeline for the zircon's age.

The oldest surviving rocks on Earth date back roughly 4 billion years. But those rocks contain zircons from an earlier epoch, and those ancient zircons have wondrous things to tell us. From sites including South Africa, India and the Jack Hills Formation in Western Australia, zircons have been dated to 4.38 billion years, nearly as old as the Earth itself (at 4.38 billion years, the Earth's crust was just stabilizing, and its atmosphere was evolving out of its reduced—lacking oxygen—atmosphere). Because zircons form in granite-like stone, which arose from the world's basalts, their birth occurred after the planet had begun to differentiate. This means that differentiation was taking place early in the Earth's development. Another bit of news from the primordial front is that the zircons hold chemical signatures of oxygen, implying that liquid water was already in play 4.38 billion years ago. Our oceans are nearly as old as the Earth itself. And finally, researchers find minuscule specks of carbon within the zircon's matrix. These graphite flecks can come from several sources, but they may be the fingerprint of the arrival of life on Earth (for more on biogenesis, see Chapter 5).

Rise of the continents

It was all Wegener's fault. Before him, the continents seemed so quiescent and organized. They built up with volcanic activity and uplift from forces beneath, and then eroded away. Simply put, the continents moved up and down. Scottish naturalist James Hutton championed this vertical picture of land-forms. Hutton was inspired by the great vertical span of sediments at Siccar Point and other sites in Scotland. For over a century, geologists embraced Hutton's concept of upright movement of the Earth's surface. But along came German meteorologist Alfred Wegener, who had noticed, even as a child, how the eastern point of Brazil fit along the west coast of Africa, while Canada seemed to snuggle nicely against the Sahara.

Wegener was not the first to suspect that continents "drift". The first to make the connection between coastlines of the eastern and western continents may have been Dutch cartographer Abraham Ortelius,[28] in 1596. In the 1830s, Scottish geologist Charles Lyell asserted that, "Continents...although permanent for whole geological epochs, shift their positions entirely in the course of ages."[29] American geologist Frank Bursley Taylor also proposed such a crazy idea in 1908. Taylor portrayed moving continents as the cause of mountain systems. The leading edges of sliding continents, he submitted, caused the ocean crust ahead of them to buckle downward, leaving troughs where sediments built up. As the continents continued to move, they uplifted the sediments, resulting in mountain chains. And what was the mechanism for all that movement? Taylor blamed the Moon. At the time, the leading theory of lunar origin was that the Moon was gravitationally captured, possibly during the Cretaceous Period. From this violent event, Taylor suggested, tidal forces arose to tear the continents asunder.

A few years later, American geologist H. Baker proposed that a supercontinent abruptly ripped apart, forming the Atlantic and Arctic Oceans. He laid the cause for this geologic catastrophe at the feet of the planet Venus, whose orbit came close enough to Earth to yank a chunk out of the continental crust. This excised planetary blob came from the Pacific Ocean, he reasoned, as the continental margins on the Atlantic side appear to fit together.

Armed with detailed maps of the world's coastlines, Wegener published his idea of global continental drift in 1915. It was not universally accepted. In fact, it spurred a worldwide, fierce debate. Wegener's strength as a meteorologist and climatologist counted against him in the area where his ideas must be tried: the halls of geology. Wegener was not yet an established geologist (that

---

[28] Ortelius is credited with the first modern atlas, the *Theatrum Orbis Terrarum*.
[29] Principles of Geology by Charles Lyell, 1874

came later), and the geologists' society was a fairly closed club. Even the most open-minded of the scientists still had to contend with Wegener's "earth-shaking" concepts. One problem his ideas faced: how could the Earth's heavy continental masses hurtle *sideways* across great expanses of the globe? Wegener's wandering continents were especially unpopular among the well-established geologists of Europe and North America. Geologists in the world's southern realms were quicker to accept Wegener's theory. In addition to seeing the jigsaw-fit of the coastlines across the Atlantic, these geologists were familiar with the matching rock types and fossils on both sides of the ocean (Figure 2.19). The fossilized leaves of the ancient Glossopteris, for example, appear in both South Africa and South America. Fossils of the wolf-sized Cynognathus, an extinct mammal-like reptile, inhabited what is now South America and South Africa. The land creature could not have made it across

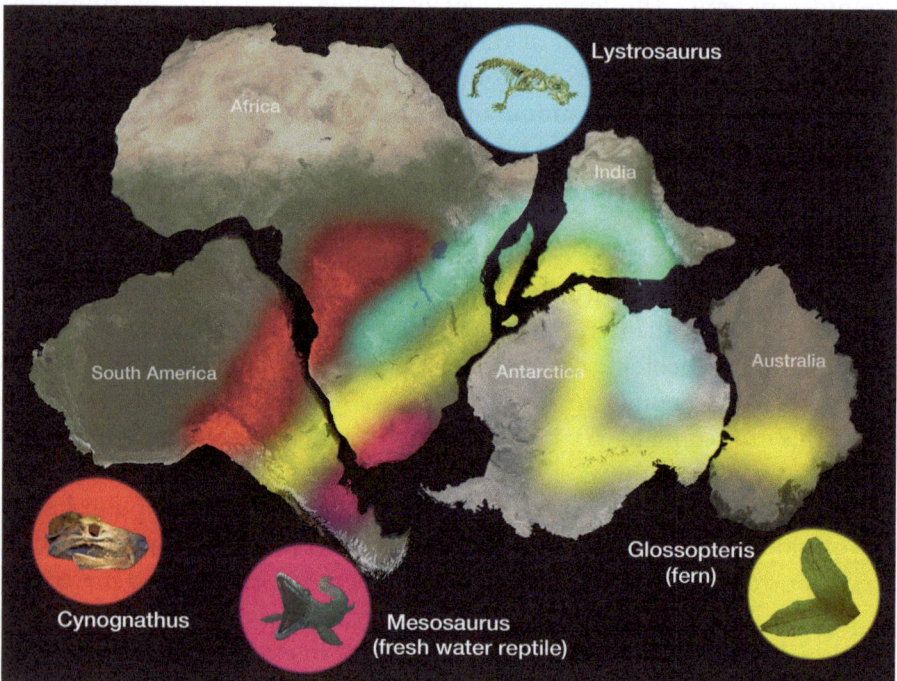

Figure 2.19 An assortment of fossils bridges the modern territory that once made up Gondwana. Common fossils have been located in Antarctica, South America, Australia, India and Africa, confirming that those landmasses were at one time merged into a supercontinent. (art by the author, based on work by Snider-Pellegrini and Wegener) (Lystrosaurus skeleton photo by Rama, Wikimedia Commons, Cc-by-sa-2.0-fr) (Cynognathus skull https://commons.wikimedia.org/wiki/File:E_-_Cynognathus_-_2.jpg)

the Atlantic from one hemisphere to the other. The idea of land bridges between the continents seemed a stretch, especially across all that seawater.

The answer was waiting in an unlikely place, several thousand fathoms beneath the ocean waves. It came from an unlikely activity, too: hunting for submarines. During World War II, the navies of several countries began to use sonar to detect submarines. But the sonar went beyond the level of lurking enemy craft, all the way down to the sea bottom. Sonar reflections ("returns") revealed canyons and mountain ranges on the ocean floor. In the 1950s, American researchers Marie Thorp and Bruce Heezen collated the sonar from around the world, and created the first global map of the ocean floor (Figure 2.20). Their groundbreaking work revealed a great underwater mountain chain, now known as the mid-Atlantic ridge. The longest mountain chain in the world, this ridge of peaks stretches from Iceland in the north all the way down the Atlantic to Antarctica, a stretch of 16,000 km. The submarine mountains rise some 3 km off the sea floor and extend up to 1500 km wide in some places. The chain rises above the surface of the Atlantic as a series of volcanic islands. From north to south, these include Jan Mayen (Norway), Iceland, the Azores (Portugal), Ascension Island (UK), Tristan da Cunha (UK) and, at the south end, Bouvet Island (Norway). The mid-Atlantic ridge connects to a series of sea-floor ridges at the edges of the planet's plates,

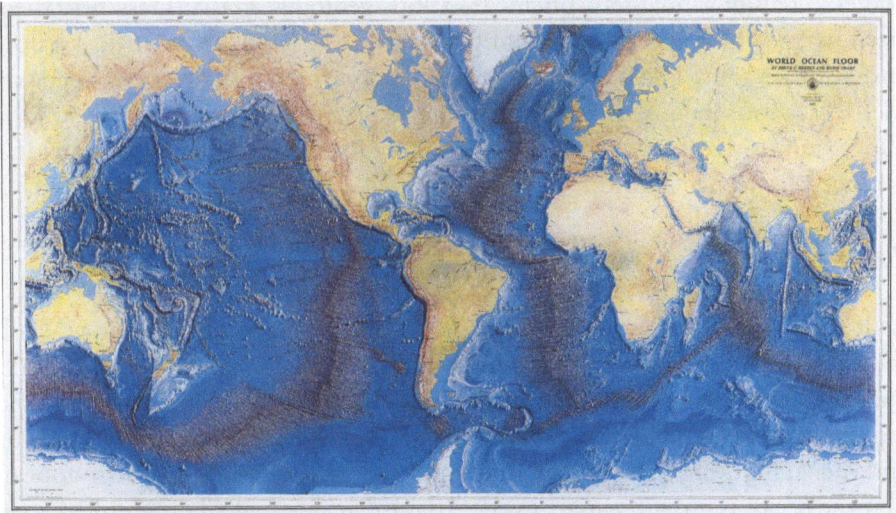

Figure 2.20 The historic seafloor discovery map by Marie Thorp and Bruce Heezen. Their work brought to light the nature of sea floor spreading and its relationship to our planet's plate tectonics. (courtesy of Fiona Yacopino)

adding up to a combined network of mountain ranges 65,000 km long. The network of linked oceanic ridges is even longer, adding up to 80,000 km long.

Even before the revelations of Thorp and Heezen's work, geologists had come to realize that earthquakes had an odd distribution, with many congregating along mid-ocean ridges and in what we now recognize as subduction zones.

The basaltic stone of the seafloor spreads out from a central ridge, both eastward and westward, creating new ocean floor material at the rate of 2.5 cm/year. Over the past 100 million years, the Atlantic Ocean has expanded by 2500 km. The realization that the mid ocean ridges constantly generated new ground provided Wegener with a solution to his problem of sideways-moving rock, and the concept of plate tectonics was born.

Magnetic studies of the sea floor bolstered the new idea. A map of magnetism in oceanic basalts along the mid-Atlantic ridge shows a curious pattern of shifting directions. The Earth's magnetic field reverses itself periodically, and these reversals are recorded in tiny crystals within the sea floor stone. The curious feature of the pattern is that a mirror image of magnetic shifts echoes the motif on both sides of the ridge (Figure 2.21). As the oceanic crust spreads apart, alternating magnetic stripes replicate on both sides, repeated in the opposite order from one side to the other: the pattern on one side is a mirror image of that on the opposite side. This mirror image confirms that the crust is expanding outward on both sides, dutifully recording the history of the planet's changing magnetosphere. In 1960, geologist Harry Hess first theorized that seafloor spreading[30] might be taking place on the Atlantic sea floor and at other sites. His theory was confirmed with the discovery of the symmetric seafloor magnetic anomalies, first discovered by geophysicist John Vine, Cambridge University's Drummond Hoyle Matthews, and Canadian geologist Lawrence Morley.

Plate tectonics and continental "drift"

As it turns out, not all of the Earth's plates mark regions of spreading. While the Atlantic expands outward from the mid-Atlantic ridge, the margins of the Pacific Ocean basin are marked by subduction, the process of one plate sinking under another. Where these slabs of crust meet and descend, rock heats up and melts, marking lines of volcanoes and erupting magma. While it appears that the ocean is growing apart in the Atlantic, it is actually being pulled apart

---

[30] a term coined in 1961 by Robert Dietz

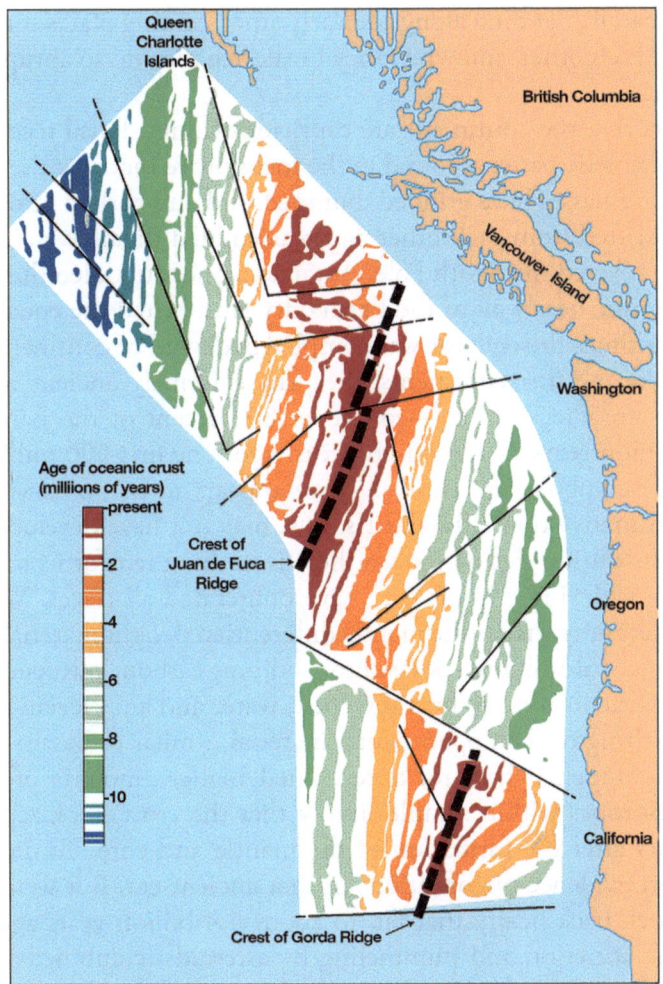

Figure 2.21 A 1991 geomagnetic map of the magnetic anomalies on the ocean floor along the Juan de Fuca and Gorda ridges just west of the North American coast, where sea floor material is spreading, leaving a mirror image pattern in the geomagnetic record. (USGS public domain map by W. Jacquelyne Kious and Robert I. Tilling via https://commons.wikimedia.org/wiki/File:Magnetic_anomalies_off_west_coast_of_North_America.gif)

by the force of sinking plates on the other side of the world. Plates sometimes ram into each other, thrusting surface material up into mountain ranges. The soaring Himalayas mark the collision and uplift between the northward-moving Indo-Australian plate and the Eurasian plate, beginning about 50 million years ago (the process is on-going). The Andes Mountains are another example of a mountain range born at the edge of colliding plates. Some plates migrate in parallel, gliding alongside each other. Sliding plates sometimes get hung up on each other, and when they break free they do so abruptly, causing earthquakes.

It appears that the continents are capricious things. Fossil records and the changing magnetic patterns found in the rocks prove that over the past billion years, the landmasses have grinded, bumped, slid and bashed into each other. They have drifted from each other and reassembled themselves. The heart of Africa once covered the South Pole. Antarctica once straddled the equator. In fact, at one time nearly all of the Earth's dry land belted the equatorial zone, and at other times the equator was completely devoid of continents.

What were the first continents like? The extremely ancient rocks of the Cryptozoic—or Precambrian—eon cover 80 percent of the Earth's history, from 4.6 billion years to 600 million years ago. The first 500 million years of our planet's scrapbook are elusive. There appears to be no surviving record from this formative time. The Earth's crust may not have developed any permanent traits during this period, so there is nothing left for us to see, or our planet's overachieving erosion may have obliterated all traces. We do know with some certainty that as the mantle differentiated, light materials including silicon, oxygen, aluminum, potassium, sodium, carbon, nitrogen, hydrogen and other elements rose to form the crust, water and air. Oceanic crust came from the melting of the upper mantle material, similar to basalts. The continental crust, much lighter rock, contained higher amounts of radioactive materials. Isotopes within samples imply that the crust stabilized some 200 million years after the formation of the mantle and core. To date, we have only indirect evidence of conditions in that ancient era. But we can run our geologic clock back nearly that far, to about 4.4 billion years ago. The heat from Earth's accretion and pummeling by asteroids, combined with energy from radiogenic (radioactive) elements like uranium and thorium, melted iron and nickel, both of which settled to form the planet's core. At 4 billion years, the first crustal rocks had formed from lighter material that rose to the surface. Exterior temperatures cooled off, and the water cycle began in earnest. At 3.9 to 3.8 billion years ago, the meteor storms abated, and small protocontinents began to merge into major, stable continents. At 3.6 billion years, we see the first hints of life engaging with the environment.

The most ancient of the Cryptozoic rock formations holds perhaps half of all the world's crustal metals. Unique to the primeval rock are banded iron formations, which record a time in our history before free oxygen was plentiful. The Cryptozoic layer marks an era devoid of most fossil traces. Fossils are rare and deformed by great pressures from later rock layers, but within the oldest appear the remnants of the earliest microfossils (Figure 2.22).

The Theia/continental drift connection?

The cataclysm that formed the Moon may have initiated our planet's unique plate tectonics. In a report to the 2023 Lunar and Planetary Science Conference, Caltech geodynamicist Qian Yuan presented new computer simulations of the impact. The computer model shows that remnants of Theia, buried deep within the Earth's mantle, could have instigated the planet's process of subduction, where one plate slides beneath another.

Two regions within the Earth's mantle behave strangely as seismic waves pass through them. Known as the large low-shear velocity provinces, these subterranean continent-sized blocks slow seismic waves as they pass through the mantle. Yuan's team proposes that the strange assemblies could be the dense remains of Theia. The new models show that these masses, if indeed they were introduced at the time of the Theia impact, would have had a profound effect on the currents within the Earth's mantle. Sinking down to the bottom of the mantle (where the mantle meets the planet's outer core), the masses would have forced plumes of hot rock to rise and impinge upon the crust, where they would force surface rock downward, and the action of subduction would have begun. The computer data projects that plate tectonics would have gotten under way roughly 200 million years after the Moon-making impact.

Some researchers caution that we cannot yet prove that the low-shear provinces are actually remnants of the great Theia impactor. If they are not, then any connection they have to plate tectonics will require more proof.

Figure 2.22 (left) Traces of early life on Earth exist today within the ancient rocks found in Australia's Dawn of Life Trail in the Pilbara of Australia. Here, fossilized colonies of microbes called stromatolites left their mark in the stone some 3.8 billion years ago. (Right) Modern stromatolite colonies still thrive in Australia's Sharks Bay. (NASA photos courtesy Chris McKay)

Earth history divisions

Geologists divide the history of the Earth into four time periods. The Hadean is the first and most ancient. During this formative time, Earth's first solid rocks coalesced from the cooling molten planet. At the beginning of the Hadean, the Moon formed, likely from a giant impact.

The second great eon in Earth history is the Archean, beginning slightly later than 4 billion years ago and ending at about 2.5 billion years ago. As the atmosphere cleared and thinned, the oceans formed. Meteor impacts subsided. The first confirmed microfossils appear in these layers, with true cells appearing as early as 3.8 billion years ago. The first supercontinent, Kenorland, appears (for a list of history's supercontinents, see Appendix E).

The next eon is called the Proterozoic, which opens with the "Great Oxygenation Event" some 2.5 billion years ago. Microbes converted much of the Earth's carbon dioxide into oxygen, transforming our atmosphere into something closer to what we see today, with abundant free oxygen in the air. Over 650 million years ago, the Earth endures several "snowball" events of near-global glaciation (see Chapter 6). Supercontinents assemble and break up, ending with the quartet of Gondwana, Siberia, Baltica and Laurentia, paving the way for the latest, present eon.

The Phanerozoic spans from 541 million years ago to the present. The Cambrian explosion brings a diversity of life forms never seen before. The ozone layer builds, and life enters the land. Major flourishing and extinctions rise and fall. Mesozoic life—including the dinosaurs—comes and goes. Ice age giants like Mammoths and Mastodons arise, becoming extinct just a few thousand years ago. We arrive at our world today. (For a more detailed list of events within the Phanerozoic, see Chapter 5).

Parade of continents

The earliest plate tectonics continue to present geologists with a mystery, and scientists love a mystery; mysteries tend to trigger competing ideas. How did the process get started in the first place? How did we get from basaltic oceanic crust to granitic continental masses? One school of thought suggests that plumes of magma moved through the mantle and built great basalt structures. At the base of these basalt piles, rock re-melted and transformed into granite, precursor to modern plates. Other geologists proffer that as Earth's magma ocean cooled on its surface, it formed a brittle crust that then cracked. Within those cracks, fresh magma from beneath shoved its way upward, pushing slabs of cooled crust apart in the beginning of plate tectonics. Plates subducted and melted into granite, the first material of modern plates. By 3 billion years ago, plate tectonics appear to have been in full swing, and they continue in similar fashion to this day. The movement of continents has played such an important role in molding our planet that we wonder about other worlds. Alas, there is no evidence of a comparable geologic process on Mercury or Venus. But as we will see in Chapter Six, Mars may once have had such an arrangement, and we find the clearest parallels to terrestrial plate tectonics on several ice moons in the outer solar system.

Just as we got wind of the Big Bang by projecting the movement of galaxies backward in time, so we can get an idea of the Earth's continental configurations in the past by charting their current movements. The chemical character and locations of granite, the fossil record, and the arrangement of faults and folded layers in the most ancient mountain chains confirm that plate tectonics have been sculpting the face of our world for at least 2.5 billion years. Supercontinents have arisen, broken up, and come together again in new forms. This repeated cycle is called the Wilson Cycle, after Canadian geologist J. Tuzo Wilson, who first recognized it. Our planet has played host to at least half a dozen supercontinents in its past. The earliest one that we can recognize is the supercontinent Vaalbara, some 3.6 billion years in the past. The duration of Vaalbara is not known, but 2.7 billion years ago—after the fracturing of Vaalbara into smaller units—the continents had gathered again into the supercontinent Kenorland, whose core region dated back as far as 3.1 billion years. At 2.3 billion years past, the continent was already breaking up into two major sections. Its dissolution coincided with increased oxygen levels in Earth's atmosphere, and led to increased rainfall, which reduced methane levels (methane is a powerful greenhouse gas). In this way, Kenorland's demise initiated a global cooling, which may have led to a period known as the Snowball Earth (see Chapter Six). The next supercontinent, Nuna, assembled itself about 1.5 billion years ago. It splintered and came back together as the supercontinent Rodinia about half a billion years later. Rodinia appears to have split into three major continents, leading to yet another supercontinent—this one somewhat controversial—the proposed supercontinent Pannotia, which may have coalesced 600 million years ago. Current reconstructions show Pannotia as land masses at both poles with a thin land bridge connecting them across the equator. Pannotia was short-lived, lasting until about 540 million years ago, when its land bridge collapsed and its reconnecting sections became the supercontinent Gondwana, along with the much smaller continents Laurentia and Baltica. (Today, Laurentia forms the central mass of the North American continent.)

From there, the picture becomes much more clear with better-preserved geological units of the Cambrian period. Laurentia spread across the equator. A large subcontinent called Avalonia broke off from Gondwana, which was moving southward, and drifted toward Laurentia. The massive rafts of rock eventually merged, creating a region called Euramerica. As the name sounds, material from this merger would eventually form parts of Europe and the Americas. But the dance continued, with Euramerica shifting toward the

Gondwana supercontinent in the south. As the two masses collided, they formed the supercontinent Pangaea some 335 million years ago. It was within this great land that the world saw carboniferous forests and the first dinosaurs. The continent remained intact until the Middle Jurassic, 175 million years ago.

Vestiges of these supercontinent collisions remain in the form of the Appalachian mountain chain in North America, the Caledonide Mountain range of Scandinavia, and—from even earlier—the folded rock layers of the Pan-African belts of Africa and South America. 100 million years ago, the mighty Rocky Mountains lofted from the North American plains, and the Andes arose along the margin of what is now South America, but the Swiss Alps and the Himalayas were still millions of years to come. A shallow sea bisected the precursor to the North American continent, and the nascent Atlantic, a mere narrow band of water, started to widen as the mid-Atlantic rift began to spread. Within 50 million years, Australia broke free of Antarctica and the Indian subcontinent moved north to ram into Asia. Finally, the continents, subcontinents and islands of the modern world began to take shape. The Earth has had no super-continents since, but it will. Geologists project several possible future supercontinents: Pangaea Proxima, Amasia and Novopangaea. The world is restless indeed.

Biological diversity and plate tectonics

From microbes to petunias to elephants, living things may have a special relationship with plate tectonics. The migration of the continents engenders increased biodiversity, keeping populations separate over time. This enables creatures to "go their own way" in terms of adapting to different environments. Plate tectonics also serve as a sort of planetary thermostat, recycling minerals and chemistry that dampen the levels of carbon dioxide, tempering spikes in its concentration. The movement of plates and building of continents also controls sea levels, which interact with those same chemicals to keep carbon dioxide in check. Tectonics also recycle water from the oceans into the deep subsurface rock. Without the process, the Earth may well have lost all water on its surface, leaving a Venus-like, desiccated world.

Aside from colliding and subducting landscapes, other forces were shaping the embryonic terrestrial landscape. Long before the rise of the supercontinents, before Vaalbara, Gondwana and Pangaea, their protocontinent ancestors emerged, stabilized and began to wander across the juvenile Earth's face, and as they rose they continued to be worn down and pulverized by weather, torn asunder by fracturing, and blasted by asteroids and comets. That deadly rain of rock and metal took over a billion years to abate.

Crater cataclysm

The entire solar system, from the plains of Mercury to the surfaces of Kuiper Belt Objects like Pluto, has been battered and beaten. Even today, new craters form on Earth at every hour, though most are small enough to avoid detection. Mars orbiters have charted many instances of the appearance of new craters on the rusty Martian plains. During four years of study from orbit, the Lunar Reconnaissance Orbiter identified at least 25 new craters. Any solid surface not sculpted by tectonics or weather bears a record of the fusillade of rock and metal that traversed down from the planets' formative years. That nightmare meteor storm may have come in gusts. As they carefully count craters on various planets, moons and asteroids, researchers have deduced that an era of concentrated meteor impacts began about 4.1 billion years ago, early in solar system history. The intense barrage—which appears to have ebbed some 3.8 billion years ago—has come to be known as the Late Heavy Bombardment (LHB) or the Lunar Cataclysm.

We have the Moon to thank, again, for our first practical inklings about this severe cosmic battering. With the advent of lunar exploration, both human and robotic, a pattern began to emerge. It seemed that the Moon rocks had all been melted from impacts about 3.9 billion years ago. A 1974 paper proposed that an extreme cratering event resurfaced the Moon in a strong pulse of impacts lasting less than 200 million years. They called it the Lunar Cataclysm, and the idea gained ground as further studies continued. Geologists realized that if such a pulse of incoming asteroids really took place, they should be able to find many samples that had been melted at about the same time. This trend was borne out among the Apollo samples, as well as with samples from Soviet Luna sample return craft. The Apollo samples offer a range of ages, but none more ancient than 3.85 billion years, which aligns with the roughly 3.9 billion-year end to the cataclysm. Soviet Luna 16 samples came in at 3.4 to 3.5 billion years old, while Luna 20 samples measured 3.9 billion years. All samples yet assessed show ages that bolster the idea that the Moon—and by implication, the Earth—underwent widespread bombardment between 4.05 and 3.85 billion years ago.

Skeptics pointed out that the Apollo samples were biased toward the large impact basins on the Moon's near side. This was true: Apollo landings were constrained to near-equatorial sites due to communication limitations with Earth. The Soviet landing sites were also all near the equator due to limits in fuel for the return trips. Could it be that the Lunar Cataclysm was a mirage of limited sampling?

In the 1980s, researchers realized that they had a second set of Moon rocks: meteorites that had been blown from the lunar surface and ended up on Earth. Since they came from random impacts from many parts of the lunar globe, these would provide a good check against the Apollo and Luna samples. It appears that at least the majority of the Moon was resurfaced in a geologically short, furious event. But appearances can be deceiving. Some new data may point to a more gradual rise and fall of the LHB, as we'll see in the next chapter.

Making a new impact

As the lunar cataclysm tailed out, many large asteroids still lurked out in the darkness. Some circled the Sun in Earth-crossing orbits, and eventually a few had to connect. The largest collisions may have led to mass extinctions across the globe. One such event befell the Earth at the end of the Mesozoic Era, the age of the dinosaurs. Half of the Earth's inhabitants became extinct at the end of the Cretaceous Period. Some types of microplankton and all marine reptiles (like the mosasaurs and plesiosaurs) disappeared. Most notably, the ammonites—which had ruled the seas since the Devonian period—vanished from the oceans with one exception: the chambered nautilus. Many of these lines of creatures were in decline over a period of time. The population crash occurred at what is known as the Cretaceous/Paleogene (or the K-Pg) event.

A father and son team first tumbled to the clues waiting in the strata of Italy. Luis and Walter Alvarez and their team of geologists noticed a striking feature: a dark layer of sediment that lay at about the depth where the Cretaceous Period ended. The dark clay was enriched with iridium, a rare metal on Earth. While most of our planet's iridium sank to the core during differentiation, it is abundant in meteorites. For some reason, this boundary—now known as the K-Pg, or Cretaceous/Paleogene boundary—was saturated with metals usually found in asteroids and comets. What could it mean?

Evidence poured in from all over the world. Unusually high levels of iridium were appearing in samples from across the globe. A planetary event had left its record in the K-Pg layer worldwide. The planet had been encrusted by an earth-shaking cosmic event. Below the iridium layer, rich flora and fauna carpeted the Earth. Above it, some 75% of the Earth's species abruptly disappeared.

---

[31] Their hypothesis was first published in the journal *Science* in 1980.

In 1979,[31] the Alvarez team controversially proposed that a giant impact had singed the Earth 66 million years ago, vaporizing much life and exterminating all non-avian dinosaurs. Their work would confirm the theory of a Yucatan crater put forth by two geophysicists, Glen Penfield and Antonio Camargo, earlier in the decade.

In the Yucatan peninsula, more evidence was piling up in a quite different direction. As data came in, a pattern emerged: the K-Pg boundary was thicker in the Americas than it was in Europe, Africa or Asia. Anthropologists and archeologists now took their turn: across the Yucatan, in the heart of the Mayan empire, a line of limestone sinkholes inscribed a great arc. Half of that arc was under the waters of the Gulf of Mexico, but the rest wrote the demarcation of a 200- kilometer crater. The pits, called cenotes, formed a critical part of the Maya civilization, supplying fresh water and forming a network of sacred sites. So while the great asteroid impact spelled the end of the dinosaurs, it may have led to the birth of one of the world's great cultures.

Gravity maps confirm the existence of a vast crater draped across Mexico's Gulf. The scar has been christened the Chicxulub Impact crater. Its center lies near the village of Chicxulub, from which it gets its name. Experts estimate the crater's diameter at 180 km, with a depth of 20 km. Samples of the crater include shocked quartz, which occurs under great pressures like those within explosions.

The impact at the end of the Cretaceous may have been the last straw for an already dying race. Paleontologists assert that many types of dinosaurs seemed in decline, according to the fossil record. The global climate was in flux, ocean currents were changing as continents drifted into different configurations, and a spurt of violent volcanism lowered temperatures across the planet. But the dinosaurs[32] likely went out with a bang, and it would have been a horrifying one (Figure 2.23).

The giant intruder was likely a carbonaceous chondrite asteroid some 10 km across. Its impact sent 1.5 km-high tsunamis across thousands of kilometers of ocean as molten shrapnel rained from the sky. The material that fell back to Earth triggered wildfires on a planetary scale. 1000 k/h winds carried burning trees and flesh across hundreds of kilometers of landscape. 25 million

---

[32] Paleontologists differentiate between avian and non-avian dinosaurs. The non-avian dinosaurs included creatures like the duck-billed hadrosaurs, Apatosaurus, Diplodocus, Triceratops, and Stegosaurus. These had thick bones and the largest bodies of anything to walk the Earth. Modern paleontologists, armed with new data, assert that avian dinosaurs did not become extinct, but rather survived to the present as birds. This light-boned avian group included tyrannosaurs, Spinosaurus and the dromaeosaurs like Deinonychus and Velociraptor.

Figure 2.23   At the end of the Cretaceous period, a massive asteroid impacted the Earth at the edge of what is now the Yucatan peninsula, bringing to extinction 70% of all plant and animal species, including all non-avian dinosaurs. What the dinosaurs thought of the event is unknown. (painting by the author); inset: a gravity map shows the outline of Chicxulub meteor crater, straddling the coast of the Yucatan peninsula. (https://commons.wikimedia.org/wiki/File:Chicxulub-Anomaly-Grav-3.jpg)

metric tons of ash rose so high that it shrouded the entire globe in a several-year winter, causing a collapsed food chain, mass starvation and deadly acid rains.

As if that wasn't bad enough, the Chicxulub asteroid may have had an accomplice. While about 200 impact craters have been identified on Earth, a new discovery off the coast of Guinea Bissau, West Africa, may line up with the timeline for Chicxulub. The remains of what appear to be an impact crater are far smaller, about 8.5 km across. That would make the impacting body 400 meters wide, less than half that of Chicxulub's culprit. The structure—called Nadir—is buried hundreds of meters beneath the sea floor. It was discovered within seismic data,[33] and displays the raised rim, bowl shape and central peak typical of a crater.[34] At the time of impact, the Nadir site would have occurred beneath 500 meters of ocean water. Although the creation of Nadir would not have had the global fallout that Chicxulub did, the impact would have added insult to injury, a one-two punch for Mesozoic life on Earth (Figure 2.24).

DARTs in space: Saving the world from another Chicxulub

We've seen the Hollywood version: a team of brave astronauts is dispatched to the surface of an incoming asteroid, where they plant explosives to save humanity. And while the stories are fantasy, they may become reality one day. Of the 29,000 identified Near-Earth asteroids, 2224 are in orbits that could, potentially, carry them on a collision course with our world. Far from being special effects entertainment, the result could be disastrous, akin to the Chicxulub impact that ended the age of the dinosaurs. In fact, on the average of once per century, a meteor or comet large enough to cause extensive destruction comes blazing through the atmosphere and explodes overhead. In June of 1908, a probable comet disintegrated above the Tunguska region of Siberia, flattening trees and killing wildlife over a wilderness area of 2100 square kilometers. Just over a century later, another meteor airburst injured 1500 people in the rural area of Chelyabinsk and 7200 buildings.

The DART mission was designed to evaluate the effect of a spacecraft impact on an asteroid. Flight engineers chose a double asteroid to track changes in the orbit of the smaller one after their craft, the Double Asteroid Redirection Test (DART), impacted the surface. Dimorphos is a 525-foot/165 meter rock orbiting the larger Didymos (780 meters or half a mile across). The pair was chosen because Didymos' orbital length—11 hours 55 minutes—was well known, and could be meticulously tracked for changes.

(continued)

---

[33] *Science Advances*, August 19, 2022

[34] Some researchers point out that volcanic structures or salt domes sometimes share these features.

On September 26, 2022, DART bashed into Dimorphos at 14,000 mph/22,530 kph. Four observatories in Chile and South Africa timed changes in the asteroid's orbit. Their results were confirmed by radar from two separate radio observatories. The impact shortened the tiny asteroid's orbit by 32 minutes, three times the amount predicted. The change was significant. In an asteroid encounter, the most likely scenario envisions a space rock passing very near the Earth and being detoured by its gravity (direct hits are nearly impossible statistically). The asteroid would come back around and, after making a circuit around the Earth, impact. Astronomers witnessed this situation at Jupiter in 1994. Comet Shoemaker-Levy 9 broke up after a close pass to Jupiter. In July of that year, its remnant nuclei impacted the surface of Jupiter in a spectacular parade of nuclear-level blasts, pummeling the planet from July 16-22.[35]

At Earth, an asteroid's close flyby, in which the body passes through a gravitational "keyhole" to become captured, would hopefully give engineers time to mount an expedition similar to DART. A fairly small change in its orbit would prevent the asteroid from returning to impact the Earth.

The question is, if something larger came at the Earth, could we scale up a craft like DART? The mission was a surprisingly efficient way to move Dimorphos. But something bigger would require either a much larger—and heavier—spacecraft impactor or a group of DART-sized ones. As one DART team member put it, "It all comes down to statistics, and the statistics say that we probably won't ever have to deal with anything bigger than a Dimorphos until long after Captain Kirk is out there with technologies way beyond what we can play with now. By then it won't matter..." Still, nature does not always play by statistics, so many analysts suggest that humanity needs to pursue planetary defense in many forms.

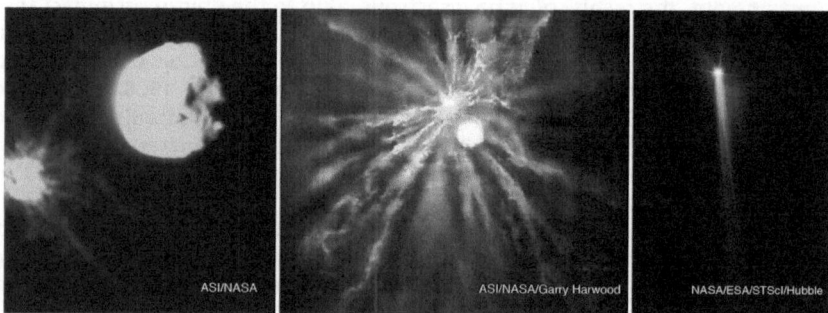

Figure 2.24 Left: NASA and the Italian Space Agency teamed up to send the DART mission on a crash course with the tiny asteroid Dimorphos (at left), a moon of the asteroid Didymos. The impact was a test for defending the Earth against incoming asteroids, but it also provided insights into the dynamics of asteroid collisions throughout the history of the solar system (ASI/NASA). Center: digitally processing brings out the details of tendrils and sheets of material blasted away during the impact (processing by Gary Harwood). Right: 285 hours after DART's collision with the asteroid, the Hubble Space Telescope imaged a trail of material streaming from Dimorphos (NASA/ESA/STScI/Hubble).

[35] Evidence of similar impacts can be seen on Jupiter's moons Callisto and Ganymede, where 16 crater chains called catenae are likely traces of comets or asteroids that fragmented before impacting.

The mass extinctions and climate collapse at the end of the Mesozoic came—at least in part—at the hands of an asteroid that survived from the planet-forming epoch of the solar system. The asteroids, Kuiper Belt Objects, and inhabitants of the Oort Cloud are all ambassadors from the birth of the planets (some even predating the birth of the Sun). And just as members of the Kuiper Belt and Oort Cloud are thought to represent primordial conditions in the outer solar system, so Mercury may embody a planetary embryo of the type seen in the second stage of planetary formation from much closer in. If so, it records an era in the early solar system typified by radical processes occurring only in the interior of the solar nebula. Mercury may provide us with insights into that violent, unruly era of Earth's formative years.

Looking back: A Brief History of...the Earth

The stage is now set for the appearance of the Earth. As we explore the details of our planet's development within the pages to come, we take a moment to survey our world's formative history, a story that is still being written.

Planet Earth coalesces from a rock pile to a planet over a period stretching roughly 4.5 billion to 4.4 billion years ago, a nightmarish epoch spanning the course of about 70 million years. To build that much mass, the rate of meteors falling to the surface must be hundreds of millions of times the rate seen today (today's rate of incoming meteors adds up to approximately 48.5 tons globally each day). That's 50 million tons of metal and rock each day in those formative years. By 4 billion years ago, the rate falls to a thousand times the present, and by 3.2 billion years past, rates approach the modern ones. The Earth's first surface is molten from all of that impact energy, a vast sea of glowing lava from horizon to horizon. Above, dust and toxic gases choke the atmosphere, an eternally gloomy and dark firmament.

The furious bombardment does more than leave craters and melted rock; larger asteroids strip copious amounts of atmosphere away. The first atmosphere is rich in hydrogen (to get a picture of Earth's first atmosphere, we need only look as far as Jupiter and Saturn: hydrogen, helium, methane, and ammonia). The giant impact that created the Moon tears an entire hemisphere of rock away as it strips the planet of most of its remaining atmosphere. A day on the Earth now lasts for only four hours.

The Earth continues to grow as asteroids and—to a lesser extent—comets add to its bulk. The planet morphs into a new, equally alien world. A crust begins to solidify across the entire globe at about the 4.3 billion year mark. We know this because we have dated zircons, the oldest minerals on Earth, to that age. Heat continues to escape as volcanoes break through, belching out a new, secondary atmosphere made of water vapor, carbon dioxide, sulfur and nitrogen. Water vapor adds so much to volcanic emissions that the new atmosphere is primarily steam and carbon dioxide, something like a wet Venus. The crust chills beneath a torrential rainfall as water vapor cools enough to condense and precipitate from the atmosphere. More water is added by asteroids and comets. Days and nights come quickly; the planet's spin now extends to five hours.

(continued)

4.2 billion years ago, the sky is clearing. Surface temperatures have dropped below the boiling point. The surface sees its first sunrises and sunsets, but that surface is a near-global sea, pummeled by a continuing hail of meteors. No continents have yet appeared. Instead, the dry land presents as volcanic islands or the raised rims of giant craters. Massive tides eat away at the rare bits of dry land, raised by a dramatically close Moon. But that Moon is drifting away. 100 million years after its birth, the Moon is halfway out to its current distance, and Earth's daylight has doubled. At 3.95 billion years ago, the magnetic field begins to seep from the molten core, but is less than half the strength of present day.

The gathering ocean waves pull carbon dioxide from the atmosphere, dissolving it in their waters. Much of this carbon dioxide will end up trapped within carbonate rocks. The air clears even more. As the Earth's interior begins to churn and fracture into plates, the continents rise from the deep seas.

As the Earth embarks upon a path to becoming the world we know today, giant impacts occasionally vaporize massive portions of the world's oceans. They continue to do so until 3.8 billion years ago. The air is still mostly carbon dioxide, but small continents have broadened and begun their parade across the planet's face, building the first mountain chains. The Sun and lightning have interacted with complex carbon molecules, seeding the oceans with novel chemistries. Sea floor volcanoes spew organics into the surrounding waters. A new force is about to change the world completely: life. At 3.5 billion years ago, the first cyanobacteria leave traces (Chapter 5). The atmosphere is now 98% nitrogen, and the day races by at 8 hours. The first supercontinents rise from the oceans at 3.3 billion years, and the first land bacteria arrives just 100 million years later. Average global temperatures now settle at about 24°C. We see the first major ice age at 2.8 billion years ago. At the 2.48 billion year mark, the Great Oxygenation Event transforms the atmosphere, dumping massive quantities of oxygen into the air. The presence of oxygen destroys atmospheric methane. Days have lengthened to 10 ½ hour periods. At 1.25 billion years past, the Earth's inner core solidifies, increasing the magnetic field. Changes in the Earth's interior and atmosphere bring a series of planetary "inventions", including a protective ozone layer 500 million years ago as oxygen levels rise. Dual ice ages cover the globe at the breakup of the supercontinent Rodinia some 720 million years ago (Chapter 6). The day is now 19 ½ hours long, and chilled temperatures make their way back up. Some 500 million years ago, life explodes across the planet's oceans and races across the land, bringing dinosaurs, Mammoths, ducks and people.

(For a brief history of Venus, see Chapter 4. Chapter 5 presents a brief history of Titan, and a brief history of Mars is posted at the end of Chapter 6.)

3

Earth=Mercury: Earth as a molten and bludgeoned world

Pick up a pencil. Grind some of the pencil "lead"—actually graphite—into the palm of your hand. Spread it around with your finger, feeling the granular texture. Smell the rich, pungent gunpowder-aroma. Now, imagine an entire world covered in the stuff, from undulating gritty hills to cratered plains to the ebony faces of wandering cliffs. This is Mercury. But it's not all black. A scattering of stars sparkle across the landscape in the brutal sunlight, mirroring the stars above in the airless sky. For Mercury is carpeted in tiny diamonds.[1] It is an exotic world, smallest of all planets. And it has a few things to teach us about the earliest days on our own planet Earth, when a vicious hail of asteroids and comets assaulted the landscape and scorched the air.

Mercury is a planetary raisin, a globe that has shrunken under the Sun's heat, leaving planet-wide wrinkles called Rupes. Despite the fact that Mercury's daytime temperatures soar to 800°F/427°C, ice encases the shadowed crater floors on its poles. In fact, Mercury's nighttime chill of -275°F/-170°C gives the planet the widest spread of temperatures of any planet (1100°F/600°C). The tiny planet should be lacking in most of the volatiles that stuck to the larger terrestrial planets. After all, it has spent most of its existence closer to the Sun, where fierce solar winds strip away atmospheres and even solid rock over time. Instead, it is rich in volatiles, perhaps more so than the early Earth or Venus. Some of those volatiles have erupted through the surface in a menagerie of volcanic forms, including vents, mounds, and collapsed pits. Mercury is a place of contradictions.

© The Author(s), under exclusive license to Springer Nature Switzerland AG 2023
M. Carroll, *Planet Earth, Past and Present*, Springer Praxis Books,
https://doi.org/10.1007/978-3-031-41360-5_3

Figure 3.1 Twilight before sunrise on Mercury brings spectacular views of the sodium cloud surrounding the planet. Distant mountains beyond the horizon cast shadows through the brightening sky. Hollows can be seen in three stages of development: a collapse feature forms at left in an area of darkened surface. At center, an active hollow generates haloes of bright material, possibly due to outgassing of vapors. At distant right, an ancient hollow has become inactive and erodes away beneath a constant drizzle of micrometeorites and solar radiation. (Painting by the author)

The planet's surface area is 75 million square kilometers, about one tenth that of Earth's. Flattened out, Mercury's combined real estate is twice the area of Asia. Mercury is the densest of the terrestrial worlds, and one of only two with an active magnetic field (the Earth is the other). Its orbital path around the Sun is eccentric in more ways than one. Its circuit is more oblong than the other terrestrials, and its distance from the Sun means that it completes two trips around the Sun for every three Hermian days. This means that in some locations, the Sun rises, turns back and sets again before rising to cross the sky (sunset is the reverse, with a double sunset).

Adding to Mercury's unique nature is its albedo: despite being closest to the Sun, Mercury is a dark world. Our first close views of Mercury came from Mariner 10 back in 1974 and 1975. Flight navigators sent the craft on a clever, looping orbit that would pass by Mercury three times.[2] Because of the

---

[1] 53rd Lunar and Planetary Science Conference, 2022.

geometry and timing of the flybys, Mariner imaged a total of just under half of the globe. Initially, engineers thought something was wrong: Mercury's surface seemed impossibly dark. But the spacecraft cameras were fine. It was Mercury that was amiss. The little planet was darker than our gunpowder-grey Moon.

Some of the Moon's dark visage comes from its iron-rich nature. But on Mercury, the surface lacks those darkening minerals. Some researchers posited that graphite might be to blame, but no proof came until years after Mariner 10, thanks to the MESSENGER (Mercury Surface, Space Environment, Geochemistry, and Ranging) Mercury orbiter. Its four-year mission[3] obtained neutron spectroscopy, data that shed light on the subject. MESSENGER scrutinized dark patches within and adjacent to craters. The dark substance, known as Low Reflectance Material (LRM), contained carbon, lots of it. That carbon was probably exhumed from a layer of graphite under the surface. Eons of meteor "rain", along with volcanic activity and the tectonic folding of Mercury's crust, likely tilled the soil, bringing the graphite up onto the sterile surface and darkening the entire alien landscape.

Some Mercury experts assert[4] that the graphite hearkens back to the primordial ages of the solar system, when magma seas covered much of the planet (the dark "seas," or maria regions, of our Moon bear witness to a similar past; they used to be oceans of molten rock). If Mercury's chemical makeup resembled what it is today, most of its minerals in that magma bath would have sunk to the bottom. The only mineral light enough to rise to the surface is graphite. That geologic movement gifts Mercury with one of the darkest surfaces in the solar system.

---

[2] En route, the craft flew by Venus, using its gravity to reach Mercury. This was the first gravity assist attempted, a technique that is now a critical part of many space missions.

[3] "…to boldly go to Mercury where no one had gone before…"

[4] See Simone Marchi's report in the July 4th 2013 issue of *Nature*

Mercury's Diamond tiara

While Mercury's stony rind may house a cache of buried mineral treasure, natural wealth of a different kind may crown the planet's crust: diamonds. Billions of years of violent impacts may have transformed nearly a third of Mercury's crust into a jeweler's paradise.

On Earth, diamonds crystallize far beneath the surface, some 150 km down. From there, volcanic eruptions or the rising of continental plates bring the gems to the surface. Diamonds form under high pressure and temperature, the same conditions present as a meteor vaporizes the ground upon impact. Carbon is the seed that sprouts a diamond, and Mercury is probably full of it. The little planet's early magma ocean held quantities of graphite, a form of carbon that can lead to the jewels. Computer modelers called upon their machines to simulate 4 billion years of impacts on a surface of graphite.[5] The models demonstrated that if Mercury was covered in 300 meters of graphite—a reasonable assumption considering its volcanic history—the prodigious pummeling would have produced 16 quadrillion tons of diamonds. That's sixteen times the amount of diamonds in the entire Earth. While some diamonds might be destroyed by later impacts, team members say that type of loss would have been "very limited."

Says Mercury researcher Ron Vervack, "Walking across Mercury, you might come across large-scale (but not global) Low Reflectance Material regions where diamonds are embedded in a bunch of [graphite]. It's kind of a rich man, poor man planet."

A different kind of ambiance

All of the terrestrial worlds (Mercury, Venus, Earth and Mars) share two phases in their common heritage: their formation within the Sun's dust-engorged nebula, and a late period of giant impacts (called the Late Heavy Bombardment, see Chapter 2). During the epoch of giant impacts, comets and asteroids introduced new chemistries into the forming planetary embryos. Some of the impacts even stripped away the crusts of planetoids.

The primordial planets harvested their initial atmospheres from the surrounding solar nebula. The skies of the early Earth probably began with a "reducing atmosphere" made up primarily of hydrogen and helium, with a smattering of water vapor, methane, and ammonia.[6] These are the gases that ruled our solar nebula as our planetary system formed, and they are still prevalent today in the outer solar system. But the reduced atmosphere of our home

---

[5] Results presented first at the March 10, 2022 Lunar and Planetary Science Conference, Woodlands, Texas.

world couldn't hold on for long; hydrogen and helium are light molecules, and the solar wind stripped the Earth of most of its ancient, first atmosphere. (Today, our atmosphere is starved of these elements compared to the cosmos around us.)

Those earliest epochs saw the cooling and solidifying of our planet's surface, and the beginning of the water cycle as the environment cooled. Darkness visited the hadean vistas as hazes and dense layers of cloud blocked sunlight from the surface. The primordial panoramas would have been an inspiration to Dante, with seas of molten rock, toxic vapors pouring from vents and mountains, lightning tearing the sky and incandescent meteors thundering through the mist, some ending in blinding explosions and a hail of stone (Figure 3.2). This little bit of Hell engendered a second atmosphere, one made up primarily of steam, before leading to a third atmosphere, one dominated by nitrogen and life-generated oxygen.

A similar hellish infancy must have visited Mercury, although the little planet never had an atmosphere as dense as ours. From the stripping away of the atmosphere and its supplanting by volcanism and asteroids, our two planets diverged.

What was that first atmosphere like? We gain insights into the atmospheres of the early planets by studying the gas giants. Jupiter and Saturn have retained many of the same gases present in the solar nebula. We can also gather clues from the study of the oldest meteorites, especially the carbonaceous chondrites. These contain carbon dioxide, carbon monoxide, hydrogen, nitrogen, and sulfur dioxide. In the case of the carbonaceous chondrites, we also find methane, nitrous oxide, and important carbon-based compounds like benzene, toluene and naphthalene. Some of these are considered as important contributors to life processes.

As with Mercury, Earth's primordial atmosphere may have thinned nearly to the point of vacuum, but the air was soon recharged with gases that volcanoes belched out: water vapor, hydrogen, water, carbon dioxide, nitrogen and methane. Some of the mix came from incoming asteroids and meteors. What the Earth did not get from these sources was its second most abundant gas: oxygen. Oxygen was quite rare early on; its presence today is the byproduct of life (for more on this subject, see Chapter 5).

---

[6] Some studies indicate higher levels of carbon dioxide, even in the first atmosphere.

Figure 3.2 A tale of two planets: In their infancy, both Mercury and the Earth endure a brutal hail of meteors. Just after sunset, the stones on the plains of Mercury (left) glow from the Sun's heat. The solar corona paints elegant curtains across the firmament, dappled with neon-pink flares. It is late in the Pre-Tolstojan period, and in the near-vacuum, ballistic volcanic eruptions sear Mercury's panorama (left horizon), and fresh craters free molten rock from beneath. At the opening of the Archean age, Earth's surface (right) thunders with even more activity as volcanism transforms the landscape into a hellish inferno. The dense atmosphere sizzles with incoming meteors and lightning. Soon, the skies will clear. (paintings by the author)

The story of Mercury's atmosphere took a wild detour from the Earth's, and to understand it we must first track the little planet back to its beginnings. One of the great mysteries of Mercury is its large core. We have seen that Earth's core is a bit oversized (see Chapter 2), probably due to the impact of the hypothesized Theia. Mercury finds itself in a similar, but more drastic, position. While the Earth's core takes up 17% of its volume, Mercury's core accounts for 57% of the planet's bulk. Its core is higher in iron than any other planetary core. Early theories suggested that Mercury formed in the inner system before the Sun became stable. In this scenario, Mercury started with twice as much mass, but as the Sun's output stabilized, temperatures on Mercury's surface soared to the vapor point of rock. Much of Mercury's outer crust would have been blown away by the infant Sun's solar wind.

A second possibility is that a massive impact or impacts stripped Mercury of its outer crust layers, leaving behind a dense globe consisting mostly of its core materials. In collisions of planetoids with core-to-total masses roughly similar to Earth, models show a single impact scattering half the total mass, mostly silicates from the outer layers. The greater part of the energy in these catastrophic collisions converts to heat, and the majority of the colliding planets and their ejected matter vaporizes, much as we saw for the Theia impact that led to Earth's Moon.

A 2004 study[7] revealed that most collisions do not result in the fusing of the two planetary embryos. Instead, fully one third are hit-and-run encounters—glancing blows in which the two impactors remain separate. Further research shows that another conceivable scenario involves a proto-Mercury skimming a larger planet and losing much of its crustal material, and then settling into an orbit closer to the Sun. The amount of material that Mercury loses in the models depends upon factors like the angle of collision and the velocity of the impact.

An intriguing possibility is that the proto-Mercury formed further out in the solar nebula, and then drifted sunward to collide with the larger planetary embryo that ultimately became Venus. Most of Mercury's ejecta would have ended up on Venus. To confirm this Venus/Mercury dance, researchers would need to know more about the composition of the Venusian crust, as well as more detailed models of terrestrial planetary collisions. A Venus/Mercury collision is a dramatic explanation for several of Mercury's mysteries, such as its large core and certain ratios found within its crust.

Something in the air...but not much

The gases in Earth's atmosphere are not what they were. The atmosphere has gone through its own evolutionary history, but it is informative to compare the air we breathe with atmospheres from other sources. The Earth's modern atmosphere contains, in descending order, nitrogen, oxygen, argon, carbon dioxide, and water vapor. It also has traces of helium, hydrogen, methane and a few inert gases, echoes of the past primordial mix of gases here.

The atmospheres of the gas giants more closely resemble the solar nebula.[8] Deep oceans of methane, ammonia, hydrogen and helium coexist within our solar system's giant worlds.

Meteorites have trapped gases from the embryonic years of our solar nebula. Laboratory tests confirm the presence of carbon dioxide, carbon monoxide, hydrogen, nitrogen, sulfur dioxide, methane and other gases.

The Earth itself contributed to its own atmosphere. Today's terrestrial volcanoes erupt water vapor (about 73%), carbon dioxide, sulfur dioxide, nitrogen, and other gases. Geysers and fumaroles vent even higher percentages of water vapor (roughly 98%), with hydrogen, methane, hydrochloric acid, carbon dioxide and ammonia rounding out the brew. All of these sources have contributed to the skies of Earth that we enjoy today.

---

[7] *Accretion Efficiency During Planetary Collisions* by Agnor and Asphaug, American Astronomical Society 2004.

[8] The Galileo probe directly sampled Jupiter's atmosphere in December of 1995.

Figure 3.3 Any volcanic eruptions in Mercury's near-vacuum would resemble the umbrella-like geysers of Jupiter's highly active moon Io. The plume over Io's northern hemisphere emanates from the volcano Tvashtar, and towers 330km into the sky. (New Horizons spacecraft photo courtesy NASA/Johns Hopkins University Applied Physics Laboratory/Southwest Research Institute)

Mercury matured into a battered, nearly airless globe. As with the other terrestrial planets, Mercury lost its initial atmosphere—what there was of it. The little planet began to replace it through the escape of interior gases. In its near-vacuum environment, Hermian volcanoes resembled the explosive geyser-like eruptions on Jupiter's moon Io today (Figure 3.3). But Mercury had a few strikes against it. The loss of Mercury's diaphanous sky came at the hands of several major causes. The first was the simple fact that Mercury has low gravity. That weak gravitational pull on the molecules making up its atmosphere could not hold on to the atmosphere for long. The escape of Mercury's atmosphere was exacerbated by heat. Heat is basically the movement of atoms. As the Sun's heat beat down on Mercury's fragile blanket of air, it triggered ever-increasing movement of the atoms of hydrogen, helium, sodium, potassium, and other gases. Speeding atoms collided and rushed in every direction, finding little resistance from Mercury's weak gravity, and bounced off into the vacuum of space. The planet has a protective, though weak, magnetosphere (about 1.1% the strength of Earth's). In the past, Mercury's magnetic "bubble" was stronger, and may have acted as a shield against the Sun's atmospheric erosion, but Mercury's proximity to the Sun exposed it to a flood of solar wind, which also stripped away the air. Soon, this battered and cooked world lost most of its gases in the face of the Sun's blistering solar wind. But not all of them. Mercury developed a rarified primeval atmosphere that still hangs on today. Its air—practically a vacuum— includes 42% oxygen, 29% sodium, 22% hydrogen, 6% helium, and trace elements of potassium, argon, neon, carbon dioxide, water, nitrogen, xenon and krypton.

"The tenuous Mercury atmosphere is complicated, and knowing what it may or may not have been doing in the past is not an easy thing," says Ron Vervack, planetary atmospheres researcher at the Johns Hopkins University's Applied Physics Lab. "I think I can pretty safely say that it never had anything like a Martian atmosphere in terms of either pressure levels or global presence."

The University of Colorado's Andrew Wilcoski agrees with Vervack's summation. Wilcoski has been studying the thin atmosphere of the Moon as it might parallel other worlds. "The surface pressure of one of these atmospheres would have been on average about 0.5 mircobar (~0.000007 psi). This is about 10,000 times thinner than the present-day surface pressure of Mars, and 2 million times thinner than Earth's surface pressure. So these atmospheres were very thin." Wilcoski draws a difference between an exosphere like that on Mercury and a "collisional atmosphere". While an exosphere is so thin that molecules do not collide with other molecules before hitting the surface or escaping to space, collisional atmospheres have enough density for their molecules to interact. "Because they were collisional, we expect these atmospheres to have had dynamics, like circulation patterns and winds," Wilcoski says. "Similar processes may have unfolded on Mercury in its past."

But Ron Vervack and others point to the possibility that Mercury may have developed more substantial, localized atmospheres in the early days when the prevalence of asteroids and comets raining down in the early solar system was high (during the Late Heavy Bombardment period). "You could imagine that a bunch of comets raining down on Mercury might have led to temporary [mini-atmospheres] at places where a comet struck Mercury. At that local position, water vapor (mixed in with carbon monoxide, carbon dioxide, and various other organics and refractory species) would form a local exosphere over the region." Transient atmospheres like this would quickly be dispersed and then lost owing to a variety of processes, but Wilcoski's work indicates that volcanic eruptions could have triggered a new localized atmosphere that lingered for up to 2500 years before disappearing. Volcanic outgassing and meteor deliveries are not the only way Mercury could have gotten its assortment of volatiles. Vervack adds that there is also the possibility of "a much larger collision with something that could have created a more substantial temporary atmosphere (e.g., the formation of Caloris Basin). But it would never be an atmosphere in the true sense, not even on Martian scales."

How to make an exosphere

Air is hard to come by on Mercury. Where does it come from? The exosphere of Mercury has three primary sources. All involve something hitting the surface. The first is called photon-stimulated desorption. In this process, the Sun's photons give surface atoms enough kick to break free of the surface. The kick is a low-energy one, so the atoms don't go high or fast enough to escape. Instead, they generally travel in a ballistic arc, returning to the surface. At the cold poles, the atoms may stick to the surface. In warmer areas, they bounce and migrate across the surface. (This process is responsible for most of the sodium in the exosphere).

The second source of Mercury's thin exosphere is called impact vaporization. Micrometeoroids hit the surface and vaporize material, ejecting it into the "air". These impacts are higher energy events than photon-stimulated desorption, casting material over 1000 km up. Atoms that reach these higher altitudes will be affected by radiation pressure: solar photons "push" the atoms in the anti-sunward direction, creating Mercury's sodium tail, which waxes and wanes with the strength of the radiation pressure as Mercury orbits the Sun. Micrometeoroid impacts add most of the calcium and magnesium into the exosphere.

The third process, sputtering, takes place when charged particles are accelerated by the magnetosphere. The planet's lines of magnetism carry particles down to the surface, where they knock atoms from the ground. There is some debate about just how much Mercury's magnetosphere contributes to the exosphere.

In contrast, the Earth's atmosphere is held to the planet by higher gravity (Mercury has only 38% the gravitational pull of Earth). Temperatures are lower, causing molecules to move more slowly and remain together, and the magnetosphere of the planet protects the atmosphere from the kind of erosion that Mercury's exosphere endures from the solar wind.

Did transient atmospheres exist on Mercury? There is one way to find out: their vapors would probably have made their way to the poles. In fact, ices have been discovered in the permanently shadowed crater floors of both Hermian poles.[9] The first hint of such ice came not from spacecraft, but rather from Earth-based radar studies. In 1992, astronomers beamed radar to Mercury using the Arecibo radio telescope in Puerto Rico.[10] The radar return,

---

[9] Since Mercury's spin axis is essentially vertical, it has no seasons, and the Sun remains directly over the equator. Deep polar canyons and craters have areas permanently sheltered from direct sunlight.

[10] Later studies also used the antennae at Goldstone and the Very Large Array.

Figure 3.4   One of the first portrayals of ice at the pole of Mercury. Here, a landslide has exposed fresh ice to Mercury's thin exosphere, essentially the vacuum of space. The 1992 painting was done for a Popular Science article that released in 1993. (painting by the author) Inset: North pole of Mercury. Red areas indicate shadowed regions. Yellow delineates ice deposits imaged by Earth-based radar studies. The pattern confirms that all ice deposits seen by radar lie in areas of persistent shadow. (NASA/Johns Hopkins University Applied Physics Laboratory/Carnegie Institution of Washington)

or "bounce", carried the distinct signature of water ice. The radar studies identified some 20 sites that were radar "bright," having the characteristics of frozen water. Later study by NASA's MESSENGER orbiter confirmed the discovery, mapping the ice regions in more detail (Figure 3.4).

MESSENGER offered three lines of evidence. First, images from the orbiter spied light areas that lined up with the radar-bright areas in the permanently shadowed spots at the poles. Added to this, MESSENGER brought to bear its neutron spectrometer, which measured hydrogen concentrations at the same radar-bright locations. Hydrogen is the component of water that the spectrometer could sense. The instrument indicated that the polar deposits exist as a layer some tens of centimeters thick, covered over by surface material 10 to 20 cm thick. This seals the water ices from the vacuum of space, which would wick the water away quickly. MESSENGER's third line of evidence came from its laser altimeter, which charted the topography of the Mercury landscape. The laser fired more than ten million times, creating detailed topographic maps. These readings confirmed the radar and spectrometer results. Mercury's ices huddle against the pole-facing crater walls and various slopes, the coldest locales within Mercury's arctic regions. And water is only one of many volatiles that Mercury holds close.

Mercury's volatile temperament

Mercury's interior is rich in radioactive material, including potassium, uranium and thorium. The heat from these materials undoubtedly helped to power Mercury's ubiquitous volcanoes. Radioactive elements may also contribute to Mercury's molten core, which generates a robust magnetic field, revealing that its interior was dynamic and active in its childhood. A 2007 radar study revealed that Mercury still has a large molten core, which generates a magnetic "bubble", or magnetosphere, around the planet. But with a core the size of Mercury's the magnetosphere is weaker than expected. Why? Mercury's core has two sections, an inner solid metallic center cocooned within an outer, molten metal core. One possibility is that the outer core is layered, or stratified. The layers could be caused by temperature gradients, or because of composition (light elements settle upward). Both of these could lead to the observed magnetic fields. Others suggest that if sodium is a major component of the core, temperatures at the upper regions of the core may drop below its melting point in a layer called the iron-snow zone. There, iron crystallizes and sinks through the core. (This is the opposite condition to that of the Earth, where crystallization first happens at the center of the core). This top-down crystallization might generate two separate regions of magnetic generation, and each field might dampen the effects of the other, weakening the overall magnetosphere.

Some of Mercury's interior elements make it to the top (via volcanism or impacts), where they meet volatiles like sodium, potassium, chlorine, and sulfur. Even hydrogen and some organics shelter at the poles. MESSENGER's geochemical studies revealed surface concentrations of volatiles similar to those found on Mars. Until recently, Mars was thought to be the most volatile-rich terrestrial planet.

The elements are not scattered evenly across Mercury's surface. Instead, they congregate in distinct regions known to geologists as terranes. Their chemical sorting may have come at the hands of different types of magmas (various lavas rich in elements of differing amounts). Geochemists recognize four distinct regions on Mercury: the Northern Terrane, Caloris Interior Plains Terrane, the High Magnesium Terrane and the Low-Fast Terrane. Each region boasts its own mix of volatiles and mineral types. For example, the Northern Terrane lacks as many craters as other regions, as it has been flooded by magmas that have left smooth plains. These plains are enriched in sodium, potassium, and chlorine. By contrast, the more heavily cratered High-Magnesium Terranes are richer in iron and are high in olivine (a greenish mineral that crystallizes out of magma). Other regions show unique mixes of volatiles as well.

Larry Nittler is a Deputy Principal Investigator on the MESSENGER science team.[11] Like other researchers, he finds Mercury's rich volatiles puzzling. "The ice and organic volatiles in polar craters are a sign of recent delivery, likely by comets, but Mercury is also relatively rich in moderately volatile rock-forming elements like sodium, potassium, and chlorine, which must reflect how the planet formed. We really don't know either why Mercury has such a large core or why it is so volatile-rich." Mercury's building blocks included materials that formed at relatively low temperatures, likely from the outer solar system beyond the "frost line". Models like the Grand Tack show much radial mixing of planetesimals in the inner system.

The polar ices can come from only two sources: meteor and comet impacts and outgassing from Mercury's interior. Comets likely have deposited ices in those deep craters over time, but Mercury's surface shows many signs of venting as well. Those gases would have migrated to the cold poles, where they would condense in the permanently shadowed crater floors. In that sense, the polar deposits may well represent the remnants of Mercury's earliest exosphere. In other words, the ancient atmosphere of Mercury is waiting for us to study it, and it has been waiting for millions or even billions of years, frozen to the floors of permanently shadowed craters.

The chemical smorgasbord of volatiles contributes to what goes on above. For someone standing on Mercury's surface today, the thin air would present a fascinating, though subtle, firmament. The rarified exosphere would be most obvious on the night side near dawn or dusk. The gases glow in various colors and shapes owing to how they are generated and which atomic species comprise them. Sodium shines with the same hue as sodium street lamps on Earth, lending an amber tint to the evening sky. A great sodium tail streams away from Mercury, carried along in the blazing solar wind. At the equator, this sodium banner streams around the planet opposite the Sun, appearing as a yellowish radiance along the horizon, brighter in the direction of the unseen

---

[11] Nittler is also a science team member on the upcoming European/Japanese Mercury mission BepiColombo.

Sun. Overhead, the sodium gleam fades away as they enter Mercury's shadow (the gases fluoresce in sunlight). The sodium sheen would undulate with changes in the solar wind.

Another element, calcium, makes its presence known by adding a purple tint to the sky. The lavender luminosity is strongest toward the dawn equator, since this is the direction of travel as Mercury rams through the interplanetary dust and solar particles on is orbital path. Combined with the sodium, shades of green may appear in the twilight just before sunrise.

Mercury's "Sleepy Hollows" and other exotic volcanoes

Mercury's interior gases should have disappeared long ago. As Arizona State University Mercury expert Mark Robinson quipped, "I think the MESSENGER mission proved that Mercury cannot exist!" Those puzzling volatiles (like water, carbon monoxide, and sodium) lead to some of the strangest features in the solar system. As vapors break through the surface of Mercury, they leave structures typical of volcanic eruptions. Investigators have identified a multitude of diffuse, bright deposits surround rimless depressions, interpreted as pyroclastic (explosive) volcanic vents (Figure 3.5).

Lava flows on Mercury are ancient. Major flows and eruptions of molten rock appear to have ceased around 3.5 billion years ago. After that time, Mercury's global contraction pinched off many of the volcanic sites. Smaller scale volcanism carried on at locations where the crust was weakened, most commonly by impacts or faulting of the crust.

In fact, the face of Mercury has been sculpted by three primary sources: tectonics (the movement of the crust and what lies beneath), volcanism and impact cratering. The deposition of Mercury's surface material seems to have been ruled by volcanism. Evidence abounds: smooth plains that embay low-lands, vent-related depressions, channels formed by the flooding of lava across the surface, and the lack of large impact basins (compared with the Moon, for example), which have been buried by lava "seas". A broad lowland region in the northern hemisphere called Borealis Planitia is a volcanic plain covering six percent of the entire planet. Within it, some fifty rimless vents—most surrounded by bright haloes of deposits—bear witness to pyroclastic volcanism. Great smooth plains cover a total of 27% of Mercury's surface, demonstrating just how important a part volcanism has played in the planet's past. The majority of volcanic forms on Mercury emerge within impact craters, but volcanic sites form at the edges of smooth plains. The relationship between

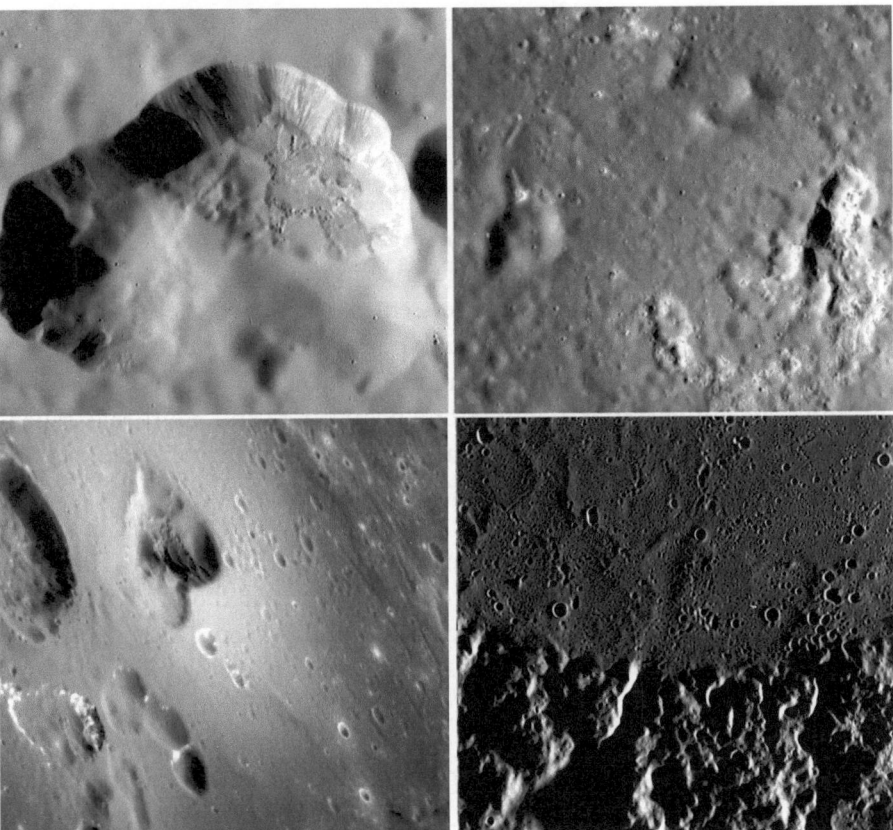

Figure 3.5 Clockwise from top left: Some surfaces of Rachmaninoff basin are blanketed by fine particles ejected explosively from the vent. Other regions, like those against the far side of the depression, are more textured, indicating a younger age. Material has slid down the walls, and a bright horizontal layer just beneath the surface can be traced along much of the northeastern rim. Top right: The floor of Mistral crater. The large irregular depressions are pyroclastic volcanic vents that have erupted bright material. The small depressions with bright halos are hollows. Bottom left: An irregular depression sits atop what may be a low-relief shield volcano (although the classification is controversial because of the gentle slopes surrounding the vent). Bottom right: In the southern portion of the Mendelssohn impact basin, smooth plains in the top half of the image stand out against the rugged, heavily cratered region that forms the basin rim across the lower part of the image. The smooth plains were created when fluid lavas flooded across the basin floor. The low Sun angle reveals subtle ridges marking the rims of impact craters buried by the lavas. (all images courtesy NASA/Johns Hopkins University Applied Physics Laboratory/Carnegie Institution of Washington)

these sites and smooth lava plains echoes that found on lunar volcanic plains,[12] where pyroclastic eruptions take place at the margins of lunar maria. The bulk of the Moon's maria regions, dark seas of basalt,[13] are estimated to have erupted between 3 and 3.5 billion years ago. They cover some 16% of the global lunar surface (considerably less than Mercury's volcanic flood plains). On the Earth-facing ("near-side") hemisphere, maria regions account for 31% of the surface. On the far side, which is far more rugged, dark maria plains cover only 2% of the surface. The difference in the two hemispheres may be due to a preponderance of radioactive elements closer to the surface on the Earth-facing side. Elements like thorium, potassium and uranium would trigger melting of subsurface rock, causing magma to rise and flood the surface. This difference in character between one hemisphere and another is called a crustal dichotomy. Crustal dichotomies are also found on Mars, Saturn's moon Iapetus, and several other moons to a lesser extent.

These heat-producing elements are part of a lunar geochemical mix called KREEP, a combination of potassium (K), rare-earth elements (REE) and phosphorus (P). The KREEP may have been concentrated on the near side of the Moon as a result of an impact on the opposite side. The impact that formed the Moon's vast South Pole–Aitken (SPA) basin would have sent a massive plume of heated magma along a shock wave through the lunar interior. That plume would have carried KREEP to the Moon's nearside. That KREEP-y concentration would have given rise to the volcanism that created the nearside volcanic flood plains that make up the maria.

While much harder to see, the Earth has similar lava plains, called Large Igneous Provinces. The provinces result from plumes of hot mantle material that convect through "hot spots", thinning regions of the crust that allow magma to surface from below. Examples include the Columbia River Basalt Group of the northwestern U.S. (which erupted only 15-17 million years ago) and India's Deccan Traps, one of the largest volcanic features on Earth. A trap is a continental region flooded by basalts. (see the box: Large Igneous Provinces (LIPs) and Earth's extinctions).

When basaltic lavas flood continents, they rise through fissures that split the continental crust into rift valleys. Magma flows to the surface through the fissures, solidifying as dikes—hard rock walls left behind as softer surrounding rock erodes away. Dikes have been located on Mercury and all the other

---

[12] Some apparent volcanic plains may also be due to tectonic forces, which leave similar features.

[13] Lunar basalts contain more iron than basalts on Earth or Mercury, but lack any minerals that have been modified by water.

terrestrial planets, along with the Moon. This process can even split entire continents, leading to ocean basins between.

Large Igneous Provinces (LIPs) and Earth's extinctions

The Deccan Traps of India hold over one million cubic kilometers of basalt. The furious volcanic activity it took to build such a province was, at one time, the leading candidate for the extinction of dinosaurs and other life at the end of the Cretaceous period. Increased volcanic eruptions contribute to reduced light and oxygen levels globally, and can dramatically change climate. In light of the potency of Large Igneous Provinces, the generation of LIPs on Earth have been provisionally tied to other major extinction events.

Other mass extinctions have occurred in close proximity to LIPs. The Permian-Triassic, Pliensbachian-Toarcian, and the Triassic/Jurassic extinction events all took place at times close to emplacement of LIPs. 252 million years ago, the Permian-Triassic event may have been triggered by the rise of the Siberian Traps LIP. This "Great Dying" marks the transition from the Permian to the Triassic periods. 69 million years later came the Toarcian Anoxic Event, when oxygen levels in the world's oceans dropped, devastating the food chain. The Karoo large igneous province in southern Africa and the Ferrar large igneous province in Antarctica are linked to this massive extinction event. The Karoo-Ferrar LIP coincided with the breakup of the supercontinent Gondwana 183 million years ago. A wide variety of Mesozoic marine life was devastated, although marine reptiles and land-based life were largely unaffected. Yet another major extinction event visited the Earth 201 million years ago, at the boundary of the Triassic and Jurassic periods. The Central Atlantic Magmatic Province, largest of all igneous zones on Earth's continents today, covers 11 million square kilometers. Its occurrence overlaps the breakup of the Pangaea supercontinent, which birthed the Atlantic Ocean. Elements of the LIP are found in the Americas as well as in southwest Europe and northwest Africa.

Why would the formation of a large igneous provinces cause mass extinctions? The large volcanic eruptions that emplaced the LIPs often release massive quantities of sulfate gases. These combine with water to form sulfuric acid. The acid in the atmosphere causes global cooling, but also damages vegetation by acidic rainfall. The change in ocean water acidity also causes vast algal blooms, whose decomposition starves the water—and anything that might be swimming in it—of oxygen. The acids also react with metals in seawater, reducing oxygen even further. The LIPs across the Earth are largely hidden by erosion, water and forest, but we can see their brethren clearly on the immense plains of Mercury. (For more on mass extinctions, see Chapter 6)

Although they have carried out extensive searches of Mercury's vast lava plains, researchers have found no classic volcanic mountains. This is an important and puzzling revelation. It means that Mercury's volcanoes are somehow different from those on the other terrestrial planets. Mercury lacks the precipitation, wind, and plate tectonics that have wiped out many traces of Earth's earliest volcanoes. Why are there no remnants of sloping cones or soaring volcanic peaks? Their rarity may be due to a difference in the nature of eruptions early on. Mercury's primordial volcanoes may have been unable to build large structures because their lavas may have been thin and fluid. Eroded lava flood channels trace the source of Mercury's vast smooth plains as magma-related. But Mercury lavas likely thickened through time, with more recent flows being higher in viscosity. As the planet cooled, eruption temperatures would have been reduced, producing molten rock with more crystals and silica. These thicker lavas could have built more classical volcanic structures.

Scientists at the Open University in the UK[14] may have spotted two candidates for volcanic mounds, but the structures may also be impact related. The team suggests that, "most eruptions on Mercury were too large and rapid" to fabricate lasting structures. Mercury's early lava spread out quickly to form the planet's smooth plains instead of building "up" as volcanic mountains, cones and mounds. (This contrasts with the Earth, where even the most ancient volcanic activity was probably robust enough to create classic volcanic domes, cones and mountains.) Smaller eruptions that might build such structures may have broken through only later in Mercury's development, when the planet had cooled and was generating less quantities of magma. Cinder cones or volcanic mountains may exist on much smaller scales. This concept is bolstered by the numerous volcanic structures seen late in MESSENGER's mission, when the spacecraft was commanded into a lower orbit (Figure 3.6). High-resolution images reveal pit craters, pyroclastic deposits, and lava flows surrounding streamlined landforms. Later volcanism seems to be related primarily to impacts, which would have broken through the crust to free magma. Mercury's "intercrater plains" predate the smooth plains regions, but are likely older versions of the smooth plains.

---

[14]Volcanic Shields on Mercury Identified at Last? by J. Wright, et al, *Lunar and Planetary Science* XLVIII (2017)

Moon volcanoes come and go

Mercury's most vigorous volcanic activity appears to have ended during its Caloris era, although eruptions and flows may have continued at subdued levels into the Mansurian. The primary reasons for cessation of volcanism were thought to be the waning of heat from accretion and impacts, the dying of radiogenic material, and Mercury's small size, which would tend to cool off more quickly than the larger terrestrial planets. Similar timelines for geologic lethargy, compressed into an even earlier scale, were thought to follow on the Moon, a smaller body whose volcanoes should have quieted even sooner in its history. But the Chinese lunar sample return mission Chang'e 5 has caused a rethinking among volcanologists and planetary geologists about how quickly the Moon cooled, and consequently how quickly Mercury did the same. Chang'e 5 made landfall on the Moon's Oceanus Procellarum in December of 2020. The site was chosen for its comparative youth: the vast lava plain is sparsely cratered. China's lunar envoy drilled a core sample and scooped soil and rock, returning nearly four pounds of the precious material to the savannah of Inner Mongolia.

Initial studies reveal that the samples are younger than any other rocks returned from the Moon by Apollo and Luna robotic missions. The material dates to 1.96 billion years back. The new studies show that our Moon was volcanically active as recently as 2 billion years ago, which doesn't fit the timeline assumed by earlier models. Planetary geologists had concluded that small bodies like the Moon or Mercury cool fairly quickly. Young worlds keep interior warmth through the heat of accretion, impacts, and radiogenic heat from elements like uranium and thorium. The smaller the body, the more quickly that heat dissipates. Models indicated that the Moon's volcanoes should have chilled out long before the time of the Oceanus Procellarum samples.

How did the Earth's small Moon remain active for so long? It may be that larger quantities of radiogenic material deep in the Moon's mantle are heating things up, but isotopes found in the Chang'e 5 rocks seem to contain about as much of those materials as the Luna and Apollo samples do. Alternatively, it may be that radiogenic elements come in different amounts and mixes than researchers thought. One new proposal involves tidal heating. In the days of the Moon's formation and active volcanism, the Earth's gravity may have pushed and pulled on our satellite, keeping its interior molten. After all, the Moon began in a much closer orbit than it has today (Chapter 2).

One thing is certain: small worlds like the Moon and Mercury may not follow the planetary cooling-off processes in ways that we currently understand. Even the smallest of planets are still full of surprises.

While vents, fissures, flows of material, irregular depressions, bright deposits and chains of collapse pits all point to past volcanic activity, strangest among the plethora of volcanic features parading across the face of Mercury are the enigmatic hollows, collapse features unique to Mercury. These puzzling formations take the form of sunken pits, often encircled by bright halos

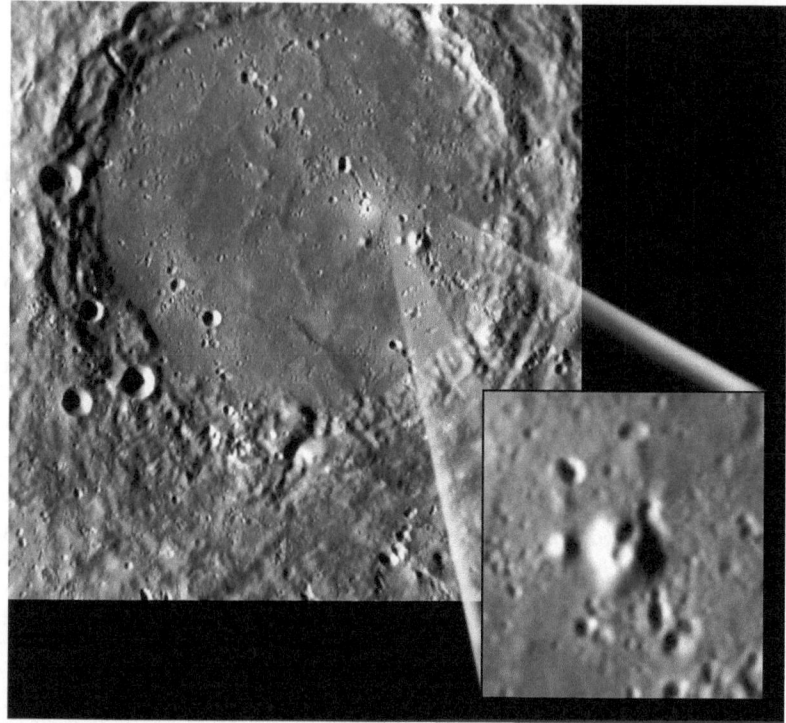

Figure 3.6 A possible cinder cone or similar volcanic structure was located late in MESSENGER's mission within Heany Crater. (NASA/Johns Hopkins University Applied Physics Laboratory/Carnegie Institution of Washington)

draped across plain and peak, glowing like pearls cast across a black velvet surface. Many appear to be eating away summits or crater rims.

Mercury's hollows were first spotted in images taken by Mariner 10 during its three flybys of the planet in 1974 and 1975. But Mariner's imaging resolution was too low to understand what Mariner scientists classified as "bright, ill-defined patches." But all that changed with the MESSENGER mission's three flybys and years-long orbital mission, from 2011 to 2015. With MESSENGER's advanced systems and long-term study, the mysterious hollows came into clearer view. A 2014 paper from a team led by Rebecca Thomas summed up the details of the hollows, describing them as "sub-kilometer scale, shallow, flat-floored, steep-sided rimless depressions typically surrounded by bright deposits and generally occurring in impact craters." Small clusters emerge on floors or along the rims of larger or complex craters, and some extensive groupings coalesce into large depressions. They form on crater walls, rims, ejecta fields, terraces and central peaks. They even arise on the hills and mountains that form central rings in vast impact craters. Hollows congregate in chains along the base of rises where crater floors meet the outer rim, or at the foot of central peaks. Occasionally, hollows can be found on their own, isolated from each other, sunken into flatlands or rolling plains. These may be associated with ancient, degraded crater rims or peaks. Some of the isolated hollows lack the typical bright haloes, instead being surrounded by dark material. They make up the darkest spots on the planet.

Thomas and other researchers point out that the hollows are quite different from volcanic collapse pits, something found plentifully across the little planet. Volcanic pits are deep, with rounded edges and irregular, rough floors. But the exotic hollows are shallow, with smooth floors, scalloped margins and often a blue coloration. Coronas of bright material surround these mysterious pits with as yet unknown substances. The depressions range from tens of meters to several kilometers across (larger hollows are usually merged collections of small ones). At some sites, such as the floors of Kertesz and Hopper craters, the hollows have uniform depth over an extended area. Because of their small size, depth measurements have been difficult, but a study of 27 clusters carried out by Thomas' team charted depths between 5 and 98 meters (an average of 47m deep). Two other studies found average depths of 37 meters and 24 meters, so the wall of a typical hollow is about the height of a seven-story building.

On the Earth and other worlds, a variety of processes can lead to irregular rimless depressions (Figure 3.7). Terrestrial pits form from subsurface wasting, the collapse of lava tubes or other cavities below the surface, volcanism, wind and water erosion, and various tectonic forces including faulting. The hollows present a different story.

On Mercury, hollows do not seem to result from secondary impact craters or classic volcanic pits. In the case of impact craters, when a large asteroid or comet tosses out ejecta during an impact, that ejecta can form chains of small secondary craters. These secondary craters sometimes are shallow with irregular contours, and usually are found in clusters within the bright rays of larger craters. The steep margins of hollows, along with their flat floors, are quite different.

Nor can they be classified as volcanic pits. Some 150 pyroclastic volcanic vents have been charted across the face of Mercury. Bright deposits (typically of reddish material) surround them. This contrasts with the blue hues of the hollows. The vents are much larger than hollows, some tens of kilometers across. Other volcanic craters, called pit-floors, are also larger than hollows, and distinctly reddish as well. While they do form chains a bit like hollows, they result from the collapse of an underground lava tube.

Figure 3.7 Depressing landscapes: a comparison of depressions on several planets. Top left: Flat-floored depressions at the edge of the Martian polar ice cap mark sites where carbon dioxide ice has sublimated. Top center: Scarps encircle flat-floored depressions called cavi in the "cantaloupe terrain" on Neptune's moon Triton. Volatiles involved are likely nitrogen or methane. Top right: Terrestrial volcanism in Death Valley, California. Several pits are associated with the pyroclastic volcanic caldera Ubehebe. Ubehebe's main crater is about one kilometer across. Bottom left: sublimation of water ice, perhaps mixed with other ices or ammonia, may be responsible for depressions on the icy satellites, such as this case on Europa. Bottom center: irregular mare patch in Mare Tranquillitatus on the Earth's Moon bears superficial resemblance to Mercury's hollows, but is formed by degraded lava flows. Bottom right: pits within the great nitrogen ice "glacier" of Pluto's Tombaugh Regio. Pits are likely the product of vaporizing nitrogen as the ices sublimate. (Mars image NASA/JPL-Caltech/University of Arizona; Triton image NASA/JPL; Ubehebe photo by the author; Europa Juno image NASA/JPL-Caltech/SwRI/MSSS; Moon: Lunar Reconnaissance Orbiter image NASA/GSFC/Arizona State University; Pluto: NASA/Johns Hopkins University Applied Physics Laboratory/Southwest Research Institute/Lunar and Planetary Institute)

The Earth's Moon has broadly similar features called lunar irregular mare patches (IMPs), but these also differ from Mercury's unique formations. IMPs are young, as their margins are well defined and they have few craters on their surfaces. Unlike hollows, they are not associated with impact crater floors, rims or mountains. They lack any halo or brightening seen in the hollows, and are the result of basaltic lava flows.

Analogues of Mercury's hollows exist in the outer solar system among the icy moons of the giant planets. On the margins of the Martian south polar cap, seasonal sublimation of carbon dioxide leaves behind irregularly shaped pits in areas called "Swiss cheese terrain". The depressions can form singly or as coalesced groups, in similar fashion to adjacent hollows. Depressions bounded by steep scarps form along nitrogen ice margins of Triton's polar cap in regions called cantaloupe terrain. The depressions, called cavi, are collapse features driven by sublimation of ices, most notably water and nitrogen ice. Europa and Ganymede both exhibit pits associated with subsurface processes. Underground melting may trigger the cavi. Enceladus features similar forms, especially near cryovolcanic sites, arranged in parallel "pit chains." The lines of depressions mark spots where loose surface debris or fluffy snow has seeped into a crack beneath (similar features have been studied in Iceland as good analogs for Enceladus). On the dwarf planet Pluto, we find intricate linear patterns of pits sunk into the ice. Fracturing of the ice, in combination with sublimation, may be responsible. Pluto's pits are hundreds of meters across.

The location of Mercury's hollows suggests that they form in the brutal light of a Hermian day, on Sun-facing slopes (Figure 3.8). This coincides with the concept that high temperatures or bright sunlight may trigger hollow formation, freeing volatiles from the subsurface. The majority of hollows congregate within Mercury's Low Reflectance Materials. A perfect example lies inside a crater on the southern rim of the Rembrandt impact basin. The northern portion of the crater is on the floor interior to Rembrandt, but the crater straddles the rim, so its southern portion lies outside of the Rembrandt. The impact basin is covered in material typical of Mercury's bright regions, while the terrain outside of Rembrandt lies within an area of low reflectance material. Inside the small crater that crosses the two terrains, hollows have formed only across the segment of the crater that overlays the low reflectance material. Something about the LRM regions must be favorable to hollows formation.

LRM is high in magnesium, calcium, and sulfur, so it may be that one or more of these elements are involved in their formation. LRM also has a greater abundance of carbon, probably in the form of graphite. Graphite, too, may be the component responsible for the creation of hollows. Hollows lie near some three-fourths of volcanic vents within impact craters. Clearly, hollows are

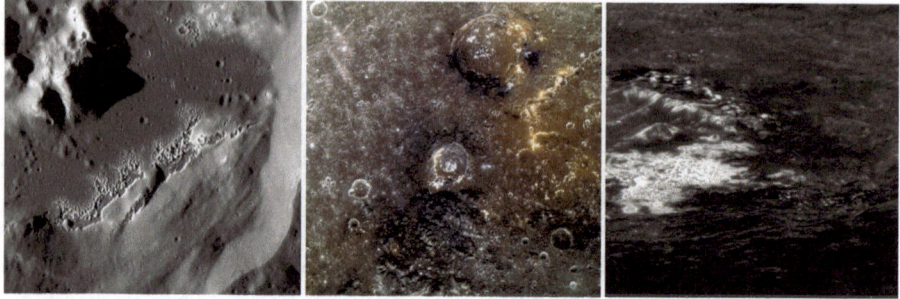

Figure 3.8 Left: Hollows form on a crater floor, along the base of the crater wall, and on the top of the central peak. Center: several craters near the eastern edge of the Caloris basin have unearthed low reflectance material. Blue-haloed hollows form across the floors of others. Reddish deposits typical of explosive eruptions spread out to the northeast, suggesting that this region may have once been the site of pyroclastic volcanism. Right: This oblique view shows the interior of Tyagaraja crater. Its central peaks, extensive hollows, and sloping terraced walls display an assortment of geologic features. (NASA/Johns Hopkins University Applied Physics Laboratory/Carnegie Institution of Washington)

somehow related to Mercury's volatiles; either they form from some type of eruptive outgassing or from collapse, or perhaps both.

The hollows appear to start as a localized darkening of the surface. A central region begins to collapse into several pits, and as it does, a brightening builds around the assemblage of depressions. This brightening is often quite dramatic, developing into a striking halo surrounding the cavities. Fine dust may be the culprit, dispersed by a process of gaseous eruption, the tearing apart of volatiles by solar radiation, or the levitation of particles by electrostatic charges, much as happens with dust on the Moon. Alternatively, the haloes may be a sort of "frost" from atoms that sublimate and then fall back to the surface, remaining for a time before solar radiation and micrometeoroids erase their elegant glow. Yet another option put forward by some researchers is that the hollows result from cometary impacts. Chunks of comet nuclei have undoubtedly impacted the surface over time, and some may have left behind bits of a particular material. Comets break up frequently when they get as close to the Sun as Mercury, so it is conceivable that Mercury could have experienced impacts from countless cometary nuclei. While most of the material probably evaporates on impact, some residual ices may survive to form the hollows. In this scenario, volatile material is trapped and preserved, forming hollows as it sublimates through the surface later.

Part of the mystery of Mercury's hollows lies in location: why do they form only in a few locations within Mercury's low reflectance material-covered regions? The LRM regions may be richer in volatiles. It may be that a hollow's growth is triggered under just the right chemical and physical settings. Variables might include surface temperatures and angle of sunlight and slope. It is likely that a concentration of volatiles must build up to a certain level before breaking through the surface. Mercury's magnetic fields may even play a role, funneling ions from the solar wind down to the surface, where they kick off the formation of hollows. If this ion barrage does play a part, then hollows may have been rare in Mercury's past, when its magnetosphere was stronger.

As the hollows age, the activity that led to the haloes tails off, and the ground darkens again. A lag deposit starved of volatiles probably builds up, and when it reaches a depth sufficient to protect the volatiles underneath, it forms a sheltering lid, shutting off the escaping gas and dust. (The ices at Mercury's poles frequently have lag deposits protecting them from the vacuum of space). Over time, the surfaces of the hollows relax. Micrometeorites and solar wind erode the walls until the hollows fade away. But meteorites could break through the surface, exposing the underlying volatiles once more, reawakening dormant hollows and starting the formation process all over again. In fact, hollows are common on steep slopes of crater walls, cliffs and peaks. It is possible that mass wasting—landslides—on these steep inclines lead to more long-lived hollows, as the moving material clears away the lag covering and exposes fresh subsurface volatiles.

Because of their locations, investigators have come to realize that the hollows are fairly recent in Mercury's geological record. The youngest craters on Mercury have crisp rims and bright rays. They were created during Mercury's most recent Kuiperian epoch (see Alien Landscapes, below). Many rayed craters play host to prominent hollows, including Balanchine, Degas and Dominici. The hollows that have been imaged in detail lack any craters overlaying them, so it is likely that they occurred after Mercury's major cratering eras. Abundant hollows have erupted at one of the youngest rayed craters on the planet, the 24-kilometer-diameter Xiao Zhao. They form bright spots on the crater's central peak, along its rim and walls, and even across the outer bright blanket of ejecta. Despite the youth of the crater itself, the hollows are even younger. In the Raditladi impact basin, hollows have formed on the basin's central peaks. The hollows are closely related to landslides that have moved down slope to bury well-preserved small craters (100-200 meters diameter). The apparent youth of these and other hollows argues that they may still be active today.

Ron Vervack envisions a fascinating vista above the hollows near twilight (Figure 3.1). "The exosphere as it exists today lends itself to some interesting

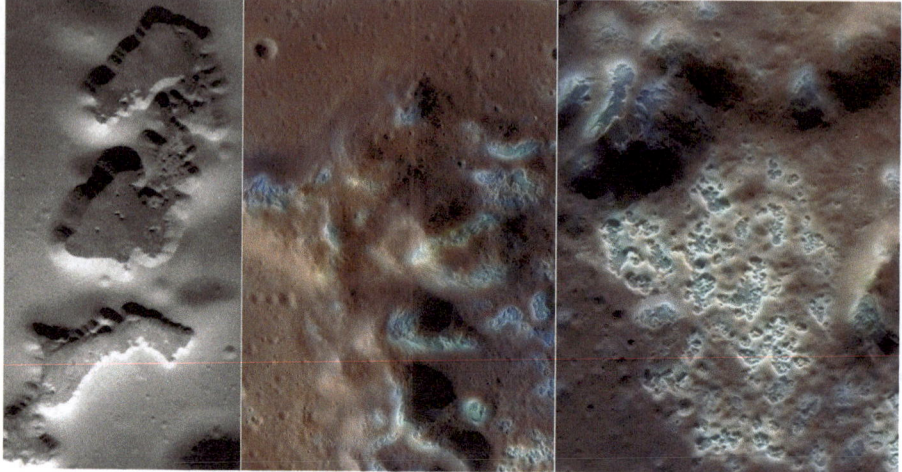

Figure 3.9 Left: An ultra-high resolution image of a series of hollows, at 3 meter/pixel resolution (illumination is coming from top of image). Center and right: MESSENGER color photos of hollows eroding the tops of hills and eating away at the cratered plains. Blue-tinted haloes of material are evident in these color-stretched images. (NASA/Johns Hopkins University Applied Physics Laboratory/Carnegie Institution of Washington)

possibilities for how it would appear to someone standing on the surface of the night side. The exosphere glows in various colors and shapes owing to how it is generated and which atomic species comprise it. Imagine being on the equator at midnight. The sodium atoms stream around from the dayside all around the terminator in the anti-sunward direction, so you would see this yellowish glow near the horizon in all directions." The sodium glow comes from a fluorescent emission, so it must be triggered by solar photons. It would not be visible in darkness. The combination of glow from various elements in Mercury's rarified air would combine to create a ghostly green and purple mist above Mercury's depressions. The display in the sky would only add to the unusual enigma of Mercury's sleepy hollows (Figure 3.9).

<u>More Alien landscapes</u>

Today, Mercury displays to us alien landscapes both elegant and desolate. Its landscapes are products of volcanism, cratering, and tectonics—the large-scale processes that change a planet's crust. All of these processes share a commonality with the Earth. Understanding the tectonic personality of a world provides a doorway through which we can comprehend its geologic evolution.

Mercury's upper crust may be a *megaregolith*, a collection of fractured bedrock left over from numerous impacts that excavated 100-km-wide craters, breaking up the surface materials. This rubble—largely a product of the Late Heavy Bombardment—may extend many kilometers down. But Mercury's volcanism may have wiped out some of the earliest marks of the LHB epoch.

Among Mercury's most intriguing and dominant geologic features are its long, arcuate escarpments—rupes—that rise above the cratered plains in sinuous lines. The dramatic cliffs span for hundreds of kilometers and tower over a mile high in places. They form a global web of precipices across the face of the dark planet. The scarps mark the sites of thrust faults, caused by fracturing of the crust. One block of the crust shoves up over another as the planet shrinks. The scarps record Mercury's planetary contraction; scientists estimate that the planet's radius may have shrunk by as much as 7 km. It may still be contracting even today.

Mariner 10 first revealed Mercury's cliffs during its three encounters in the mid-1970s. MESSENGER sighted many more. The scarps crisscross the entire planet, confirming that much of Mercury's surface is dominated by features related to the planet's contraction. The planet's shrinking is not unique: all of the terrestrial worlds have done so as they have cooled over the 4.6 billion year history of the solar system. As they cool, they shrink, and often this cracks the surface crust.

Both the Moon and Mercury have preserved a record of this global shriveling. On Venus, Earth and Mars, such records are rare because of erosion (see the box **Crags Compared**). On our Moon, dense basaltic magma has flooded across crust made of much lighter anorthosite. As the basalt filled in lowlands, its increased weight caused the crust to sag, compressing the underlying surface and deforming the basalt blanket into ridges. On Mars, we find other similar ridges. Near the great Tharsis bulge (Chapter Six), linear features score the landscape for hundreds of kilometers. As on Mercury, these scarps result from compression of the surface. With heights up to 300 meters, these uplifts may have been lubricated by underground water or carbon dioxide ice. We see ridges on the Earth as well. A classic example is the Sawtooth Range of Montana, where the Rocky Mountain chain uplifts the rock of the adjacent plains. While their morphology is similar to the rupes on Mercury, their cause is quite different, having nothing to do with the planet's global shrinkage. Another North American escarpment snakes across southwestern Utah. The Kaibab monocline has a similar profile—if projected backward to account for erosion—to lobate scarps on Mercury, the Moon and Mars. Geologists cite other sites on Earth as analogs to Mercury's lobate scarps, including locales in Algeria, Australia, and the Solomon Islands.

Mercury's remarkable crags rear up in a multitude of forms. High-relief ridges show more abrupt margins than other types of escarpments, and are symmetrical in cross-section. Wrinkle ridges have lower relief with steep sides along both front and back, as they have been compressed from two directions. They are often flat-topped. Lobate scarps tend to have a gentle rise on one side and a steep drop-off on the other. They follow a curving arc across the landscape. But the gang of precipices are ill-behaved; they often do not slave to any specific classification. Some wrinkle ridges have only one steep side, while other scarps become muted or sharper at various portions along their wandering facades. Too add to our confusion, some even split, branching into two types of structures. Mercury's longest lobate scarp, Enterprise Rupes, crosses the rim of the great Rembrandt impact basin. The segment of the cliff within the basin has a classic profile of a gentle rise on one side and a cliff on the other. But as the rupes moves beyond the confines of the basin, it forms steep scarps on both front and back, much like a wrinkle ridge (Figure 3.10).

Crags compared

1. The ridge named Dorsun Heim looms above the lunar plains of Mare Imbrium. While the slope away from us is gentle, the ridge drops off precipitously on the facing side. For scale, the crater at left—C. Herschel—is about 15 km across. The image was taken by Apollo 17 astronauts. (NASA/JSC)
2. West of the Moon's Montes Teneriffe lies another wrinkle ridge. Boulders rest precariously on its flanks and summit. This Lunar Reconnaissance Orbiter image covers an area less than 2 km across. (NASA/GSFC/Arizona State University)
3. The Mars Reconnaissance Orbiter captured this view of a wandering wrinkle ridge crossing the region of Oxia Planum, a 3.6-4 billion year old site rich in clays. (NASA/JPL/University of Arizona)
4. Even asteroids host wrinkle ridges. This view of the asteroid 433 Eros reveals Hinks Dorsum (toward the right), an 18 km-long crest resembling some of the ridges on Mercury. The ridge was probably formed by compression, but unlike the Hermian ridges, this one was triggered by stresses due to the impact of the adjacent crater, whose rim is visible at left.
5. Ridges similar to those on Mercury are rare on Earth. One such rarity is the East Kaibab Monocline, which breaks across the surface of the Colorado Plateau. Compression from the rise of the Rocky Mountains created the dramatic uplift. (Wikipedia, https://commons.wikimedia.org/wiki/File:East_Kaibab_Monocline_north-of-Paria_Utah.jpg)

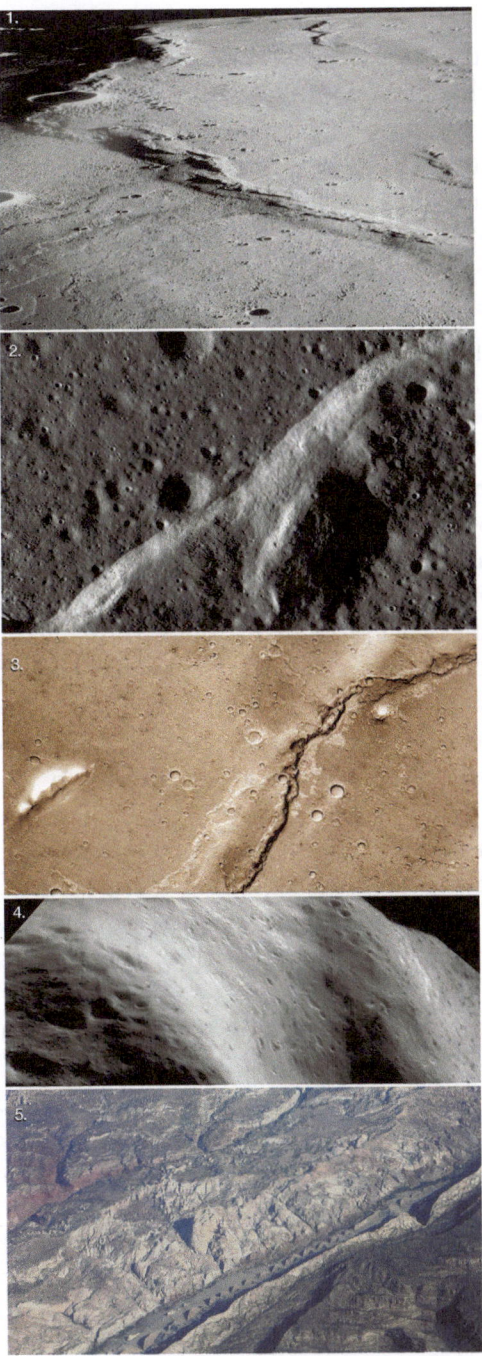

Figure 3.10 The faults, wrinkle ridges, and escarpments found throughout the terrestrial planets and the Moon display similarities in form, but not all mark planetary compression. Upwelling of magma causes some. Others rise as a result of tectonic forces quite different from those found on Mercury (such as the Earth's unique plate tectonics or Venus' strange vertical "columnar" tectonics). Here, we explore a few archetypal examples.

If Mercury's global contraction lingers to this day, the planet should continue to manufacture scarps. In fact, researchers have found evidence that the planet continued that process until very recently, and may be carrying on even now. In the latter, low-altitude phase of MESSENGER's mission during the craft's last 18 months of flight, the spacecraft imaged mini-scarps, cliffs orders of magnitude smaller than the lobate scarps seen earlier. Rather than stretching for hundreds of kilometers, these ridges extend for only a few. Their height reaches tens of meters rather than the kilometer rise seen on some rupes and other escarpments. A paper[15] in the journal *Nature Geoscience* reported that their "small-scale, pristine appearance, crosscutting of impact craters and association with small graben all indicate an age of less than 50 Myr." In contrast, some of the larger, more eroded rupes may date back 3.5 billion years (Figure 3.12). The small-scale escarpments must be young, because a constant drizzle of micrometeoroids on planetary surfaces wipes out land forms of this size over billions of years. Similar-sized lobate scarps have been found on the Moon. They are thought to be no more than 800 million years old, and could be even younger than 50 million years.

The authors of the paper propose that these scarps are the smallest members of a family of thrust fault scarps on Mercury that range from small to vast, with their size related to their age. If the small scarps are as young as they appear, their presence intimates that Mercury is tectonically active today. This means that the little planet's interior is in the midst of a prolonged, slow cooling.

Some researchers advise caution on the true meaning of the small faults. They point out that we have not yet gained a deep understanding of how the small scarps relate to the larger, older ones. Scarps of this scale could also be produced by flowing surface material, they assert, or from recent impact shock waves.

A different kind of escarpment shows up in many crater floors. Within "ghost craters", craters that have been almost completely flooded and erased by volcanic floes, ridges and troughs inscribe a pattern across crater floor and rim. The ridge/trough patterns result from expansion and contraction of thick lavas draping across the crater. Several craters within the Goethe impact basin exhibit such marks (Figure 3.11). In some places graben play a part in the complex crustal interplay. Here, blocks of the surface sink between two rising sections of crust, which bound the edge of the blocks in fault cliffs. Many ghost craters display crosscut patterns that sometimes cross the crater rim.

[15] *Recent tectonic activity on Mercury revealed by small thrust fault scarps*, by Watters, et al, *Nature Geoscience,* September 26, 2016.

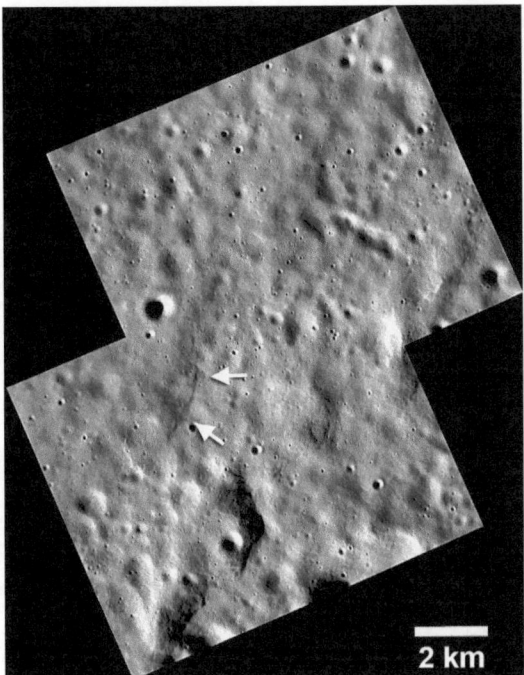

Figure 3.11 Taken in the latter part of MESSENGER's low-orbit campaign, images like this show small, fresh scarps cutting through older craters. (NASA/Johns Hopkins University Applied Physics Laboratory/Carnegie Institution of Washington)

A cavalcade of craters

In addition to its tectonic and volcanic narrative, Mercury's surface also displays a record of the battering it has taken under a downpour of rock, metal and ice. Meteors, asteroids and comets have left dramatic ray craters and colossal impact scars. The largest is the Caloris basin, a 1525 km wide wound surrounded by mile-high mountains. Like ripples in a frozen pond, concentric mountain chains define vast multi-ringed basins like Caloris. Several dozen of the bulls-eye impact features have been identified in MESSENGER orbital and flyby images.

Craters come in several distinct configurations. Their form is dependent, for the most part, on the size of the object that impacted the surface (although the makeup of the surface also comes into play). The smallest impact features are called simple craters. These are bowl-shaped, often with slightly raised rims. Larger meteors leave a complex crater. Similar to a simple crater, complex craters typically have central peaks and flat floors. As the meteor impacts the surface, its energy converts rock into molten magma fluid. That fluid

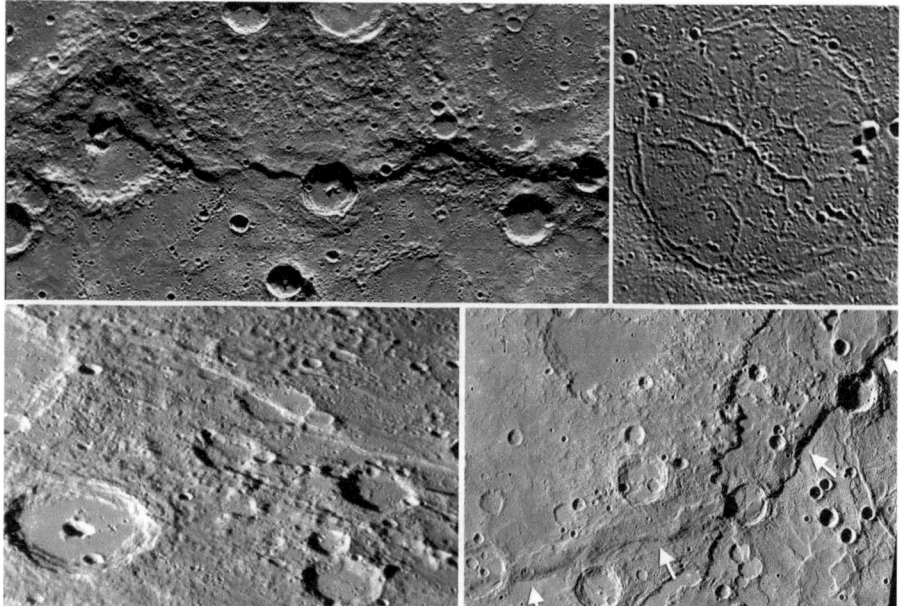

Figure 3.12 Top left: This image mosaic shows a section of Victoria Rupes, one of the famous scarps on Mercury. Wrinkle ridges like this are interpreted as tectonic in origin and are usually only found on volcanic plains. Top right: troughs score the floor of a ghost crater in Goethe basin. This pattern is caused by extension and contraction of the thick lava filling the crater floor. Lower left: A large tectonic scarp—formed when the planet's interior cooled and contracted—runs through a crater near the center of the image. In the bottom left corner of the image, Lessing crater lacks the typical central peak found in a complex crater on Mercury. Instead, Lessing hosts a central pit, likely formed by volcanic activity. Lower right: Enterprise Rupes, (white arrows) stretches for 1000 km/mi and towers over 3 km high. MESSENGER has confirmed that Mercury's shrinking has resulted in a global network of lobate scarps, tectonic landforms that are the outcome of thrust faults. (all images courtesy NASA/Johns Hopkins University Applied Physics Laboratory/Carnegie Institution of Washington)

rebounds as the crater forms, splashing up at center to leave a mountain as the rock cools and solidifies. Large complex craters develop terraced, "stair-step" walls, and streams of ejecta fan out from many. The freshest of these are called ray craters. Their ejecta is brighter (or, more rarely, darker) than the surrounding landscape, leaving rays of material scattered around like a starburst with the crater at center. Some complex craters, especially on Mars, have lobate ejecta blankets surrounding them, caused by water or other ices within the impacted surface. Craters greater than about 485 km across are in a new class, that of impact basins. These huge features often have a series of ring-shaped mountain chains within. The extent of their depth sometimes breaches the crust, freeing lavas to flood their interiors. Impact basins frequently produce associated volcanism.

Furious asteroid and comet bombardment from the surrounding solar nebula, along with fierce volcanism, wracked Mercury from its earliest days. The geologic record of this period, beginning toward the end of the Late Heavy Bombardment some 3.8 billion years ago, is partially wiped out by tectonic features like those lobate scarps, left behind by the shrinking of the planet as it cooled. At the same time, the Earth was suffering its own mauling. Our home world is far more geologically active than Mercury, so forces like plate tectonics and volcanism obliterated many of our impact scars. What craters didn't disappear beneath sliding plates or flowing lavas were worn away by our relatively dense atmosphere and its complex weather systems. Temperature swings, wind, liquid water and water ice all contributed to the breakdown of the rocky structures that delineated those ancient craters. Today, geologists have identified 190 surviving impact structures across our globe.[16] But for its part, Mercury has more craters per square inch than any other terrestrial planet, and many are pristinely preserved.

Mercury transitioned through several major epochs of cratering, volcanism and erosion: the pre-Tolstojan terrains exhibit some of Mercury's oldest landscapes. Smooth volcanic plains formed during the Tolstojan (beginning with the formation of the Tolstojan impact basin), and the Calorian (beginning with the creation of the Caloris impact basin) eras, when volcanic flows were common. Volcanism began to wane at the end of the Calorian, but pyroclastic eruptions continued into the Mansurian period, and perhaps into the Kuiperian. Many of Mercury's famous hollows, which we will visit in a moment, formed during the Kuiperian age, which also records the most recent impacts. The Kuiperian period stretches into the present (Figure 3.13).

Mercury's exosphere provided little protection from the blitz of meteors and comets that battered the terrestrial planets throughout the solar system's early history. Impactors continued their assault as the surface of the planet cooled and solidified. Although Mercury's early crust was substantially set down by flows or eruptions of magma from within—much like its terrestrial siblings—crystalline material rose through the melted rock to form a scum on the top of the global magma seas. Heavy rock sank, and light material like graphite came to the surface. Small areas of the earliest crust may still be preserved, although any ancient crust has undoubtedly been modified by fracturing and impacts. The majority of Mercury's face is dominated by what

---

[16] This number is according to the Earth Impact Database, which was last updated in 2017.

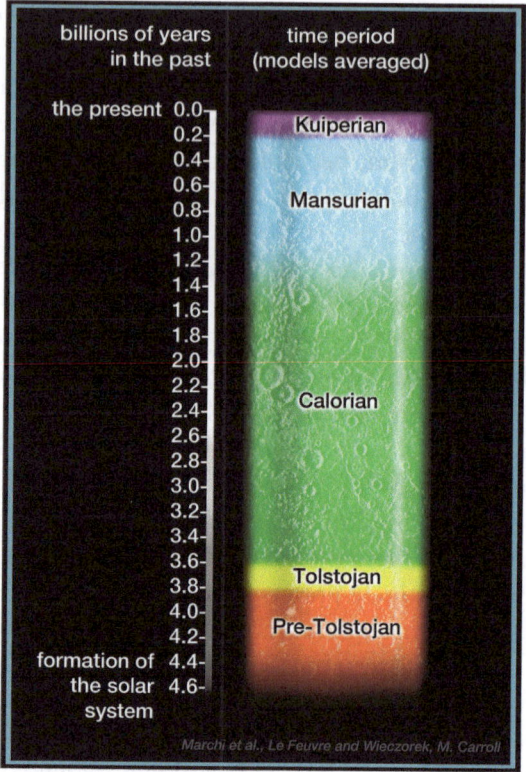

Figure 3.13 Researchers look at the density of craters as a yardstick for Mercury's age and evolution. Several leading models have come up with differing timelines. In this diagram, estimates from two leading studies are averaged. The widest variation in models is the transition between the Mansurian and Calorian periods. (art by author)

geologists call intercrater- and smooth- plains. The intercrater plains regions date back to Pre-Tolstojan and Tolstojan ages, while the smooth plains are younger (Tolstojan and Calorian). Both of these types of terrain arose through a combination of extensive volcanism and impact cratering. The volcanic flows raged across the primordial landscapes, but declined in the Calorian age. Pyroclastic volcanoes continued to spew lavas at least through the Mansurian Period. As volcanism became rare in the Kuiperian period, it was replaced by continuing impacts and the formation of hollows, a process that may continue today (Figure 3.14).

It would seem, at first blush, that the cloud of meteors and comets within the solar nebula should have tailed out gradually as the planets and moons swept them up during the age of planetary formation. But crater counts on the terrestrial planets and the asteroids tell another tale. Some indications are

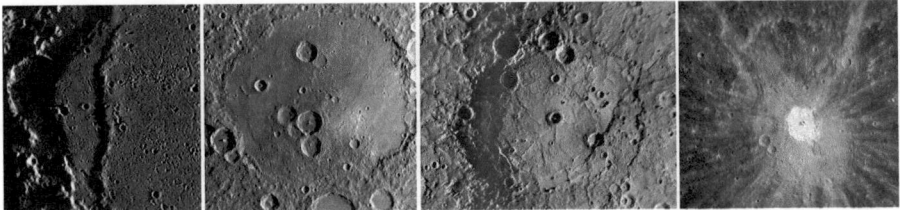

Figure 3.14 Craters from four eras (l to r): The edge of Goethe basin. Goethe is a very ancient, pre-Tolstojan impact feature; Raphael was formed during the Tolstojan era, and includes areas that may exhibit remnants of explosive volcanic outgassing (bright area, center right); Rembrandt, from the Caloris period, is the second largest impact basin on Mercury, and may be as old as 3.9 billion years; the young crater Kuiper, from which the Kuiperian period gets its name. Kuiper's dazzling interior and extensive ray system combine to create the brightest province on the planet's surface. (MESSENGER photos, NASA/Johns Hopkins University Applied Physics Laboratory/Smithsonian Institution/Carnegie Institution of Washington, public domain, accessed via https://en.wikipedia.org/wiki/Rembrandt_(crater)#/media/File:Rembrandt_crater_mosaic.jpg)

that a vicious flurry of impacts gusted through the solar system late in the era of planetary formation. Called the Late Heavy Bombardment or the Lunar Cataclysm, this terrifying age of falling rock and metal began—according to some lines of evidence—about 4.1 billion years ago, and continued until about 3.8 billion years ago, at the end of Earth's Hadean Era. The Hadean Era was typified by extensive volcanism and a high influx of meteors.

All the other planets and moons may have suffered similarly at the hands of the Late Heavy Bombardment, and their diaries of the event, inscribed across their cratered faces, give us an even clearer picture of that terrifying epoch. Mercury is one such faithful recorder; asteroids carved out Caloris and Rembrandt basins about 3.9 billion years ago, at the time when the LHB raged. Mercury got a heavier beating than the Earth/Moon system, perhaps by a factor of 4, because of the planet's proximity to the Sun. The Sun's gravity pulls in more material, and at greater speeds, than out at the Earth's distance. Titanic scars from the LHB chronicle its events on Mars and the Moon as well. On Mars, the giant impact basins of Hellas and Argyre, both in the heavily cratered southern hemisphere, appeared between 3.8 and 4.1 billion years ago. The protoplanet that left behind the Imbrium basin, one of the largest impact features on the Moon,[17] did so roughly 3.9 billion years ago, at the height of the LHB event. The age of the younger Orientale impact basin still falls within the window for the LHB. And we find traces of the deadly rock

---

[17] In fact, Imbrium is one of the larger impact basins among all the terrestrial planets.

fall from the asteroid belt, too. Meteorites collected on Earth that hearken from the asteroids are riddled with evidence of shockwaves. The shocked features form under great pressure from high-velocity impacts. No such evidence of ancient cratering exists on Venus, because the planet appears to have been resurfaced long after the Late Heavy Bombardment (see Chapter 4).

Although the initial models for the Late Heavy Bombardment suggested that the event began and ended abruptly, the increase in impacts may have been more gradual. A steady stream of asteroids and comets may have rained down from the very beginning of the solar system, rising gradually up to 3.5 billion years ago. At that time, the inventory of asteroids would have been starved enough that the drop-off in impacts was rapid. The geologic record shows impact melt crystals (due to high velocity impacts) back to about 4.2 billion years ago. But some researchers argue that the reason we see few examples of impact melt before 4.2 billion years is simply because older material was eroded by later impacts, destroying evidence of earlier impacts.

A wet reboot for LHB?

The skeptics of the Late Heavy Bombardment model have a new arrow in their quiver, they claim. Evidence from recent work seems to indicate that the early Earth may have had standing water on its surface before the proposed LHB. UC Boulder astronomer Steven Mojzsis points to evidence that the peak in bombardment may have come much earlier, at approximately 4.48 billion years ago.[18] If true, this early bombardment would leave plenty of time for the Earth to cool and life to begin when it did. Mojzsis and his team have studied dozens of meteorites, dating the moment when the rocks had been melted, effectively restarting their geological clocks. This melting would have occurred during the massive asteroid bombardment of the planet. They could find no uptick at the 3.9 billion year mark. Instead of a violent spike of incoming debris, Mojzsis says, the new work indicates a more gradual tapering off. Mojzsis team finds that the migrations of the planets must have taken place at an earlier time to match the results.

But David Kring, geoscientist and impact expert at the Lunar and Planetary Institute in Houston, points out that, "Based on my own group's work and integrating what others have learned, I think we can say (a) that bombardment was intense in the first billion years of solar system history, (b) there

---

[18] Stephen J. Mojzsis, et al., "Onset of Giant planet Migration before 4480 Million Years Ago", August 12, 2019 *Astrophysical Journal*

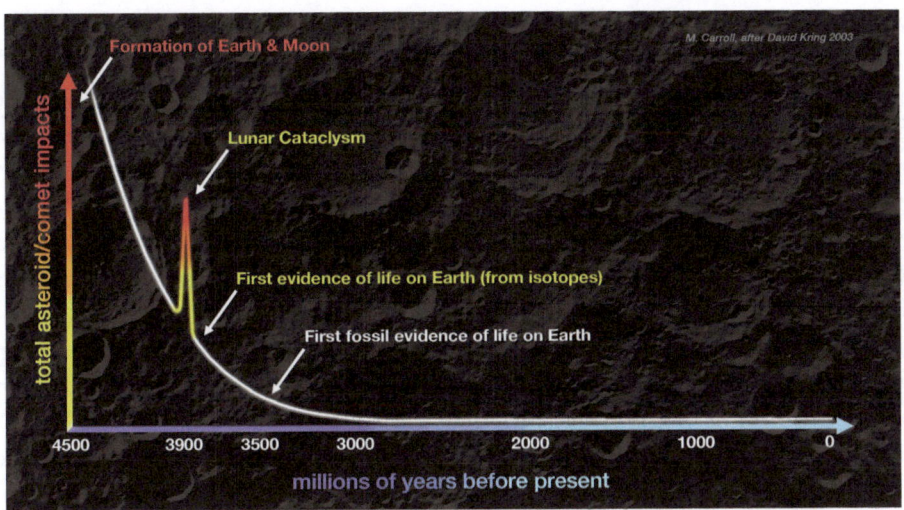

Figure 3.15 Like Mercury, the Earth's pummeling by meteors, asteroids and comets began as a fierce barrage during the formation of the solar system. According to one school of thought, the incidence of impacts dropped off precipitously until about 4 billion years ago, when a spike in impacts marked the Late Heavy Bombardment. (Art by the author after Kring, 2003; Background courtesy NASA/ASU/Lunar Reconnaissance Orbiter)

were significant impacts in a short period of time (e.g., Imbrium, Orientale, Schrodinger), representing a pretty intense influx of debris 3.8-4.1 Ga that one can call an impact cataclysm, although (c) the magnitude and duration of that bombardment is still uncertain, and – finally – that (d) Schrodinger and Orientale represent the end of the basin-forming epoch on the Moon." (Figure 3.15) Because the Schrodinger impact basin formed at the end of the Late Heavy Bombardment, Kring advocates sending sample return missions (perhaps as part of the human return to the Moon in the Artemis program), from Schrodinger. There, samples of the oldest basin material from nearby South Pole Aitken basin rest alongside samples of the youngest, Shrodinger. These bookend samples could lock down the timeline for the Late Heavy Bombardment. And while some critics of the classic Late Heavy Bombardment model assert that cratering records on Mercury and Mars fail to tally with the timeline, Kring points out that a fraction of the craters in the highlands of Mercury and Mars have been obliterated over time, so their records are not as complete and well-preserved as those of the Moon. The more pristine nature of the lunar highlands makes them a much more informative sampling. "Not only is the cratering record better preserved on the Moon," Kring adds, "but we have samples to calibrate that cratering record. Samples and sample ages provide a better picture of early Solar System history, not a less-informed one."

What would the effect of a Late Heavy Bombardment have been upon the Earth? Based on cratering rates on the Moon and on other worlds like Mercury, investigators conclude that the fall of meteors onto the surface of the Earth created more than 22,000 impact craters larger than 20 km/ 12 mi in diameter. The fusillade left at least 40 vast impact basins greater than a thousand km/620mi across the globe, and a handful of extinction-level impact scars some 5,000 km/3100 mi across. Asteroid material has been dated at 4.6 billion years, at a time when scientists believe the first solid objects were forming within the solar nebula. But no terrestrial rocks have been found older than about 3.8 billion years. It appears that the LHB heated most of the Earth's surface to the melting point, destroying the older rocks. No wonder the time period is known as the Hadean. Isotopic ratios within rare zircons from Canada and Australia[19] indicate that the Earth before the LHB had a solid surface with acidic seas. Plate tectonics had not yet commenced, so any continental rock rising from the seas came in the form of crater rims, or was uplifted by volcanism.

Biologists point out that the earliest evidence of active life on Earth, found by isotopic studies of rock, shows biological changes in 3.8 billion year old strata. These biogenic markers appear in the record immediately after the end of the posited LHB. How did the Late Heavy Bombardment affect life's progress on this planet?

What if the LHB never happened?

The Late Heavy Bombardment clearly poses intriguing problems for biologists. Could life have started beneath such a punishing rain of meteors? Now, geologists are asking a provocative question: what if it never happened? Using crater counts to determine relative ages of surfaces, researchers estimated that three large impact basins on the Moon, Imbrium, Serenitatus and Nectaris, all dated to about 3.95 billion years old. But more detailed imaging by NASA's Lunar Reconnaissance Orbiter spotted streams of material scattered across Serenitatus and Nectaris from the great Imbrium basin. Apollo landing sites had been chosen to sample material from different epochs, so their landing sites were spread out. But it now appears that samples from various landing sites may have all been seeded with material from Imbrium. Despite coming from widespread

(continued)

---

[19] The 4.4 billion year old zircon from the Jack Hills of Western Australia—mentioned above—is the oldest yet found, and predates the LHB by only a few million years. The zircon is thought to be a scrap of pre-LHB crust encased in a younger, 3.8 billion year old matrix of rock.

landing sites, the majority of Apollo samples may represent samples from Imbrium, calling into doubt the dating of other sites. A 2010 study showed that samples thought to have originated in Nectaris were geologically and chemically similar to Imbrium samples. And further investigations using newer, advanced techniques indicate ages of 4.2 billion years. This suggests that large impact basins were forming long before the purported spike of the LHB. In light of the new data, some impact experts now propose a more gradual Late Heavy Bombardment beginning as early as 4.2 billion years ago, occurring after a lull in infalling debris. This scenario is consistent with radiometric dating of meteorites and with the most ancient of Earth's geologic samples. Arguments against a Late Heavy Bombardment also come from the cratered surfaces of Mercury, Mars and the moons of the outer solar system, LHB skeptics assert. They maintain that none of these planets and moons evidence an increase in the rate of bombardment. The controversy rages on.

If the LHB is an illusion within the data, the emergence of life may be easier to explain. The earliest of Earth's microfossils, from about 3.5 billion years ago, are complex, with segmented forms and outer sheaths. This complexity argues for a much earlier biogenesis. If biogenesis began earlier than the LHB, the first life would have had to survive the onslaught of meteors. A gradual, more gentle LHB provides one possible solution to the problem. Still, some advocates of an abrupt LHB point to evidence that the earliest microbes were microorganisms that thrived in high temperatures (known as hyperthermophiles). The heat and pressure from frequent impacts would be less of a problem to life forms such as these.

Circling Mercury: the MESSENGER and BepiColumbo orbiters

MESSENGER, the Mercury Surface, Space Environment, Geochemistry and Ranging spacecraft, made three flybys of Mercury before settling into orbit in the spring of 2011. Its journey began with its launch from Cape Canaveral Air Force Station in Florida on August 3, 2004. The massive spacecraft took a circuitous route to Mercury. This was due, in part, to the design of the mission: MESSENGER needed to approach Mercury gradually enough to settle into orbit. The spacecraft carried out a gravity-assist flyby of Earth and two of Venus before carrying out three Mercury flybys on January 14, October 6 of 2008, and September 29, 2009. During its three flybys, MESSENGER was able to map nearly the entire planet. Finally, the craft rocketed into orbit March 18, 2011. During MESSENGER's primary mission, the orbiter's observations began the long process of untangling the major mysteries that its designers set out to solve: why does Mercury's have such an abundance of metals? What is the general outline of the planet's history? What is the nature of Mercury's magnetic field and the structure of its core? What are Mercury's polar deposits made of, and what might their source be? Finally, why is the surface of the little planet so rich in volatiles?

MESSENGER confirmed that Mercury's polar craters are filled with water ices. MESSENGER went on to an extended mission, in which it operated during a more active part of the Sun's eleven-year cycle. Researchers searched for surface effects from the Sun's increased energy outflow. The mission also included a variety of targeted imaging of interesting places noted in the primary mission. The orbiter swooped lower for more detailed observations. MESSENGER continued to operate into a second extended mission with even more detailed imaging and other observations from a nerve-wracking 5–37 km. The flight was a tribute not only to mission planners but also to the navigators who had to carefully control the spacecraft's path. MESSENGER's advanced instruments enabled a broad spectrum of data in varied scientific disciplines.

The next Mercury orbiter, ESA's BepiColombo, will be able to "see" the distinct signature of diamond light glinting off Mercury's surfaces. It is scheduled to make six flybys of Mercury before orbit insertion in 2025 (Figure 3.16). The BepiColumbo mission is made up of two orbiters: ESA's Mercury Planetary Orbiter and JAXA's Mercury Magnetospheric Orbiter. A suite of

Figure 3.16 The E-type asteroid 2867 Steins, imaged by the European Space Agency's comet orbiter Rosetta. Was this space rock once part of Mercury's interior? (ESA 2008 MPS for OSIRIS Team MPS/UPD/LAM/IAA/RSSD/INTA/UPM/DASP/IDA; processing by T. Stryk.)

Figure 3.17   One of BepiColumbo's second Mercury encounter photos, taken on June 23, 2022.The spacecraft was 920 km from Mercury. The overexposed magnetometer boom of the Mercury Planetary Orbiter (ESAs portion of the dual craft) is seen at left, and part of the medium gain antenna is also visible. The dramatic 200-kilometer-long Challenger Rupes parallels the magnetometer boom. (copyright ESA/BepiColombo/MTM, CC BY-SA 3.0 IGO)

eleven sophisticated instruments will survey Mercury over several years. After a seven-year cruise, the craft will split into its two separate orbiters, with one following a polar orbit. And as with all exploration, BepiColumbo will answer some questions and reveal others. The solution to many of Mercury's great mysteries will undoubtedly have to await visits from landers. With each passing discovery we make about this enigmatic, tiny world, we gain more insight into the Earth's early life, when volcanoes ruled the land and meteors ruled the skies.

Drops of Mercury in her hair?

Getting samples from asteroids and comets is one thing, but getting chunks of Mercury back to the Earth is an entirely different challenge. Any sample return vehicle would have to overcome Mercury's gravity to get back, along with the uphill battle of the Sun's gravity, requiring prodigious amounts of energy (the microgravity environment of the asteroids makes a return trip to Earth relatively easy). But what if we already have samples of Mercury lurking in dusty meteorite collections around the world?

Some of the rarest meteorites are the aubrites (named after the site of the first discovery, in the French village of Aubres). Aubrites are an unusual pale yellow, and they contain traces of metal. They are also low in oxygen, and appear to have formed in a magma ocean. These characteristics line up with models of Mercury's primordial ages, when the planet was awash in molten rock. If a massive protoplanet smashed into the young Mercury, up to a third of the planet's mass would have been cast into space, and some of that debris would have been pushed outward by the Sun's solar wind, ending up in the asteroid belt. This debris would have become E-type asteroids, which are likely the source of the aubrites. Most aubrites reside within the Hungaria group, an inner region of the asteroid belt (see Figure 2.13). There, the planetary fragments would have orbited the Sun for billions of years, soaking up the solar radiation that has left its long-term traces on the aubrites. Over time, some of the asteroids would have been perturbed gravitationally by passing comets or other asteroids, and would have coasted toward the inner solar system. Eventually, some would have fallen to Earth as aubrite meteorites.

Adding weight to the scenario, tests on aubrites have revealed low levels of cobalt and nickel, similar to estimates of levels in Mercury's shallow regions of the mantle. MESSENGER chemical reconnaissance also bolsters the parallels between an embryonic Mercury's inner composition and these strange, rare meteorites. In September of 2008, en route to the comet 67P, ESA's Rosetta spacecraft flew by the E-type asteroid Steins. During its encounter, the craft confirmed the iron-poor nature of the asteroid, and found minerals including feldspar, olivine and enstatite (pyroxene) (Figure 3.17).

Looking Back: A Brief History of...Mercury

Condensing out of the solar nebula close in to the Sun, Mercury begins to differentiate. Heavy material sinks toward the core while lighter matter rises to form the mantle and crust. The majority of the lighter matter is graphite, still present today. Mercury's core is huge in comparison to its overall size. One or more giant impacts may have stripped off the lighter outer layers of the planet.

Violent bombardment from the surrounding solar nebula and its leftovers, along with raging volcanism, wrack the planet from its earliest days. The geologic record of this period, beginning toward the end of the Late Heavy Bombardment, is partially wiped out by tectonic features like faults and scarps, left by the shrinking of Mercury as it cools. Mercury transitions through several

(continued)

major epochs of cratering, volcanism, tectonics and erosion: the pre-Tolstojan terrains exhibit some of Mercury's oldest landscapes. Smooth plains form during the Tolstojan (beginning with the formation of the Tolstojan impact basin), and the Calorian (beginning with the establishment of the Caloris impact basin) eras, when volcanic flows are common. Volcanism begins to wane at the end of the Caloris, but explosive (*pyroclastic*) eruptions continue into the Mansurian period, and perhaps into the Kuiperian. Many of Mercury's famous hollows form during the Kuiperian age, which also records the most recent impacts.

Mercury's interior is rich in radioactive material, including potassium, uranium and thorium. The heat from these materials helps to power Mercury's ubiquitous volcanoes. The elements may also contribute to Mercury's molten core, which generates a robust magnetic field, revealing that its interior is dynamic and active in its childhood. Its first tenuous atmosphere comes directly from the solar nebula, containing mostly hydrogen and helium. But volcanic eruptions add transient atmosphere as the planet matures. The magnetosphere and weak gravity cannot hold on to the thin air, and soon only a trace remains. Today, the planet stands as a battered world offering us a rich historic chronicle of the early solar system.

(For comparison, find a brief history of Earth in Chapter 2, and Venus at the end of Chapter 4. Chapter 5 offers a brief history of Titan, and Mars has a brief history at the end of Chapter 6)

4

Earth=Venus: Our planet as a Dante-esque oven

Venus and the Earth both orbit the Sun within the "habitable zone",[1] that golden distance whose conditions allow for liquid water on the surface. The two planets are practically twins in mass and size, both rocky, terrestrial worlds. Like the Earth, Venus has high mountains, plains, and great basins, although its variations tend to be more subdued than the lofty peaks[2] and ocean basins of our home planet. But Venusian conditions are nightmarish: pressures at the surface are 93 times that of Earth at sea level (roughly the pressure of the ocean at a depth of 3000 feet/900 meters), and temperatures are hot enough to melt lead. Despite the fact that Venus is closer to the Sun by a third, its distance alone cannot explain the dramatic differences between the two worlds. This singular fact stands as one of the great mysteries of our solar system. And although Venus gets a bad rap as Earth's "evil twin", it is an absolutely fascinating world. Its extraordinarily massive ocean of air and radical environment make it not only interesting in its own right, but also critical to our understanding of such issues as climate change and extreme weather.

---

[1] Mars may be on the outer edge of the Sun's habitable zone, barely.

[2] Although Venusian topography tends to be flatter, on average, than the Earth's, its highest peak is taller than Mount Everest. Skadi Mons, part of the Maxwell Montes mountain chain, rises 37,795 feet/11,520 m above the planetary mean (Venus' established "sea level").

M. Carroll, *Planet Earth, Past and Present*, Springer Praxis Books, https://doi.org/10.1007/978-3-031-41360-5_4

Figure 4.1 Artist's impression of Akna Montes, a Venusian peak that may be covered by condensed iron pyrite ("fool's gold"). Painting by the author

The childhood of two siblings

After solar wind cleared the early Earth of its initial atmosphere, another atmosphere arose. Both Venus and Earth experienced the efficient outgassing of water, carbon dioxide, hydrogen sulfide and other volatiles when the last global magma ocean solidified (see Chapter Three). As the Earth's molten surface congealed and cooled, the air above it remained well above the boiling point of water. This heated canopy of gas may have been 100 times as dense as today's. Atmospheric physicists call this a "steam atmosphere", a precursor to the planet's more stable second atmosphere. When temperatures dropped, the steam was finally able to condense, raining down on the hot surface. Surface lakes and seas may have begun to form as early as 4.4 billion years ago, just a few million years after the Earth became a differentiating globe. The planet's guts still had some settling to do.

The Earth's secondary atmosphere continued to form as heat from the interior thrust carbon dioxide, carbon monoxide, nitrogen and water vapor out through volcanoes. Our second atmosphere got an extra boost from the Late Heavy Bombardment, that fusillade of meteors, comets and asteroids, which seeded the air with water and other constituents (see Chapter Three). The heavy dose of carbon dioxide in this atmosphere draws parallels to the Venus of today.

The exception in our analogy with Venus lies in one important ingredient: water. The Earth was blessed with it, while Venus became desiccated. Water is what separates the Earth from every other world in the solar system. Water is abundant in the moons of the outer solar system, in the form of frozen crust and mantle and in the form of subsurface oceans. Mars hides vast quantities of it beneath its surface and in its polar caps. But our world is the only one with great seas of liquid water on its surface.

Heat drove water vapor and gases like nitrogen and carbon dioxide from Earth's interior. The dense, hot atmosphere formed by this outgassing was far denser than the one we currently have, perhaps a haunting parallel to Venus today. Some of the atmosphere's carbon dioxide reacted with stone and water, ending up locked into limestone. Today, Earth's mantle still contains more water than its oceans. The water within the Earth's interior may enable the action of plate tectonics, as we will see.

Our first seascapes were probably alien indeed. Before oxygen invaded our world to react with surface rocks, the continents would have been mostly black lava rock. At some stages, the oceans would have been a deep green because of iron suspended in the water. Even the skies above would seem a little off to a time traveler, as the air was likely tinted orange by organic material generated by the interaction of methane and sunlight. These floating hydrocarbons are the same compounds that give Saturn's moon Titan its distinctive orange color.

Our deep blue sea

To understand the source of our great oceans—which is somewhat mysterious—we must travel back some 4.6 billion years, to a time when the Earth was still condensing out of the solar nebula. As we saw in Chapter One, this was a violent epoch. Meteors, comets and asteroids crashed together in a process known as accretion. Earth grew as its increasing gravity pulled rocks and ice to itself. Scientists who study such ancient times have come to realize that our water could have come from three sources. The first source may have been the water-rich meteors (such as carbonaceous chondrites, a common type of meteorite thought to be emblematic of the most ancient, undifferentiated material, precursors to the terrestrial planets) that helped to build our planet. The second source may have been comets, which are basically cosmic icebergs. The third source might have been the rocky material that makes up the bulk of the planet (Figure 4.2).

But how can we figure out which source led to our oceanic globe? We find clues within the make-up of seawater. Our salty brine contains a mix of hydrogen and oxygen (that's how you make water, after all), but the hydrogen provides an interesting yardstick that we can use to compare our water to other sources. The hydrogen in our oceans comes in two flavors, ordinary hydrogen and a heavy isotope of hydrogen called deuterium. Ocean water contains a specific ratio of ordinary hydrogen to deuterium. We have been able to chart the deuterium in only eleven comets, including Comet Halley, Comet 67P/Churyumov-Gerasimenko and Comet Hyakutake. Only one, Comet 103P/Hartley 2, came close to matching the deuterium levels in Earth's oceans. Comet 67P/Churyumov-Gerasimenko was the most recently visited, studied by the European Space Agency's Rosetta comet orbiter. Rosetta found nearly three times the deuterium in Comet 67P/Churyumov-Gerasimenko compared to Earth ocean water (Figure 4.3).

Figure 4.2   Earth's early atmosphere was likely replaced by volcanism and the influx of meteors and comets. Even shortly after the formation of the Moon, lightning and solar radiation were already converting gases into complex chains of carbon, precursors of life. As continental uplift had not yet begun, most dry land was the result of volcanoes or raised crater rims. (painting by the author)

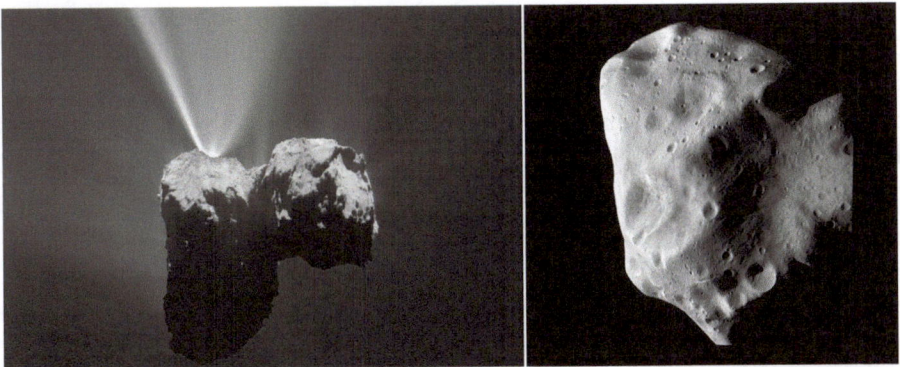

Figure 4.3   Left: Jets of gas and dust erupt from the surface of Comet 67/P. Comets like Comet 67P/Churyumov-Gerasimenko may not account for the water in Earth's oceans. (ESA/Rosetta/MPS for OSIRIS Team MPS/UPD/LAM/IAA/SSO/INTA/UPM/DASP/IDA). Right: Asteroid Lutetia and similar space rocks contain a different kind of water, but still do not fit the model for Earth's ocean brines. (ESA 2010 MPS for OSIRIS Team)

If all comets have similar ratios of hydrogen to deuterium, it turns out that cometary ice holds, on average, more than twice as much deuterium. It appears that at least the majority of our water did not come from comets.

For a decade, this seemed to be the established situation, but new data returned from ESA's Herschel Space Observatory has caused analysts to revisit the argument for cometary water as a source of terrestrial seas. Herschel encountered Comet Hartley 2, and found that its water vapor had "the same chemical signature" as Earth's oceans. Recent work now intimates that more than ten percent of terrestrial water came from comets, and perhaps a lot more, depending on how many comets have similar makeups to Hartley 2.

What about meteors and asteroids? It turns out that stony meteors—especially ones called carbonaceous chondrites—contain deuterium levels much closer to the Earthly range. And just as deuterium gives us a yardstick for measuring water, the inert gas xenon provides such a measure in meteorites. All meteorites have it.[3] If meteors were responsible for Earth's oceans, our atmosphere would have ten times the amount of xenon that it does.

This was the direction that research was pointing…until recently. But in 2022, a discovery muddied the waters. The spoiler came in the form of a soccer ball-sized meteor that fragmented along a path that included the skies over Winchcombe, a historic village in northern England. The pristine carbonaceous chondrite, pieces of which ended up in someone's driveway, came from an asteroid that used to hang out in the neighborhood of Jupiter's orbit. The coal-black stones—scattered from Winchcombe to Scotland—hold an estimated eleven percent water in the form of hydrated minerals. That water contains levels of deuterium that closely fit those for terrestrial seas.

The Winchcombe meteor was not the final nail in the coffin, some researchers assert, and the mystery continues. Even a mix of meteors and comets doesn't do the trick, because the two would still result in much higher concentrations of deuterium than the Earth contains…Which leaves us with the rocky material within the Earth itself. Computer models suggest that if water escaped from the Earth's interior rock, and then mixed with some comet ices, the resulting xenon levels in our atmosphere would match. During accretion,

---

[3] Studies indicate that cometary ice does not trap as much xenon as stony meteors do when they form.

the energy of all those colliding asteroids and comets transformed into heat. The infant planet simmered at temperatures so high that ice could not contribute much water. Instead, the majority of today's water was cocooned in clays or trapped as hydrocarbons (hydrogen) and iron oxides (oxygen), later combining to form water.

The scenario also matches the Earth's deuterium levels *if* some water was picked up from the surrounding solar nebula. It appears that volatiles (compounds like water that vaporizes at low temperatures) issued from the rock and metal as the Earth accreted. In a sense, our oceans grew as our planet did. As the planet settled down after differentiation (see Chapter One), heat from the interior escaped in the form of volcanic eruptions. Volcanoes released water vapor and carbon dioxide. Rocks locked the carbon dioxide down in a mineralogical prison, changing it into carbonate stone. The rocky plates also absorbed water. But plate tectonics, pushing and pulling at the crust of the planet, pulled some of the trapped carbon dioxide and water down into the mantle. There, the rock melted, releasing the carbon dioxide through volcanoes. This carbon carousel continued to recycle the atmosphere, and continues to this day.

Yet another theory[4] suggests that a great deal of Earth's water came from Theia, that vagabond planetoid that struck Earth and led to the existence of our large moon. In this scenario, Theia originated in the outer solar system, perhaps in the Kuiper Belt, and migrated in with its rich supply of ices. If this is the case, our oceans came from water like that found on Pluto and Triton, a reminder of the outer solar system in our own seas.

In those primordial times, terrestrial shores were barren and sterile beneath an angry young Sun. Early estimates put the temperature of Davy Jones' Locker at nearly the boiling point. More recently, new data suggests a more temperate clime. Isotope studies of South Africa's Baberton Greenstone belt include some of the most ancient rock layers on our planet. The isotopes indicate that the stone formed under cool conditions far below the boiling point. 3.5 billion year old gypsum formed in deep, cool seawater. Even the Earth's magnetic field provides a glimpse of conditions: paleomagnetic study reveals that layered rocks forming in cold standing water ("varved" sediments) formed

---

[4] Budde, Gerrit; Burkhardt, Christoph; Kleine, Thorsten (2019-05-20). "Molybdenum isotopic evidence for the late accretion of outer Solar System material to Earth". *Nature Astronomy.*

near tropical latitudes. Even Earth's early seas were not so alien as we might think.

Alien Seas, greenhouse effects and very long days

But what of our sister planet? Did frothy surf and ocean waves ever grace the sandy beaches of Venus? How could two planets end up so differently? Perhaps it had to do with where they formed, their distance from the Sun. This idea was popular for a while, but it has a major problem: nature tends toward balance. It blends things together to bring that balance. The Earth's weather is caused by nature's attempt to bring balance to temperature (evening out day warmth and night chill) or humidity (air currents often carry moist air to dry regions). This attempt at equilibrium forges ocean currents and airy gales. Rather than creating differences between neighboring planets forming in the solar nebula, nature tends to even things out. So Venus and Earth must have come from roughly the same building blocks. As astrobiologist David Grinspoon puts it, "Unless some tremendously efficient process dried Venus out completely, it must have started with more water, perhaps even an Earth's oceans worth of water—perhaps ten times this much."

The idea of oceans on other worlds came long before the age of space exploration. The ancients believed that the dark areas on the Moon were seas, and this led to our naming of lunar features like the "Ocean of Storms", the "Sea (Mare) of Tranquility", the "Lake (Sinus) of Sorrow" and the "Marsh (Palus) of Decay" (Figure 4.4).

In the early days of telescopic observation, some astronomers assumed that clouds obscured the surface of Venus. They reasoned that foggy sky hid vast marshes, something like they envisioned the Earth's Carboniferous Period. For science at the time, it made sense. Venus was our twin in size, only slightly closer to the Sun (so a little warmer), with a thick, cloudy atmosphere. Venus was the tropics! Beneath that warm fog, what was waiting within that swampy realm?

As observations progressed, astronomers found that Venus was cloaked in carbon dioxide—a lot of it. But observers still strained to glimpse the surface and chart its movements. If its rotation could be locked down, its length of day would be the same—an important bit of information for understanding a planet. It had been easy for Mars. The little planet's relatively clear, thin atmosphere plainly displayed what was underneath, and surface patterns could be mapped in detail. Of course, some of those details turned out to be wrong (i.e. the famous canals and oases), but astronomers spied the polar caps, clouds and even dust storms. Certain obvious features appeared at the

 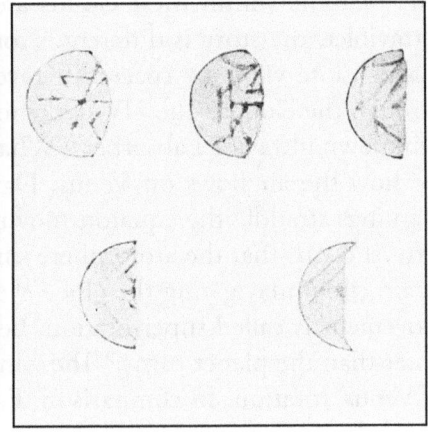

Figure 4.4   Left: Early observers named features on the Moon after bodies of water. In this view, we see Palus Putredinis, the Marsh of Decay, which lies at the edge of the Sea of Rains (Mare Imbrium). (NASA/Lunar Reconnaissance Orbiter). Right: some observers thought they could see surface features on the disk of Venus, as this 1897 sketch by Percival Lowell shows. As it turns out, Lowell's Venus drawings were a sort of self-portrait: against the bright, plain disk of Venus he was likely seeing the pattern of veins on his own retina. (public domain, Gutenberg Project)

planet's edge and marched across its face in predictable lengths of time.[5] The Martian day turned out to be just a few minutes longer than the Earth's. Could Venus observations reveal similar results?

The blindingly white disk of Venus seemed to repeat faint patterns. Many of the best observers reported seeing bright polar caps. Astronomers assumed these to be made of ice, like those on Mars and Earth. Bright spots seemed to appear here and there, apparently snow-capped peaks or plumes of cloud from volcanoes.[6]

Many attempted to discover the length of day on Venus, with estimates ranging from 23 hours and 21 minutes[7] to 225 Earth days.[8] (Today, we know that nothing on Venus lines up with either of these estimates, and the planet actually spins backwards). But it became apparent that the surface was not visible; Earthbound observers were looking at an impenetrable layer of cloud.

---

[5] One of the earliest distinct Martian features to be recorded at the telescope was Syrtis Major, drawn by Christian Huygens in 1659.

[6] Even to this day, transient bright clouds appear. Their cause is unknown, but may be due to meteorological effects of the high-altitude plumes of volcanic eruptions.

[7] Giovanni Cassini, 1666

[8] Giovanni Schiaparelli, 1877

Any patterns within these clouds are difficult to see in visible light. In the ultraviolet, the story is different. From Earth, vague, dark patterns cross the globe. Close views by spacecraft reveal swirls, whorls and chevrons moving through the clouds. The UV darkening is quite mysterious, and is called the "unknown ultraviolet absorber". Whatever is darkening the clouds helps us to see how the air flows on Venus. Dark "c"-shaped and sideways "y"-shaped markings straddle the equator, moving at 190 mph (300km/hr). Those patterns show us that the atmosphere surges across the planet in a great tsunami of air, circumnavigating the globe every four Earth days. This remarkable air movement is called superrotation, because the atmosphere circles the planet faster than the planet turns.[9] The winds are moving some 60 times the speed of Venus' rotation. In comparison, Earth's top winds move at 10 to 20% the speed that our planet turns.

With the invention of radar, strange microwave readings using Earth's first radio telescopes hinted that something might not be so pleasant down there. Some of the earliest observations came in 1956 from US Naval Research Laboratory in Washington, D.C. Turning their 50-foot dish toward Venus, the antenna picked up radiation coming from the planet. The returns seemed almost ridiculous: temperatures at least 570°F (300°C), far too hot for swamps, seas, or sea serpents. The readings were puzzling. Venus received about twice the solar energy that Earth does, far less than would explain such blistering temperatures. And even if the Venus atmosphere were much thicker than Earth's, it seemed unreasonable that it could trap enough heat to raise temperatures that amount. Perhaps, instead, the microwaves came from electrical discharges within Venus' ionosphere, high above the cloud tops. This theory also seemed unsatisfying. What else could it be? In the final analysis, the baffling "anomalous microwave radiation" seemed as though it just might be due to a very hot landscape below all those inviting clouds. It was time to take a closer look.

The Mariner 2 Venus spacecraft carried out the first successful planetary flyby in 1962. Its radiometer revealed surface temperatures nearing 900°F (~482°C). Even before Mariner 2 confirmed Venus' extreme environment, scientists had been hard at work figuring out how a dense atmosphere could retain high temperatures. Their computations centered on the famous greenhouse effect. As visible light from the Sun passes through the air, it shifts down the spectrum, becoming redder (as we see at sunset). But it continues to

---

[9] We see superrotation on another world cloaked in a dense atmosphere: Saturn's huge moon Titan.

shift, falling into the infrared, or heat, part of the spectrum. Infrared light cannot pass through air as efficiently as visible light can, so the light—which has become heat—becomes trapped inside the air, unable to escape back out into space. Just how much heat was trapped? Scientists used several variables in their models, trying out different amounts of pressure, water vapor and carbon dioxide. They assumed the Venusian atmosphere contained no more than 75% carbon dioxide. But it was difficult to visualize an atmosphere that could so efficiently hold enough heat to cook the surface the way Venus did.

The Earth's clear atmosphere allows some heat to escape. But instead of a thin nitrogen/oxygen atmosphere like ours, a dense canopy of almost pure carbon dioxide blankets the planet next door. Its atmosphere captures and holds more of the heat from the sun. Just as car windows let in sunlight, convert it to heat energy, and trap it inside long enough to melt your chocolate bar to the dashboard, Venus' dense atmosphere acts as a heat-trapping shell, ensnaring the heat in a self-reinforcing greenhouse cycle. As carbon dioxide levels steadily rose during the Venusian formative years, water vapor molecules leaked into space or chemically combined with other gases to make a nasty brew of sulfuric acid in the clouds. If we could free all the carbon dioxide within the limestone on Earth, our atmosphere would thicken to a pressure equivalent to Venus' atmosphere.[10] In other words, the carbon dioxide in the Venusian atmosphere is roughly equal to the amount of carbon in Earth's air and in its carbonate rocks (like limestone). So, Venus has the kind of atmosphere that our world would if our carbon had not been shunted away into the rocks by its interaction with our oceans and air. The atmosphere on Venus holds 200,000 times the carbon dioxide that Earth's does (Figure 4.5).

On Earth, the greenhouse effect is important. Earth's thin atmosphere traps just enough heat to raise average temperatures by about 35 degrees. The Venusian air is pumped up to conditions of industrial ovens. No briny sea could survive those conditions, but evidence has grown in support of ancient seas on Venus. Much of it is circumstantial, but some of the data points strongly to a young, watery world.

Nearly two decades after Mariner 2, the Pioneer Venus orbiter detected huge amounts of deuterium in the Venusian atmosphere. Sound familiar? Many researchers contend that such a vast inventory of deuterium is a telltale sign of ancient oceans. Another bit of evidence comes from the Soviet Venera

---

[10] No wonder our environmentalists—not eager to turn Earth into another Venus—are very interested in reducing humankind's contributions of carbon dioxide into the air.

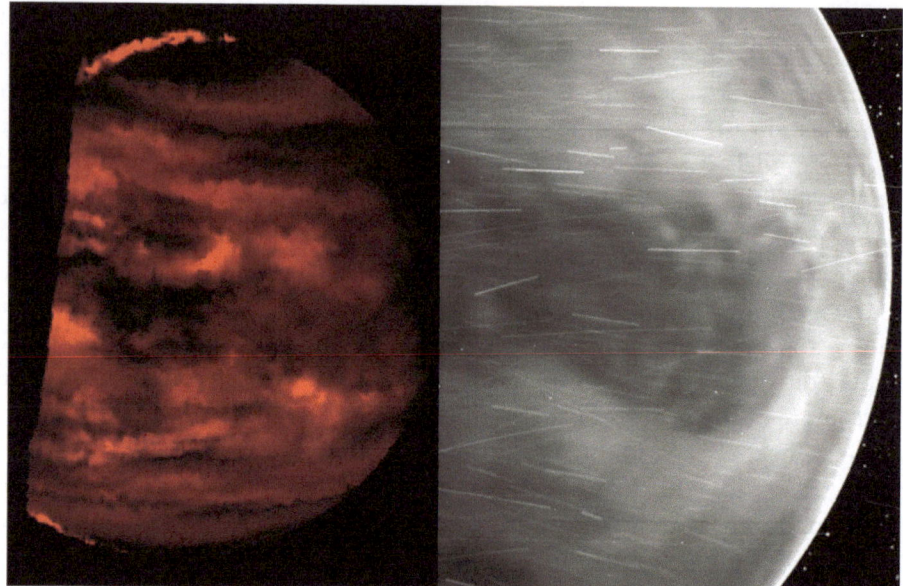

Figure 4.5 Left: En route to Jupiter, the Galileo spacecraft mapped the night side of Venus in near-infrared light (heat). In false color, this image shows heat leaking through the turbulent, mid-level clouds floating about 50 km above the simmering surface. The dark clouds are made of sulfuric acid mist. (NASA/JPL) Right: a more recent view of Venusian heat from NASA's Parker Solar Probe reveals features simmering on the surface. The image shows light visible to the human eye. The dark region is Venus' South America-sized continent Aphrodite Terra. Streaks are from solar radiation and cosmic rays. (NASA/APL/NRL)

8 lander, which touched down on the plains of Navka in 1972. The spacecraft carried a gamma ray spectrometer, which it used to take a series of measurements during descent, along with two precious readings from the surface. The instrument detected potassium, uranium and thorium in a recipe perfect for granite. What's so great about granite? It forms in the presence of water, another hint at ancient seas. Venus is completely parched today, but as its oceans escaped, they left behind the telltale sign of more temperate days. Those days may have been idyllic.

Dysfunctional adolescence

Taking a dim view: the Faint Young Sun Paradox

While Venus is closer to the Sun than the Earth is by a third, the Sun itself began as a much dimmer star. In Venus' childhood, our star shone only 60% as brightly, so temperatures on oceanic Venus would have been similar to Earth's today. In fact, many models of stellar evolution demonstrate that global temperatures on both Mars and Earth could have plunged far below the freezing point of water, and the Venusian outlook would have been more balmy than blazing. But things didn't work out that way. This is the source of the Faint Young Sun Paradox. Evidence locked within terrestrial rocks tells us that primeval Mars and Earth both hosted environments that allowed for liquid water on their surfaces. The young Sun's luminosity would have increased only gradually over a billion years. Why did these primordial seas not dive into deep freeze?

Researchers consider several explanations. The leading theory is that increased mixtures of greenhouse gases—particularly carbon monoxide, ammonia and methane—would have raised temperatures above water's melting point. Impacts could introduce and release concentrations of atmospheric greenhouse gases, especially carbon dioxide. Studies show that trapped gases from impacts could linger for some time, bolstering warming greenhouse gases for extended periods. This would have been advantageous on Earth and Mars, where temperatures under a faint young Sun would otherwise plummet below freezing. But on the already toasty Venus, greenhouse gases would add insult to injury.

A more radical concept proposes that the young Sun was more massive than today, and lost copious amounts of its material through "enhanced ionized wind".[11] In either case, conditions on Earth and Mars somehow remained warm enough for liquid water. Conditions on Venus may have benefited from the young Sun's lower luminosity, with reduced temperatures allowing for surface lakes and seas.

In the most optimistic models, Venus may have held on to oceans for a billion years or more. But as the Sun matured, its light and heat increased, warming the seas. Evaporating water infused the skies with vapor. Water vapor is a strong greenhouse "gas", and its ascendancy would have warmed the air more and more, causing increasing evaporation. (A natural cycle like this is known as a *feedback loop*.)

Some studies suggest that if Venus did have great seas, they would not have survived much longer than 100 million years after the planet formed. Other research seems to indicate that water on primordial Venus never did condense to form seas, but rather remained in vapor form.

Whether Venusian water played upon shorelines or drifted as hot vapor, sunlight eventually split it into its two constituents, hydrogen and oxygen.

---

[11] *Stability of Earth-Like N₂ Atmospheres: Implications for Habitability*; Helmut Lammer, et al; chapter 4 of *Early Evolution of the Atmospheres of Terrestrial Planets*; Springer Publishers 2013.

Much of the hydrogen drifted away into space; whatever oxygen that didn't follow was chemically locked in the rocks as rust. In a sense, Mars is red because the planet has rusted. Much of its oxygen is stashed within its rocks, where some combined with iron to form iron oxide (rust). As we have seen, a similar process worked to sequester Earth's carbon dioxide within its limestone rock.

Venus did not give up its water easily. It waged a fierce battle of chemistry and weather. Rocky terrestrial worlds like Earth have a built-in thermostat that regulates climate, forcing temperatures to remain in a range for liquid water to remain. As temperatures drop, carbon dioxide from volcanoes builds up in the air, warming the planet again. As temperatures rise, the rate at which water dissolves carbon dioxide increases. Water also reacts with silicate rock, changing it into carbonate rock. This lowers carbon dioxide and cools the planet. It's a cycle—a feedback loop—that has worked beautifully for eons, keeping Earth's climate stable.

The same process probably operated on the young Venus as well. But a little water in the air can have a lot of ramifications. Water keeps the heat in, and as temperatures rose, more surface water evaporated. As the deadly cycle continued, new carbonate rocks could not form quickly enough to offset the increasing carbon dioxide and the heating onslaught. The "thermostat" of Venus failed, and no planetary air conditioning could come to the rescue. This deadly runaway cycle vaporized any hypothetical Venusian oceans, pumped up temperatures, and ultimately led to the inferno that is Earth's "evil twin."

The Earth has oceans, and it has water vapor. What is keeping us from the same greenhouse effect that Venus suffered? Our planet has yet another feedback loop, and it involves clouds. The hotter the atmosphere becomes, the more water evaporates, just as it probably did on early Venus. But on Earth, this causes clouds to form, and those clouds are bright. The clouds reflect solar radiation back up into space, away from the planet, cooling things down once again.

How long did seas wash the face of our planetary sibling? Simulations at NASA's Goddard Spaceflight Center provide some insights. Michael Way and Anthony Del Genio ran five complex simulations using different variables. All five results showed that Venus might have been habitable for two to three billion years. That's a long time. Within that same time frame, life took hold on Earth and spread energetically across the globe.

Three of the simulations assumed that the ancient topography was similar to that found on Venus today, and included a deep (~310 meter) ocean along with a shallow sea (~10 meters deep). A fourth scenario incorporated landforms similar to Earth's with the same deep (~310 m) ocean. The fifth simulation assumed a globe-soaking ocean 158 meters deep. In all of the Goddard computer models, Venus could have held on to its oceans over billions of years, a more optimistic view than other studies. The early Venus would have had abundant carbon dioxide that became locked within rocks over time. In the various models, temperatures during this clement era ranged from room temperature (68°F/20°C) to highs exceeding a sultry 122°F/50°C. But the Goddard studies showed that something shifted, and temperatures soared.

About 700 to 750 million years ago, a planet-wide event resurfaced the landscape of Venus, wiping out old craters, filling in canyons, and releasing a vast amount of carbon dioxide that had been biding its time beneath the surface. The Goddard team proposes one possible explanation: tsunamis of magma flowing beneath and across the surface freed carbon dioxide trapped within the rock. The blanket of magma could have served as a seal preventing the carbon dioxide in the atmosphere from being soaked up by chemical reactions in the surface rock beneath the dried lava cap.

Twins: close relatives?

Some scenarios offered by planetary atmosphere experts suggest that Earth began with a formidable atmosphere which was then lost to space, replaced later by one more like the air we breathe today. Venus may well have paralleled Earth's beginnings, but something caused our two planets to part ways in remarkable fashion. This divergence has puzzled researchers for decades.

A University of Arizona team may have discovered the culprit. Their work, published in the October 2022 issue of *Planetary Science Journal*, casts a spotlight on impacts during the solar system's formative years. In the early days of planetary accretion, infant planetoids bashed into each other with one of three results. A slow impact resulted in the two protoplanets sticking together, growing into a larger planet. If the impact were fast enough, the two would obliterate each other, leaving behind clouds of debris. Yet a third possibility is that of a glancing blow, a sort of hit-and-run fender-bender in which both planetoids go on their merry way. These glancing collisions cause the two objects to assume orbits that will keep bringing them together until they eventually merge. But the Tucson team ran 4000 computer simulations to see what would happen if a Mars-sized globe grazed the early Earth. In half of the simulated smashups, the grazing Mars-sized object continued on to end up at Venus. The Earth impact slowed the wandering planet enough that it arrived at Venus more slowly, sticking to the planet upon impact.

The grazing impact would result in significant differences between the compositions of the two worlds. While Venus would gather to itself the material from the iron-rich cores of the impactors, Earth would have ended up with the lighter crustal material and upper mantle. The new story might further explain some of the most significant contrasts in the twin planet's natures. Why does Venus have no magnetosphere? No moon? Why does it spin so slowly? The only way to confirm the new scenario is to lock down the interior structure and chemistry of Venus. This comparison will reveal whether or not the planetary divergence arose through a game of cosmic bumper-car.

Planet Volcano

Whether the scenarios from the Goddard team are correct in the details, the fact remains that Venus has been sculpted by lava flows and violent volcanism for a long time. The most recent global map of Venusian surface reveals 85,000 volcanoes, with sizes down to the limit of the Magellan radar orbiter. More may be found with upcoming missions (see Appendix D). Formations on the Venusian surface generally fall into three categories, each according to how they formed. The first is volcanic-related land forms (including cones, calderas, mountains and lava flows). The second is impact craters. The third is tectonic activity (such as faulting or uplift of the crust). By far the most prevalent are the volcanic features.

Venus is "planet volcano". Venusian territory sports more volcanic shields, domes, cones, and flows per square mile than any other real estate in our solar system. Thanks to Soviet Venera landers, we have actual, in situ measurements from several surface locations. Using spectrometers and drills, the landers revealed that much Venusian rock is basaltic, nearly identical to terrestrial volcanic rock. Gamma ray spectrometers detected levels of radioactive material (such as uranium, thorium and potassium) consistent with that found in Earth's volcanic rock.

Basalt is the most common stone within the Earth's crust. Because of its low melting point, the basalt migrated up toward the surface as the Earth condensed from the solar nebula and differentiated. Most of the material in lava plains on the Moon, Venus, Mercury and Mars likely consists of similar rock.

Preparing for Hell

The Soviet Veneras had a tough assignment: descend through a sulfuric acid-laden sky, land on unknown terrain, radio back to Earth complex scientific results from a host of experiments, all before succumbing to temperatures higher than any domestic kitchen oven. Each landing was a race against time: the lander was chilled to below the freezing point just before entry into the atmosphere. By the time the craft reached the surface, its interior had warmed to comfortable room temperature. But the brutal pressure and searing outside air inexorably cooked the delicate electronics inside. Engineers had to design hardware made with special lubricants and composites. Various connections and interfaces needed to take into account clearances that changed as metals expanded in the heat. Laboratory testing took place in huge ovens that could reach temperatures similar to those found on Venus. Though some instruments failed, the survival and spectacular results from the Veneras are a testament to spacecraft engineering and creativity.

Were it not for the gloomy, acid-laden sky, many of its volcanic landscapes would seem familiar to any Hawaiian resident. Other volcanic forms are as alien as the stuff of science fiction. Their bizarre nature may stem from the hostile environment of Venus. In addition to high temperatures near the ground, air pressure at the surface is 90 times that of Earth at sea level, similar to the depths of Earth's oceans at a depth of 3000 feet (900 m). Sulfuric acid drizzles in the lightning-laced clouds above.[12] It's not a great place for a picnic.

Many Venusian volcanoes are in pristine condition, looking like they formed only yesterday. And they're everywhere; at least 85% of Earth's twin is wrapped in volcanic structures. The preservation of Venusian features may be the result of muted erosion; winds at the surface seldom break 4 mph/6 kph.

The surface of our cloud-covered celestial neighbor was revealed, a frustrating bit at a time, through the radar eyes of Russian Venera and U.S. Pioneer and Magellan orbiters. As the fog lifted, formations both familiar and unfamiliar appeared upon the planet next door. Earth's ocean basins find their counterpart in the lowlands of Venus. The Venusian ocean basins—once devoid of water—gained another kind of flowing liquid: surges of molten basalt broke upon Venusian "coastlines". Floods of extended lava flows fanned out across smooth plains. Relics of these hadean days show themselves in long, sinuous riverbeds that snake their way over the stony landscapes. Some of them are thousands of kilometers long. The Baltis Vallis may be the longest river valley in the solar system, 2 km wide at places and over 6800 km long. Newer magma floods have obscured both ends of the dried-up lava river; it must have been considerably longer.

Some of these lava "rivers" leave behind features reminiscent of terrestrial waterways: tributaries, flood plains, bars, oxbows, and meanders. Some fan out in the end, just as classic deltas do. In some places we find evidence of shifting in the rivers' flow and migration of flood plains. These features point to extended, long-term processes rather than flash floods or single event eruption flows (commonly seen on Mars). The long-lived flows puzzle geologists. As we saw in Chapter 2, our own world once had great courses of molten rock, but on Earth, lava rivers cannot remain liquid for long. But it appears that something in Venusian chemistry allows lava to flow over long periods and long distances. Some types of sulfur-rich rock can stay molten and thin for very long periods of time. Ultramafic lavas, such as those that may be present on Io, are another possibility, as they are very low viscosity and can flow

---

[12] Lightning on Venus appears to travel from cloud to cloud, seldom striking the surface.

across lengthy expanses. Another candidate material, suggested by David Grinspoon and other Venus experts, is called carbonatite. Carbonatite volcanism is rare on Earth. It involves carbon-rich rock with a very low melting temperature, 914°F (490°C). This is close to the Venusian daily temperature. Some researchers even envision a carbonatite aquifer stored below the surface, serving a similar role to Earth's underground water table. For now, the long-lived lava rivers of Venus remain a mystery.

Coronae

While the shield volcanoes of Venus may conjure thoughts of tropical, palm-covered beaches, other sites have no Earthly parallel at all. Perhaps most remarkable are the coronae (Figure 4.6). Coronae blemish the surface in circular or oblong tracts of fractures and wrinkles, disfigurements on the Venusian crust spanning hundreds of kilometers. A trough usually surrounds the ringed fracture system like a moat around a castle. We have found nothing like them on any other known planet or moon.

The quintessential corona is called Artemis. While most coronae are several hundred kilometers across, Artemis' formidable ridges would stretch from Los Angeles to Seattle. Complex ridges border Artemis' moat, bounded by a steep

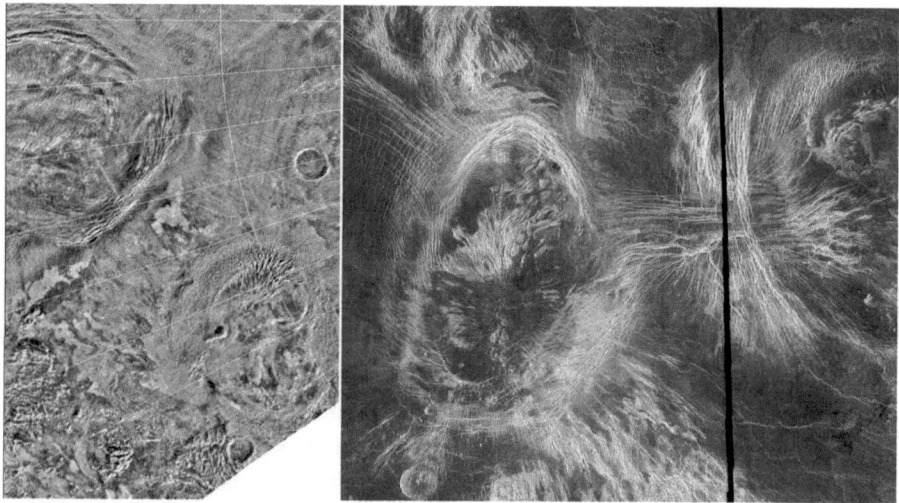

Figure 4.6 Our first glimpse of coronae came from Venera 15 and 16 radar images (left). At right is a later Magellan radar image of Bahet and Onatah Coronae. Black stripe is from missing data. (Venera image Russian Academy of Sciences, courtesy NSSDC; Magellan image NASA/JPL-Caltech)

trough. This surrounding structure is about 120 kilometers across, and is as deep as 2.5 km, deeper than the Grand Canyon of the Colorado.

Our first glimpse of the bizarre geological puzzlers came from the Russian *Venera 15* and *16* radar mappers. At first glance, the coronae appeared to be the ghosts of immense impact basins. But it soon became evident that the Venusian surface was geologically far too young to have preserved the basins from the kind of giant impacts that ended around 3.9 billion years ago. What were they? And why were they so varied, ranging from clear concentric loops to ghostly disks? A wide spread of age may account for the diversity of coronae: coronae have been forming on Venus for a very long time, and may be forming even today. The remarkable features may result from the upwelling of magma. As the magma rises from the hot interior of the planet, it shoves up against the crust, forming a dome. The heated surface softens, spreads out under its own weight and flattens. As the entire area cools[13]. It sinks and cracks, forming a circular ring around a depression. Similar vertical plumes are responsible for the Hawaiian Island chain, but any tourist will tell you that the results are quite different.

The Earth's crust is in constant motion, stirring beneath the surface and heaving mountain chains aloft. Our world's crust is split into jigsaw puzzle pieces called plates. The Earth's plates slide under, bash up against, or uplift each other in a planet-wide conveyor belt called plate tectonics. Venus doesn't seem to have plates similar to Earth's at all. Instead, the coronae may be the telltale sign of a new and alien type of tectonics. Surface features seem to be created at the hands of localized powers below them. The vertical forces that rule the crust of Venus today are known as "columnar tectonics". Rather than a planet broken into moving plates, the heat from within Venus rises through the mantle and crust in vertical plumes, or columns.

While Venus might have a thick, fairly immobile crust, recent work at North Carolina State University conflicts with that narrative. Venus may have a gentler version of plate tectonics after all, say researchers.[14] Within Venus' extensive lowland terrain, Paul Byrne and colleagues identified large blocks of the crust that have drifted, pulled apart or rotated in ways resembling the rafting motion seen in pack ice on Earth's polar oceans. This style of tectonism is known as block tectonics. Unlike terrestrial tectonics, the movement of Venusian blocks does not raise mountain chains, but the sections of moving crust have deformed the plains. This may be an indication that Venus is

---

[13] "Cooling" is a relative term, as the coolest surfaces on Venus are as hot as an industrial pizza oven.
[14] Paul K. Byrne *et al.* "A globally fragmented and mobile lithosphere on Venus." *PNAS* 118 (26), 2021

geologically active even today. Unlike the Earth, Venusian plate tectonics appear to be localized.

Recent work by planetary scientists at the Jet Propulsion Laboratory bolsters the idea of a more active lithosphere/crust. Researchers were able to deduce the Venusian crustal thickness by appraising the bending of coronae. They used altimetry data from decades-old Magellan radar orbiter records. 65 separate coronae yielded the thickness of the lithosphere at 75 locations. The measurements showed that heat flow from Venus was higher in these regions than it is on Earth, and similar to values at active tectonic areas like sites of sea floor spreading.

The researchers conclude that Venusian tectonics are driven by the subsurface movement of magma and plumes to shed heat through the crust (Figure 4.7). Similar magma plumes may well have been responsible for triggering Earth's plate tectonics, making Venus an analog to Earth's Archaean

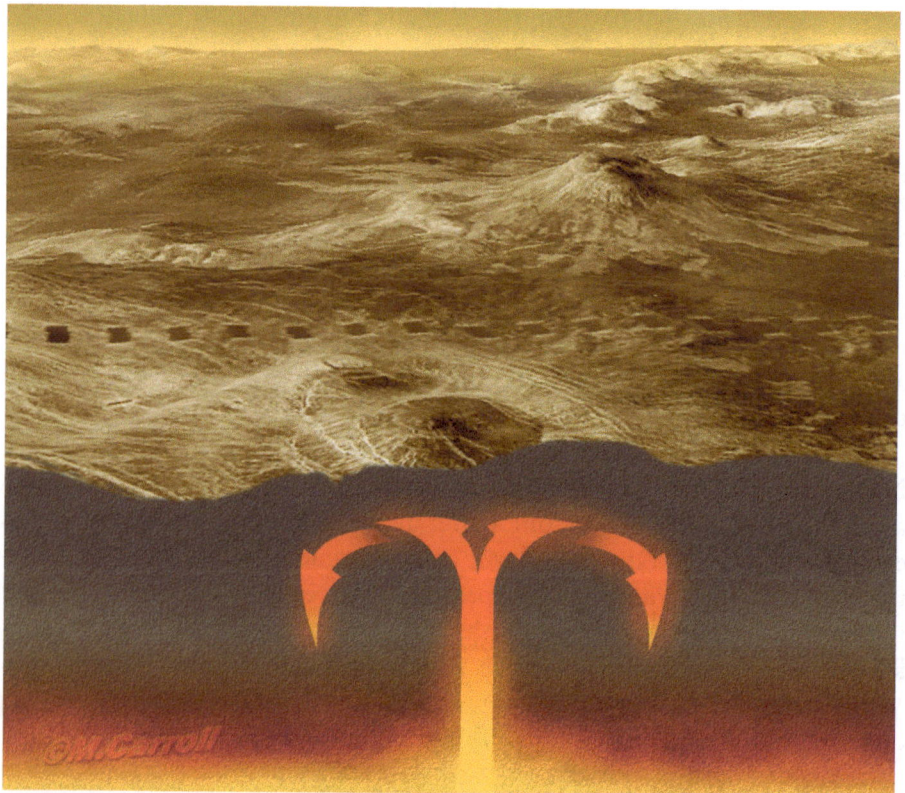

Figure 4.7 A view of the possible interior of Nagavonyi Corona. A vertical column of magma rises to form the domed region. Black squares are missing data. (Magellan image NASA/JPL/USGS, art by author)

period, which lasted from 4 to 2.5 billion years ago. During this period, when conditions on Earth were hotter, terrestrial heat flow was three times that of today.

"Venus today does not have plate tectonics," says Planetary Science Institute's astrobiologist David Grinspoon. "Was its early tectonic environment more like Earth was? A lot of us think it might have been." The planet's lack of plate tectonics today might be due to the fact that Venus dried out because of its later atmospheric evolution. As its atmosphere grew, Venus dried out. "Being that desiccated," Grinspoon says, "it's hard to maintain plate tectonics. Venus may have had a more Earth-like plate tectonic regime and then seized up when the water left. If that general picture is true, then I'd say Earth and Venus may have been much more similar than they are today."

The surface we see on Venus today is almost all young by plate tectonic standards (less than a billion years old except for a few areas that are probably much older, such as the tesserae). Any kind of global pattern might have well been erased.

Just how active is the Venusian landscape? How recently have its volcanoes, chasms and coronae formed? New work by a team from the University of Maryland and the Institute of Geophysics in Zurich, Switzerland sheds light on the question.[15] The team found that coronae differ in age. Some are subtle and have been inactive for a long period, while others are well preserved. The researchers identified 37 coronae as "active", with "extremely recent activity".

The mantle plumes occur across the globe. It may be that the underground plumes have resurfaced the entire planet piece by piece, region by region. Some coronae have sunken almost beyond recognition, fading into indistinct patterns with almost no vertical relief. Others are so fresh that they have ruptured, spilling melted stone across the Venusian landscape. Coronae seem to congregate in certain regions. The larger features have a propensity to gather along rises. The majority cluster at the edge of the canyons (chasmata) or parade along fractures in a belt across the southern hemisphere of the planet.[16] Their locations hint at variations in the planet's upper layers; they may form where the crust is thinnest. Since Venus lacks the kind of plate tectonics present on Earth, the coronae may serve as a sort of pressure valve for the energy generated by Venus' subterranean inferno. "Maybe hot spot volcanics plays a

---

[15] See "Corona structures driven by plume–lithosphere interactions and evidence for ongoing plume activity on Venus" by Gülcher, A. J. P., Gerya, T. V, Montési, L. G. J. & Munch, J. *Nature Geoscience*, July 20, 2020

[16] Most of the coronae-infested fractures crisscross the Beta-Atla-Themis zone, a vast area with young geologic features.

bigger role," Grinspoon suggests. "Maybe there's a variation on plate tectonics where things are being delaminated beneath the surface, so there are areas of subduction and uplift that you don't see on the surface. There are a number of ideas. Clearly the planet doesn't operate today, tectonically in a global sense, like the Earth does." Whatever form the Venusian version of tectonics takes, it is truly an alien affair.

Venusian Vesuvius

While the composition of the lavas on Venus remains unknown, evidence of eruption is everywhere. Among the volcanoes most familiar to us Earth-dwellers are the gently sloping shield volcanoes such as those of the Hawaiian Island chain. Scattered across the Venusian lowlands, most are a dozen kilometers or less across. Like the Hawaiian shields, these flattened mounts are the consequence of many recurring flows of thin lava. Eons of gradual buildup mark these low-profile summits.

In many areas, the volcanoes congregate in groupings of what are likely cinder cones, what researchers call "shield fields". These clusters average roughly 60 kilometers across. The small volcanic vents and cones could be seen down to the limit of Magellan's resolution from orbit. The vast ocean plains of Venus may play host to hundreds of thousands of small shield volcanoes, a number similar to those found on the Earth's restless sea floor.

Some Venusian volcanoes have grown to titanic size, towering 8 km into the sulfuric acid haze. They dot the plains and rest atop the highlands. One exemplary giant is Gula Mons, a steep mount rising in the northwest region of Eistla Regio, a 6000 km long east-west trending upland area that abuts the great continent of Aphrodite Terra.

Volcanoes like Gula are similar to terrestrial stratovolcanoes. More violent and unpredictable than gentle shield volcanoes, these erupters include such celebrities as Mt St Helens and Mt Vesuvius. Gula Mons towers 5 kilometers above the plains of Eistla Regio. Jagged scarps and troughs slash its northern face, radiating out from the central summit crater (caldera). Magellan radar images reveal that the northwestern face is very rough. The area may represent the most recent of lava flows, perhaps similar to rough aa flows on Earth. A rift valley spreads from its base, suggesting that the Eistla region has been subjected not only to volcanism but also to faulting in a geologically active past.

Evidence for geologically recent volcanic activity is ubiquitous on Venus. The coronae surround some of the strangest territory on the sweltering world. Geologists and cartographers have assigned creative and whimsical names to a bizarre collection of volcanic formations found there (Figure 4.8).

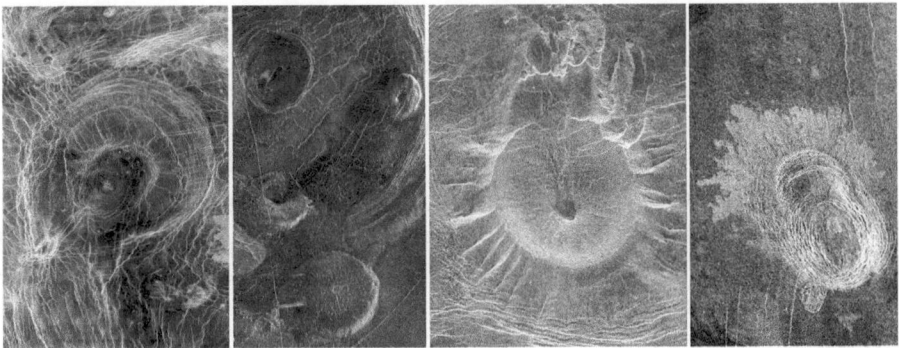

Figure 4.8   Assorted Venusian volcanoes. (l to r): Arachnoids, pancake domes, ticks and anemones. (Magellan images courtesy NASA/JPL-Caltech)

Foundational to this assortment are the pancake domes. These eerie volcanoes are roughly disk-shaped. Their summits are flat to slightly curved (domed) with fractures or summit calderas. They average 25 km in diameter and 770 meters in height. Their radar-bright, steep sides are rugged, cut by ravines and collapsed slopes. It is likely that the domes are the result of very thick lava flows. Radar images have charted over 150. The Earth has some similar formations, known as domes. We find examples of domes in many volcanic areas across the terrestrial globe, including Lassen Peak in California or Greece's Nea Kameni. But their Venusian counterparts are ten to one hundred times larger.

An excellent example of pancake domes stands at the eastern edge of Alpha Regio, where seven domes loom above the plains. Complex fractures etch the top of each. It appears that lava cooled on top, and then cracked as new lava expanded underneath—or drained away from inside—breaking or collapsing the crust. Fractures also stretch across the plains below. Some of these domes cover the canyons, while others are scored by them, providing evidence of ongoing tectonic movement before, during, and after formation of the pancakes themselves.

Some of the pancake domes seem to give rise to other forms. One such volcanic form is known as a "tick". Ticks appear to be eroded pancake domes. Often associated with rift zones, the Ticks are bounded by landslides and ridges radiating outward from the dome itself. The ridges terminate in sharp points, giving the appearance of the legs of a tick. At times, dark flows have sprung from a summit caldera, streaming along lava channels. Their summits tend to be concave. Some fifty have been discovered so far. Ticks bear a strong resemblance to the Earth's seamounts, undersea volcanoes whose flanks have avalanched off into radial ridges.

Anemones are yet another member of Venus' volcano club. These volcanoes exhibit lava flows arranged in overlapping sheets that extend outward in flower-like patterns. The lava patterns are usually related to fissure eruptions, with a series of elongated vents slicing into the summit.

Arachnoids join the volcanic dome clan, surrounded by a cobweb of fractures and crests. Russian scientists first named these features after seeing great concentric fissures spreading out from volcanic sources in Venera radar images. Their sizes range from 50 to 230 kilometers. Upwelling magma may crack the surroundings, leaving ridges and canyons radiating beyond the arachnoids themselves.

Eruptions today?

Venus' apparently fresh lava flows, crisp volcanic slopes and well-preserved calderas inspired scientists to search for changes in volcanic structures during the Magellan mission. It wasn't easy: in most locations, Magellan overflew the same territory only twice, roughly an Earth-year apart. Careful study of the data[17] revealed changes in a volcano called Maat Mons. On the flanks of the volcano, a 2 square kilometer vent changed shape over the course of an eight-month period. The caldera's margins appear to elongate and spread to the northeast. Volcanic flows downslope from the site may also be new, although the flows may not have been visible in the earlier radar imaging, which had a slightly different illumination direction. Some researchers interpret the changes as indicative of active volcanism today.

There is other evidence, and it comes from even older data radioed home by NASA's Pioneer orbiter. In the late 1970s, Pioneer's ultraviolet spectrometer clocked a steady decrease in the amount of sulfur dioxide in the Venusian upper atmosphere. The decrease continued for several years. What could have caused it? If the rate continued unabated, Venus would run out of sulfur fast, and we'd be inviting Earth tourists in no time. But rather than a sky-clearing development, some researchers propose that shortly before Pioneer arrived at Venus, a massive volcanic eruption sent billowing clouds of sulfur into the atmosphere. It's a common occurrence on Earth, but in our terrestrial air the sulfur combines with water to break down fairly quickly. On Venus it would remain for some time. If there are active volcanoes on Venus today, our radar

---

[17] For more, see "Surface Changes Observed on a Venusian Volcano During the Magellan Mission" by Herrick and Hensley, *Science*, March 15, 2023

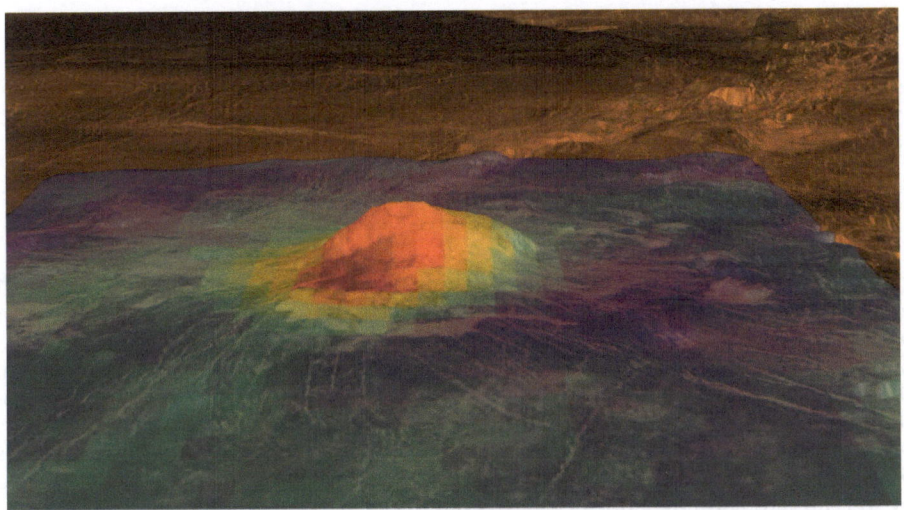

Figure 4.9 The volcano Idunn Mons appears in exaggerated height (30x). Topographic Magellan data forms the base, while colored data from ESA's Venus Express shows the heat patterns. Red-orange is the warmest temperature, with purple the coolest. The warmest area is centered on the summit. Bright flows—likely lava fields—stream down the flanks. (ESA/NASA/JPL)

images may have given us glimpses of them, and Pioneer's spectrometer has shown us the smoking gun evidence.

We have even seen "hot spots" from orbit. In 2010, the European Space Agency's Venus Express Orbiter spied "infrared anomalies" near the summit and on the eastern flank of Idunn Mons, a 2.5 km-high volcano in Venus' Imdr Regio province (Figure 4.9). The large mountain fans out across 200 km of Venus terrain. Heightened temperatures hint at fresh lava flows in the region, which line up with radar images of the mountainsides. Venus Express also observed slower winds in the area, which researchers suggest are related to heat from very recent—perhaps ongoing—lava flows. Laboratory experiments show that the rock on the surface of Venusian volcanoes may change, chemically, more rapidly than once thought. Previous work indicated that flows like those on Idunn Mons were as young as 250,000 years old. The new analysis suggests that the flanks of the volcano could have been blanketed by new lava anywhere from 10,000 years ago to yesterday.

Venus Express revealed other lines of evidence for current activity. The ESA spacecraft charted temperatures elevated by several hundred degrees in several areas. The toasty locales follow a series of fractures in the Ganiki Chasma region. Ganiki is geologically young, and associated with volcanic formations including the corona Sapas Mons. Some of the warm sources are small (~1

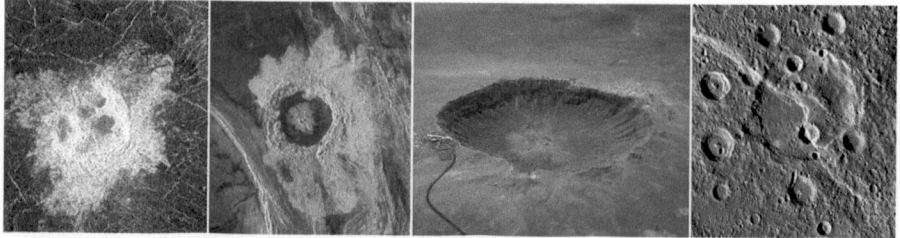

Figure 4.10 Craters compared: Far left: This irregular crater in the Lavinia Region of Venus is actually a cluster of craters, probably the result of a meteor breaking up just before impact. The crater grouping spans roughly 14 kilometers across. (NASA/JPL-Caltech, Magellan mission) Left center: With its central peaks and raised, terraced rim, the crater Dickinson is a complex Venusian crater. (NASA/JPL-Caltech, Magellan mission) Right center: Barringer Meteor Crater in Arizona is roughly ten thousand years old. Earth's weathering has already made its mark in the rounded crater rim and gullies down the sides. (courtesy Dan Durda). Far right: Mercury's Duccio crater is interrupted by the cliff-like Carnegie Rupes. Note the difference between the smaller complex craters near it and Venus' Dickinson Crater. (NASA/Johns Hopkins University Applied Physics Laboratory/Carnegie Institution of Washington)

km), while others extend over 200 square km. Close inspection of the infrared data supports the prospect that the heated areas are likely active lava flows flooding the rocky landscape.[18] Researchers have discovered more heated sites near the volcano Maat Mons. Four transient hotspots, perhaps as hot as 825 °C may be kilometers-long lava flows, bands of cinder cones, or even molten lava lakes.

One trademark of active volcanoes is the presence of lightning. Four Soviet Venera landers sensed lightning as they descended beneath the cloud decks, but their instruments could not lock down specific locations. Japan's Venus Climate Orbiter, Akatsuki, has carried out a multi-year search for lightning, but has not identified any from its orbital perch. In the course of 16.8 total hours of observations, "no flashes attributable to lightning"[19] could be detected.

A Swiss Cheese World

Craters on Venus supply us with a critical indicator of the planet's young age. There aren't very many, and the few that survive are large (Figure 4.10). With

---

[18] See "Active Volcanism on Venus in the Ganis Chasma rift zone" by Shalygin, et al; 06/17/2015 *Geophysical Letters* 42.

[19] *Constraints on Venus Lightning From Akatsuki's First Three Years in Orbit*, Lorenz, et. al. 03 July 2019 https://doi.org/10.1029/2019GL083311

Venus' dense atmosphere, this makes sense. Most meteors burn up in the atmosphere on their way down, or are so slowed by atmospheric friction that they leave a subdued scar upon impact. But the remaining craters reside on the extreme ends of the erosion spectrum, either fairly pristine (newly made) or almost completely buried. There are not many in a state of deterioration somewhere in between. And the craters aren't as old as the ones we find on Mercury, the Moon, or Mars. Venus has no evidence of the Late Heavy Bombardment. Either Venus somehow escaped this hail of stone and metal—unlike any of the other planets—or the resurfacing events of the planet's past wiped out the pockmarks from the early solar system's formation. That resurfacing must have come to an end fairly recently: all the Venusian craters are relatively young. Observed craters and other evidence indicate a surface age of less than 500 million years (±200 million).

As for those larger craters on Venus today, 96% of those larger than 15 kilometers across are *complex*. While simple craters are bowl-shaped, complex craters have raised circular rims, a central mound or peak, terraced walls leading to a flat floor inside, and extensive ejecta. To make this type of crater, the asteroid must survive its searing voyage through Venus' thick atmosphere, and must be intact as it reaches the ground.

Six tenths of craters smaller than the complex ones tend to be irregular (Figure 4.11). These impact wounds have irregular or broken rims and hummocky, rough floors. Irregular craters are the result of asteroids that have been crushed or torn apart by their trip through the atmosphere. Their ejecta patterns are as irregular as the rest of their form, typically with asymmetric material scattered around the crater. Since Venus' atmosphere is so deep and dense, smaller meteors shatter as they descend. The resulting impact leaves a cluster of overlapping craters, something rarely seen on other worlds.

Venusian reboot

Lava flows cover parts of some Venus' impact features, but if volcanoes wiped out most of the Venusian craters, we would expect to see many partially buried ones. This is not the case. It appears that something has obliterated the craters without a trace. Venus researchers have put forth several theories. One possibility is that the craters we see today are all recent impacts from a fairly limited, single event. But this theory doesn't sit well, because all evidence throughout the terrestrial planets points to the fact that there has *not* been a recent episode of cratering in the inner solar system.

Another intriguing proposal is that a Venusian moon recently fell out of orbit, showering the planet with new craters. The moon scenario is hard to disprove,

(continued)

but until other data comes to light to support it, it seems a bit forced. Additionally, modern science abhors theories evoking special events or situations (although nature seems to be full of them).

Perhaps the most convincing theory asserts that half a billion years ago, something triggered extensive volcanism after a period of geologic calm. This planet-wide paroxysm re-set the crater counts on the surface of Venus. During this period of time, the coronae began to form across the planet. From a 700 million-year-old resurfaced globe, craters began to appear at the current, unhurried rate, leaving the numbers we see today. There is precedent for this on Earth. At the end of the Mesozoic—the age of the dinosaurs—massive volcanic eruptions in a region of India called the Deccan Traps ejected so much ash and gas into the Earth's atmosphere that the climate was significantly changed for an extended period of time (see Chapter 3). If the crust of Venus is as thick as it appears, it may be that Venus periodically suffers from global, climate-changing eruptions involving thousands of volcanoes at a time.

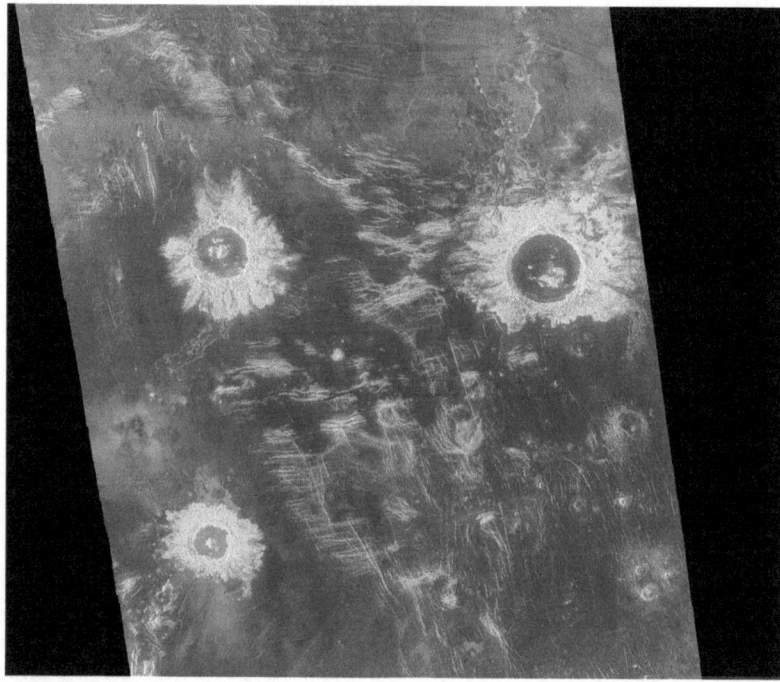

Figure 4.11 Three large impact craters scar the plains of Lavinia. Their diameters range from 37-50 kilometers. Smaller domes accompany the complex craters to the south. Several domes have central pits typical of volcanic cones or shields. (NASA/ JPL-Caltech)

Beneath acid-filled skies

Venus' dense, sluggish air pulls at the landscape with gentle winds at the surface, but hurricane-force gales at higher altitudes. The ground wind, formidable even at those low velocities, softens craters and tears across the face of mountains. It builds sand dunes and carves rocks and boulders. Sulfuric acid rains erode the highest ground.[20] Where tectonics have lifted or fractured the crust, edges and slopes appear sharper, steeper, less eroded than similar features on Earth, even the areas that seem to be quite ancient. The reason may be the lack of water in the surface and air. Erosion on Earth is driven by such things as rainfall and ice (which gets into rock cracks and breaks boulders apart), seasonal changes (Venus has none[21]), and day/night temperature variations (Venus day temperatures are almost identical to night ones). Terrestrial craters erode quickly on Earth, thanks to our weather and tectonics. The odd preservation of Venusian craters and the weathering of mountains and canyons tell us that erosion in Venus' dense atmosphere operates differently than it does on Earth.

In Venus' planetary inferno, metals may condense as "snow" on some high spots. Radar imaging of the planet reveals a strange pattern of brightening on high ground, beginning at about 3 ½ kilometers above the planetary "sea level". In radar imagery, bright reflections usually mean a rough surface, but something else is going on here. The glimmering patina blankets everything from rugged mountains to high plateaus. A variety of metals could explain the strange gleaming highlands. At Venus' drastic pressures, specific metals called halides and chalcogenides can exist as vapor. Chlorine, fluorine and sulfur, all common on the planet, would have radar signatures that appear as radar-bright as well. Other intriguing candidates include lead sulfide, bismuth, the chrome-like metal tellurium, or iron pyrite ("fool's gold"). On Venus, as on Earth, temperatures drop with altitude. Low-lying plains on Venus simmer at a blistering 467°C, while the radiant, higher elevations cool down to a lovely 387°C. Volatile metals may vaporize in the lowlands and migrate to higher terrain, where they condense again as they cool. Imagine a mountain peak crowned by shimmering fool's gold! (Figure 4.1)

---

[20] Although most sulfuric acid "rain" is actually virga, evaporating before making it to the ground.

[21] The tilt of a planet's axis determines its seasons. The Earth tilts at 23.4°, while Venus is nearly upright at 2.7°.

Some of the highest terrain displays strange dark spots devoid of the radar-bright blanket. Just as the metal vapor appears to condense above 7900 feet (2400 m), it may vaporize again at altitudes near 15,400 feet (4700 m). But the baffling dark spots are inconsistent, and not found on some high ground, such as the summit of Maxwell Montes.

Metallic mist could also explain a mystery concerning planetary exploration. Within the acidic virga falling from Venusian skies, something strange is happening. The phenomenon—called the Pioneer Venus 12.5 kilometer anomaly—assaulted all four probes in the Pioneer Venus multiprobe mission. At an altitude of about 12.5 kilometers, a power spike surged through all four probes, despite the fact that they were thousands of kilometers from each other, some in daylight and two in regions experiencing night. As each probe reached the 12 km mark, they transmitted bizarre readings of temperature and pressure. Many of their instruments failed completely, losing valuable data.

In 1995, three Washington University geochemists proposed a fascinating explanation[22]: they suggested that the cause is metallic fog in the Venusian highlands. The candidate metals vaporize at about the same altitude as the Pioneer Venus anomaly (Figure 4.12). A haze of metallic vapor could explain the mysterious 12.5 kilometer failures, and would certainly add to the alien nature of Venusian meteorology.

But a second possibility is completely independent of Venusian weather. Instead, say some engineers, the power spikes center on the hardware design of the probes themselves.[23] A 1993 NASA study found that, "All anomalies were attributed to an insulation failure of the external harness." Protective Kapton tape had been tested in chambers simulating Venus conditions, and it was deemed adequate for the wiring on the probes' external sensors. But technicians decided to reinforce the external connections by using Kynar, a type of shrink tubing not tested in the same manner as the Kapton had been. When this

[22] See *Volatile transport on Venus and implications for surface geochemistry and geology* by Brackett, Fegley and Arvidson, *Journal of Geophysical Research*, pp1553-1563, January 25, 1995

[23] Pioneer Venus 12.5 km Anomaly Workshop Report, Volume 1; NASA Conference Publication 3303

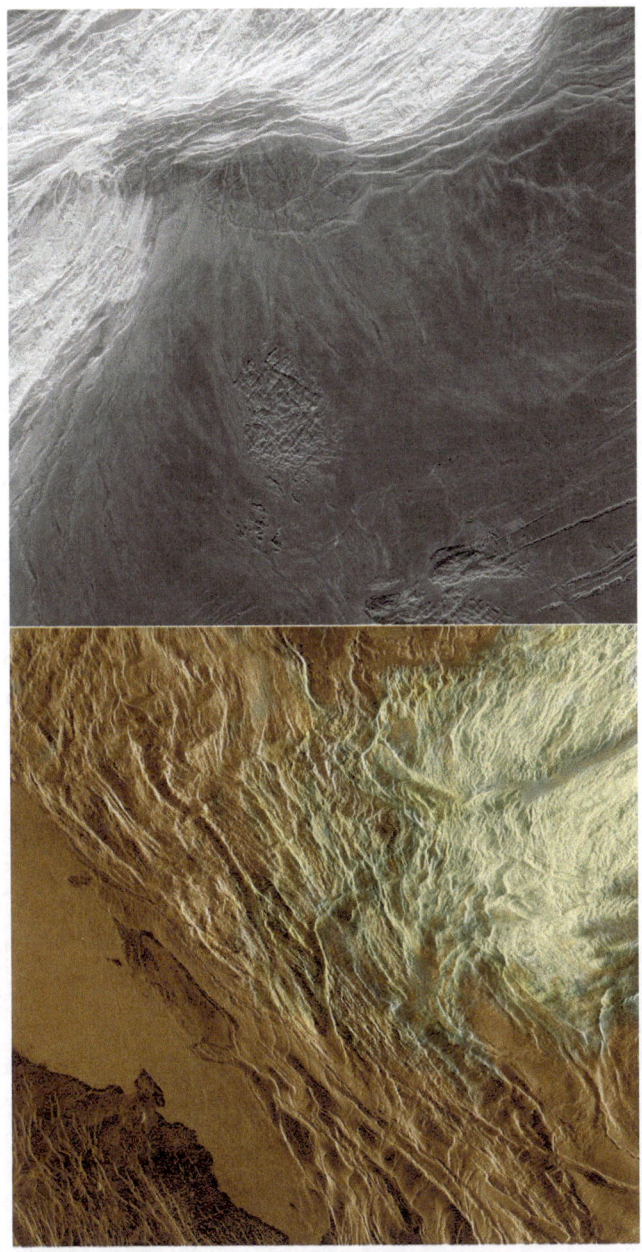

Figure 4.12 (top) Looking down on a behemoth: the Maxwell Montes are the highest mountains on Venus. Maxwell Montes itself rises to a summit 11 kilometers above the surrounding plains of Lakshmi. Its higher elevations appear bright in radar. Such radar returns are common on Venusian highlands. The phenomenon may be the result of radar-reflective minerals such as iron pyrite. The highest regions on Maxwell are devoid of the material, suggesting that whatever creates the bright areas condenses at a specific altitude range. (Bottom) Akna Montes is another mountain with a radar-bright blanket at the top. This colorized version of a grayscale radar image demonstrates the remarkable feature, which may be pure "fool's gold". (Magellan images courtesy NASA/JPL-Caltech. Akna image colorized by the author)

tubing reached 620°F (327°C) it released corrosive hydrogen fluoride vapors which dissolved the Kapton insulation, shorting out the connections.

But a hardware failure common to the Pioneer probes does not solve another aspect of the phenomenon: Pioneer Venus probes were not alone in their malfunctions. The Soviet probes Venera 11, 12, 13 and 14 all experienced similar power spikes at about the same altitude. Years later, the VEGA 1 Venus lander may have prematurely triggered its landing sequence at about the same altitude. Its sensors told it that it had landed…some 18 kilometers above the surface! (Strangely, the sister craft, VeGa 2, had no ill effects at similar altitudes). It is possible that a haze or mist of metallic compounds condensed on the surfaces of all these probes, triggering a host of breakdowns and temporary failures. Future missions will attempt to solve the great mystery of the Pioneer 12.5 km anomaly.

The creative engineers who took it upon themselves to replace the laboratory-tested Pioneer Venus materials with "better" ones have a counterpart in the Soviet Venus program. When engineers outfitted Venera 13 and 14 with color cameras, designers realized that the cameras would need a color bar, a sort of multi-colored scale so that photos could be calibrated to show true color (Figure 4.13). Venusian temperatures required very special paint

Figure 4.13 The Soviet Venera 13 and 14 gave us the first color images from the surface of Venus. Horizon is at upper right. Note the subdued colors on the calibration strip, and the crescent-shaped lens cap on the ground. Venera 14's lens cap is directly beneath its soil science arm. The angled rock in the center foreground may have been disturbed by Venera 14's landing. (image © Ted Stryk, from Venera data courtesy Russian Academy of Sciences)

that could survive without fading or shifting in color. But in the complexities of space missions, scientists, engineers, and technicians all have different priorities. Sometimes, divergent goals can breed conflicting solutions to problems. American planetary scientist Bill Hartmann noticed that the color strips on the Venera images seemed subdued. He asked a Soviet colleague why the colors seemed so neutral.

Hartmann's scientist friend described how—in those days—the Russian system was divided. A project first went to the expert engineers who actually did the research and design, and then it went to a different ministry when it came time to fabricate the hardware. Soviet researchers developed advanced paints that could retain their color even in the 900-degree heat. But when the project was transferred to the manufacturing side, a new group of workers came into the picture. Hartmann was told how the new industrial plants had "suggestion boxes" where workers were encouraged to make suggestions about ongoing projects. According to the story, a factory worker said, "This is ridiculous. We're paying incredible amounts for these paint chips and cheaper paint chips are available." Commercial paint was substituted, the worker received an award, and the paint almost immediately lost its color in the Venusian furnace. The human variable is an unpredictable and entertaining part of space exploration, no matter which country or agency is doing it.

Life in a Venusian Sea?

We find meteorites from Mars in Antarctica, and meteorites from the asteroid Vesta in Africa.[24] These rocks came to us courtesy of massive impacts that tossed them from one planet to another. Such cosmic events happened with much greater frequency when the planets were young. Venus, Earth and Mars frequently tossed bits of themselves toward each other. We know that simple forms of life (like bacteria or viruses) can survive extreme temperatures, pressure changes, and radiation. These "extremophiles" could conceivably take an interplanetary tour and make it to the other side if they were lodged within the interior of rocks. If life formed on any one of this triad of worlds, meteor impacts could have spread microbes to any of the others. The early oceans of Venus, Earth and Mars were not quarantined from each other. They were tied together in a hail of meteors that were formerly pieces of planets.

On Venus, the problem lies in what happened next. The planet heated up and lost its rivers, lakes and oceans. Then it heated up some more, to the point

---

[24] In fact, we find meteorites on every continent of the world.

where temperatures can tear apart any organic matter. We have seen that life on Earth inhabits nearly every niche and corner it can get into. If Martian landscapes were awash with ancient life-filled seas that eventually froze and dried, life may well have moved underground. But where could life flee on a hellish place like Venus?

If it could not go down, perhaps it went up, where the oceans fled, to clouds in the sky. At altitudes of 48 to 60 kilometers above the scorched, acid-laden landscape, air pressure is similar to Earth's at sea level, and clouds drift in benign temperatures. Within this cool sweet spot, temperatures hit lows around freezing, reaching highs of 93°C. Scientists at MIT, Cardiff University, and several other institutions reported indirect evidence that may signal life in these temperate clouds.[25]

The team's spectral studies revealed the signature of phosphine. An earlier MIT study demonstrated that on rocky planets, living organisms are the most likely cause of its presence. Reanalysis of the data later showed less phosphine than originally projected, and other work disputes the finding. But a new analysis of results from the 1978 Pioneer Venus atmospheric probes confirms the phosphine readings. More studies are scheduled.

Other investigators have attempted to understand the "unknown UV absorber", dark regions that appear in ultraviolet images of Venus' clouds. Many explanations have been offered for the mysterious darkening: particulate substances within the clouds, a compound dissolved by the ubiquitous droplets of sulfuric acid, or perhaps some crystalline form like water ice. An August 2019 study[26] based its findings on data from Japan's Akatsuki Venus orbiter, ESA's Venus Express, the MESSENGER Venus-Mercury mission and the Hubble Space Telescope. Researchers reported a long-term pattern of UV absorption, a darkening that might be caused by unknown chemical processes or high-altitude colonies of microorganisms in the clouds.

Visiting Venus: a detailed summary from Venera 9 through Akatsuki

The picture we have of Venus today was hard-won. Centuries of telescopic observation, laboratory and computer modeling, and difficult space missions all have contributed to our knowledge of our sister world. The humble

---

[25] *Nature Astronomy*, September 14, 2020.

[26] *Lee, Yeon Joo; et al. (26 August 2019). "Long-term Variations of Venus's 365 nm Albedo Observed by Venus Express, Akatsuki, MESSENGER, and the Hubble Space Telescope". The Astronomical Journal. **158**: 126-152. arXiv:1907.09683.*

Figure 4.14   The Venera 9 landed on a rocky slope. It returned the first images from the surface of another world. Note the cooling pipes on the right, used to circulate chilled liquid into the lander minutes before it made it to the ground. (painting by the author)

Mariner 2 spacecraft kicked off Venus exploration in the space age.[27] It was followed by a parade of Soviet and American explorers. European and Japanese probes added to the treasure trove. Our knowledge has built, incrementally.

The grand succession of spacecraft that entered the atmosphere, landed, orbited or flew by Venus all led up to the first views of the Venusian landscape. The Soviet Venera 9 (Figure 4.14) returned history's first images from the surface of another planet. Engineers had made some improvements in Venera 9 and its subsequent sisters. Flight controllers chose a pathway into the atmosphere that was shallower than before, reducing G loads[28] from ~500 to 150-180 G. The landers that preceded Venera 9 were ball-shaped, but the advanced design called for a spherical pressure vessel beneath a flat disk. The disk served as an aerobrake, in place of a parachute, which would be jettisoned

---

[27] The US followed it by another successful flyby mission, Mariner 5, in 1967. Mariner 10 flew by Venus in 1974, gaining a gravity assist for three encounters of Mercury in 1974-5.

[28] One G is one Earth gravity.

at higher altitude (earlier landers remained attached to their parachutes all the way down). Beneath the lander, a donut-shaped ring acted as shock absorber. Cameras mounted just under the aerobrake imaged a thin strip from one horizon down to the front of the lander and back up to the opposite horizon.

Venera 9 arrived on October 20, 1975. It survived the customary fiery entry. 50 kilometers up, Venera 9 cut it parachutes loose and coasted all the way to the ground. The dense atmosphere slowed it to 20 km/hour. Venera carried floodlights, because many scientists predicted a very dim vista beneath all those acidic clouds. But light levels turned out to be something akin to a heavily overcast afternoon on Earth. The craft came to rest on the side of a hill or mountain (perhaps a small volcano). The landscape was rugged, covered in sharp rocks. Their chemical composition matched that of basalt—they were volcanic. A full 180° panorama made it back to ground stations in the Soviet Union, but a second panorama in the opposite direction was foiled by a jammed lens cap. Still, it was a spectacular mission.

Venera 10 followed five days later. The weather? Cloudy skies. Barometric pressure: 92 atmospheres. Temperature: a balmy 465°C. Winds: slight. Venera 10 also suffered a stuck lens cap, but one of the two cameras worked beautifully, radioing back a haunting landscape of rolling, sandy plains broken by outcrops of eroded, flat rock. The craft leaned back after landing on a three-meter/yard slab of stone, which gave it a better view of the distant horizon.

Orbiters accompanied both landers, beaming back measurements of atmosphere, cloud particles, magnetic fields and global temperatures.

The Soviet Institute for Cosmic Research (IKI) dispatched Veneras 11 and 12 in 1978. The probes transmitted the first detailed data on Venus' triple cloud deck, but Venera 11 landed so hard that its soil sample device was damaged. Both carried lens caps with new seals. The seals worked too well: all camera covers jammed on, so no images were returned. The mother ships for this mission flew by without stopping.[29] The two landers also carried the first sound equipment on another world. They detected thunder and wind. The Russian scientists counted 1200 lightning strikes from Venera 12 alone. During descent, both probes encountered the 12.5 km anomaly, with their readings going off the charts accompanied by sudden electrical discharges. This was the first encounter of the famous Pioneer 12.5 km anomaly.

At the same Venus launch opportunity, the US mounted its highly successful Pioneer Venus missions, dispatching four atmospheric probes and an

---

[29] Some launch opportunities are better than others. In the case of Venera 11 and 12, it took more fuel to get to Venus than at other times, so the mother ships were unable to carry enough heavy fuel to settle into orbit.

orbiter that carried a radar altimeter. The orbiter's resolution was less than 75 kilometers, but good enough to hint at raised continents, plains, lowland basins, and even possible mountain chains. This orbiter detected the deuterium so critical to understanding the Venusian past. It also detected diminishing sulfur, offering the possibility that the craft had arrived just after a volcanic eruption, in time to watch its effects wane over the course of the mission.

In 1981, Venera 13 and 14 returned long-awaited color images of the surface. All camera systems operated well, and multiple panoramas were returned from each site using several filters, enabling investigators to reconstruct the images in approximate color. Venera 13 landed in a landscape of hilly, ancient rock predicted to be granite. Soil interspersed between platy rock. Venera's soil sampler determined the makeup of its surroundings to actually be basaltic with alkaline salts. This type of rock is found on Earth's ocean floors, and is thought to make up 2/3 of Venus.

Venera 14 set down on younger terrain. Its dramatic vista revealed a lava plain described by Moscow radio as "wrinkled brownish slate like sandstone," with less loose soil than at Venera 13's site. It, too, drilled the surface, also detecting elements consistent with basalt.

Venera 15 and 16 orbited Venus, mapping the northern hemisphere in detail from October 1983 to March of 1985. Their radar was more sophisticated than that of Pioneer Venus, and was able to discover large continent areas, mountains, craters and forms such as arachnoids and coronae.

The next Soviet Venus visitors were given new names and a new mission. Called VeGa, the name was a combination of *Venus* (Venera) and Comet *Halley* (Galli). The ambitious mission would carry two Venera-style landers, along with the first interplanetary balloons to study the middle layers of the atmosphere. The mother ships would carry out science during their flyby, and use the planet's gravity to assist them in encountering Comet Halley. The mission designers tried something else: inviting international participation at a level never seen in past Soviet missions. European, Japanese, Soviet and American scientists all worked together to make the complex mission a success.

French aerospace master Jacques Blamont originally designed the Venus balloons. When the balloon probes were scaled down, the jovial Vachislav Linkin of IKI took the reins for the Soviet side. Blamont conscripted 20 radio observatories across the world to chart the path of his VeGa balloons.

The balloons deployed from the VeGa landers during descent, at an altitude of 54 kilometers. Both landers deployed their balloons on the night side, but winds would soon carry them toward dawn. Both floating probes suffered from high winds. At one point, VeGa 2's aerostat dropped nearly 3 kilometers in half an hour. It burst after about 26 hours in flight. VeGa' 1's balloon fared a little better, traveling 9000 km in 46 hours. When it reached sunlight, its white balloon burst and contact was lost.[30] Ground observatories clocked the speeds of both balloons at 55 kilometers per hour, with winds reaching 70m/second. The balloons monitored temperature, pressure and light levels and the makeup of the air around them.

Soviet authorities were slow in releasing the first VeGa lander's results. During its descent, the craft took numerous readings, helping to paint a clearer portrait of the Venusian clouds. But for surface science, there was only official silence. A few years later, the truth came out. Something had triggered VeGa 1's landing sequence while it was still 18 km up. Its soil sample device diligently drilled the air and other experiments attempted surface investigations, but the only device that returned good data from the surface was a gamma ray spectrometer that tested the surface. The sequence was triggered at about the altitude of the Pioneer 12.5 km anomaly.

VeGa 2 was unaffected during its descent, and was able to directly sample the clouds. It landed on the night side. Its drill brought a sample into an internal laboratory, where it found that the basaltic rocks there had lower levels of iron than other sites, but were rich in aluminum and silicon, making them similar to the rocks brought back by the Apollo astronauts from the lunar highlands. They also bear a similarity to the Italian Apennines. VeGa 2 had, in fact, uncovered exciting evidence about the Venusian past. As Valeri Barsukov of the Vernadsky Institute put it, "It can be concluded that a more water-saturated atmosphere and even hydrosphere may have existed in Venus geologic past."

Many scientists and policy-makers realized that Venus was an important doorway through which we could understand some deep truths about our own planet. In many ways, Venus seemed to be the poster child of comparative planetology, the art of learning about the Earth by studying other planets. But Mars beckoned, and took the lion's share of planetary exploration funding for the US space program. It would be another decade before a NASA mission would fly to Earth's foggy sister. That mission would be an orbiter equipped with the best radar imaging system ever to visit the planet. It was called Magellan.

Magellan inserted itself into Venus orbit on August 10, 1990. Its circuit took it close to both poles so that the spacecraft could map nearly the entire

---

[30] Its energy packs were due to expire at about that time, so the mission may have ended with exhausted batteries.

globe. Its advanced Synthetic Aperture Radar system mapped the broiling world down to resolutions of better than 325 feet (100 m), covering 98% of the globe. In some areas, Magellan was able to chart the same features multiple times, searching for large-scale changes. While few were found, Magellan gave us a much clearer understanding of the world next door. Investigators were able to make gravity maps from variations in Magellan's orbit, gaining new insights into its interior.

Three spacecraft made flybys of Venus over the next decades, although none were specifically designed to study Venus. The Jupiter-bound Galileo spacecraft used its experiment suite to scrutinize the atmosphere as it flew within 16,000 kilometers of the planet. The probe imaged the mid-deck and lowest layer of clouds in the infrared (heat). Galileo encountered Venus twice, using the planet as a boost for its long voyage to Jupiter. (Figure 4.5)

The Saturn Cassini Orbiter/Huygens probe was next up, skimming past Venus twice, in April of 1998 and June of 1999. During its first flyby, the school-bus-sized craft coasted past the planet at a nerve-wracking 285 kilometers. Its second encounter took it to within 3600 km. Cassini trained its atmospheric instruments on the planet, monitoring clouds and charting radiation.

The MESSENGER spacecraft, a NASA/Johns Hopkins Applied Physics Laboratory project, used Venus as a springboard to its ultimate goal of achieving orbit around Mercury. MESSENGER carried out two flybys, but the first returned no science as Venus was on the side of the Sun opposite Earth. During the second flyby, MESSENGER scanned the globe with its full complement of science instruments, focusing on the upper atmosphere.

MESSENGER encountered Venus after the European Space Agency mounted its ambitious Venus expedition, a mission involving an orbiter called Venus Express. The spacecraft dropped into polar orbit in 2006, where it trained seven sophisticated instruments on the Venusian atmosphere. Among its many discoveries, Venus Express confirmed the presence of lightning, found more evidence for past oceans, and discovered a gigantic vortex in the clouds over the south pole.

The heroic mission of Japan's Akatsuki Venus orbiter seemed over before it started (Figure 4.15). Launched on May 20, 2010, the craft was due to make orbit around Venus the following December. But during a planned twelve-minute rocket burn, its main engine failed after three minutes. The dead engine stranded Akatsuki in a long, leisurely orbit around the Sun. The press called the mission a failure, but clever flight engineers at JAXA (Japan Aerospace Exploration Agency) did not give up. They put the spacecraft into hibernation to prolong the life of its systems. After five years of planning and waiting, Akatsuki returned to Venus, where controllers awakened the probe

Figure 4.15 A beautiful portrait of Venus in ultraviolet light, taken by JAXA's Akatsuki. (photo credit https://commons.wikimedia.org/wiki/File:Venus_-_October_24_2018.png)

and commanded it to fire its tiny attitude control rockets. The craft slowed by 540 mph/243 meters per second, just enough to enter into an extremely elliptical orbit ranging as far away as 100,000 kilometers, dipping down to a perigee (close approach) of a few thousand kilometers of the surface. Although two instruments failed (their design life was 4.5 years), three imaging systems continue to work as of this writing (2023). Akatsuki discovered a vast permanent wave in the clouds above the Aphrodite Terra continent, mapped wind patterns across the globe, searched for lightning, and enabled scientists to construct 3-D maps of the complex Venusian cloud decks.

The joint JAXA and ESA Mercury mission, BepiColumbo, rounds out our current state of Venus exploration. On its way to an orbital mission at the closest planet to the Sun, the spacecraft encountered Venus twice, in October 2020 and Aug 2021. Bringing to bear its full quota of science instruments, BepiColumbo took readings of the environment around Venus, passing within 8000 kilometers. Fortuitously, the joint ESA/NASA Solar Probe spacecraft had passed by the planet the day before. For the first time, scientists were able to observe the environment of a planet from multiple viewpoints.

For a preview of upcoming planetary missions, see Chapter 7.

Venus and Earth, a summary

Although the two planets differ in dramatic ways, Venus and Earth may be looked upon not as evil twins but as sisters. Venus is the closest planet to the Earth, not only in distance but also in size. It is slightly less massive, but some of its surface geology is hauntingly similar to Earth's, especially in terra firma's volcanic regions. Most of Earth's erupting mounts straggle along the edges of plates. Venus has no plate tectonics, probably due to its thicker crust and lack of lubricating water. Still, internal pressures there mold the surface, as they do here. Thanks to Soviet Venera landers, we know that most surface rocks on Venus are basaltic, with some variation from place to place. If a Venera lander settled upon the landscape of the Canary Islands or the Kamchatka peninsula, its sensors would certainly recognize the stone as reminiscent of Venus' plains and highlands.

Carbon dioxide dominates the air on both Venus and Mars, but makes up scarcely 2% of Earth's atmosphere, making our own world an outlier in that sense.[31] Still, we have seen that at the same time life was getting started on Earth, Venus had clear skies (or perhaps partly cloudy) and perhaps vast oceans, much like ours. And while our two worlds diverged along the way, we have much family history—and even a family resemblance—to share.

The young Earth of the Hadean eon began as a hellacious world of dense, toxic air—rich in carbon dioxide, like Venus—and high temperatures. But as we move forward in time and continue our exploration of other worlds, we find an Earth beginning to cool, its skies clearing and its seas appearing across the solidifying land. In short, the Earth continues its journey from Hell to Paradise, a journey interrupted on its sibling world. As we take a last glimpse over our shoulder, Venus challenges us. Humankind is changing the Earth's environment like never before. We are making inroads in the slowing of extinctions, and we are beginning to husband our resources using recycling of materials and fuels. But much human activity involves the generation of carbon dioxide, carbon monoxide, and methane. These are all greenhouse gases.

---

[31] In fact, the air on Venus and Mars is markedly similar. Both are primarily carbon dioxide, with tiny amounts of nitrogen, carbon monoxide, water vapor, and argon thrown in. Mars also has a bit of oxygen, and Venus rounds out its atmosphere with sulfur compounds. This makes the Earth's nitrogen/oxygen atmosphere quite remarkable.

It is as if we are trying, as a global community, to transform our planet into a Venusian world. Venus has provided us with a cautionary tale. Whether climate change is primarily a natural shift or human-caused, humans can take action to curb its effects. If we do not, we may well face the kind of future described in Chapter Nine. We leave Chapter 4 with the Earth as an early Venus, an alien Earth with dense atmosphere and overcast skies above a landscape less than a billion years old. Where will we go from here?

(For comparison, see a brief history of Earth at the close of Chapter 2, and a history of Mercury at the end of Chapter 3. A brief history of Titan is found at the end of Chapter 5, and a brief history of Mars at the end of Chapter 6.)

Looking Back: A Brief History of...Venus

As Mercury and Earth coalesce out of the solar nebula of dust and gases, so Venus grows, gradually, into a planet. The heat of that initial condensation is fierce; gravity, the energy of impacts, and heat from radioactive materials like uranium all contribute to an internal planetary furnace, transforming early Venus into a molten globe. In this cosmic crucible, heavy materials sink toward the core in the process of differentiation. Molten rock and lighter metals rise to form the Venusian crust. The planet ends up lighter than its sibling, the Earth.

Venus also ends up with a vastly different day from Earth and Mars. The Earth's Moon is born from an impact by a body roughly the size of Mars. The glancing blow peels off much of the outer crust material, which later coalesces into our Moon. The impact may also have constrained the length of day on Earth. It may be that Venus is struck in similar fashion as it forms, but in a direction against the spin, leaving it with the current day, 243 times as long as a day on Earth. It also leaves the planet spinning in a retrograde motion, opposite of most planets and moons in the solar system.

As the planet cools, carbon dioxide and water vapor pour from the interior into the atmosphere. The water vapor—as steam—makes up most of that early air. This period is the first of four Venusian epochs, the Oceanic. If the Venusian water totaled something like the Earth's inventory today, but remained as steam, the surface pressure may have soared higher than 300 times the pressure of Earth at sea level (300 "bars"). High above the cloud tops, sunlight breaks apart the water molecules into hydrogen and oxygen. Most of the hydrogen drifts off into space, dragging the heavier oxygen atoms[32] with it. But some heavy hydrogen (deuterium) is left behind for the planetary forensic teams to ponder.

(continued)

---

[32] Oxygen atoms are 16 times as heavy as hydrogen atoms.

Like other young stars, the young Sun goes through an energetic stage called the T-tauri phase. During this stage in a star's development, its outflowing energy clears the inner solar system of much dust and gas (see Chapter 1). The Sun eventually settles down, but even today its solar wind erodes the atmospheres of the planets.

After a billion years of battering by the solar wind, and through the loss of molecules to space, Venus' water supply runs dry. During this period of cooling, oceans may form on the surface. But soon, the ocean beds—if there are any—wither away into desiccated lowlands. Those former seas fill with lava from thousands of active volcanoes. Eruptions vomit out gas and water vapor at rates far greater than the Earth's volcanic eruptions today, perhaps 100 times the Earth's modern rate. The interior of Venus begins to cool as the planet ages. Volcanism calms down, and whatever water is left bakes out of the crust due to the high temperatures at the surface. The volcanoes continue to pump carbon into the air, now at about the same rate as the Earth's today.[33] The atmosphere fills with stable gases like nitrogen and argon, and temperatures on the ground settle to a level similar to the Earth's tropics. On Earth, where pressures may have been 60 atmospheres during the Hadean, its majority carbon dioxide was converted to carbonates, left as chalk and limestone. But at the same moment in history, Venus' water is gone, so the planet cannot cloister its carbon dioxide within its rocks. Its atmosphere remains at the 90 bars we find today.

700 million years in the past, something dramatic happens to our sister world. Volcanism tails off abruptly after a mysterious resurfacing epoch, leaving behind the exotic hothouse world we see today.

---

[33] Volcanoes and other natural sources vent about 300 million tons of carbon dioxide into the Earth's air each year.

5

Earth=Titan: cradle of life?

<u>Living planet</u>

The universe seems to be engineered to generate the building blocks of life, at least the kind that we find on Earth (and that's the only kind we've found so far). The six main chemical elements of living systems on Earth are carbon, hydrogen, nitrogen, oxygen, phosphorus and sulfur (collectively referred to as CHNOPS). Of these, the four primary elements are carbon, hydrogen, oxygen and nitrogen, referred to as CHON. As stars are born in the hearts of nebulae, drift through space, and eventually die, they leave a trail of debris from their death-explosions (Chapter 1). Different stars leave behind different leftovers, depending on their makeup. The CHNOPS elements come from many different stars, so they vary in relative proportions. But they are common throughout the cosmos, life's puzzle-pieces waiting to be fit together.

Earth's life is fundamentally rooted in carbon. On Earth, it's a universal given. And in the primordial mix of the planets, certain other compounds were abundant. Compounds like methane and carbon dioxide may be critical, for when they are irradiated, they combine chemically to generate organic compounds called tholins.[1] Tholins and other hydrocarbons are the raw material of amino acids that power biology, and they pervade the outer solar system (Figure 5.2). They leave their telltale orange tint on the surfaces of ice

---

[1] Tholins are like over-heated spaghetti sauce, splattering everywhere with red stains that won't go away.

© The Author(s), under exclusive license to Springer Nature Switzerland AG 2023
M. Carroll, *Planet Earth, Past and Present*, Springer Praxis Books,
https://doi.org/10.1007/978-3-031-41360-5_5

Figure 5.1   Titan's exotic labyrinth terrain stands as some of the most rugged territory on Saturn's giant moon. Pictured here is the Sikun Labyrinthus overlooking the depressed region surrounding Veliko Lacuna. The integration of valleys suggests an expanding lake with underground conduits draining the fluid (in this case, methane) away through the plateau. (painting by the author)

Figure 5.2   The interaction of methane and sunlight (photochemistry) works to create organics, as evidenced by dark brownish stains (tholins) in the outer solar system. Left: orange tinted onion-layers in the atmosphere of Titan betray the presence of organic material suspended in the air; center: Pluto's moon Charon may be borrowing methane and nitrogen from Pluto to hood its polar regions in the brown sludge of hydrocarbons; right: an exceptionally smoggy evening in Panama City, Florida, has a lot in common with the chemistry of organics on other worlds. (Left: NASA/JPL/SSI; center: NASA/JHUAPL; right: courtesy Kara Szathmary)

moons, in the high-altitude hazes of the gas and ice giants, and in the atmospheres of places like Pluto, Triton and Titan. For a glimpse of related hydrocarbons, take a look at the brown haze over any large city during a smoggy afternoon.

Is life universal and ubiquitous throughout the universe? Is it a freak exception on Earth, a rare and unlikely congruence of many variables? Most importantly, how did living organisms—those capable of reproducing, flexing under changing conditions, even achieving consciousness—issue from simple organic materials? These questions have weighty implications for our place in the cosmos. Theologians and philosophers certainly have answers, as do physicists, organic chemists, and biologists. Their answers come from different directions using different assumptions and tools, reflecting only parts of the whole, and we are just beginning to understand the nature of life itself.

One thing is certain: the Earth hosts a habitat for robust and diverse living things across a variety of environments, from frigid ices to the hearts of volcanoes, from lofty mountain peaks to the great abysses at the bottom of the deepest oceans. We have seen that the Earth is not as it always has been. When life arose here, chemistry and environment were very different—in fact, quite alien—from the temperate paradise we experience today. Very little oxygen drifted through the air, which was probably a good thing. The kind of simple organic materials that life uses tend to be torn apart by oxygen. As we saw in Chapter 1, Earth's early atmosphere consisted mostly of gases belched from volcanoes: hydrogen, water, carbon dioxide, nitrogen and methane. Sunlight and cosmic radiation reacted with methane, and organic material formed (see Miller-Urey and Other Experiments, below). If only we could find a place that simulated some of those primordial conditions, we might gain insight into the rise of life on Earth!

As it turns out, we do have such a planet-sized simulator, and it orbits the planet Saturn (Figure 5.1). Titan, second largest moon in the solar system, gives us our structure for exploring the mysteries of life's genesis here on Earth.[2]

Titan's frosty nitrogen/methane atmosphere interacts with sunlight, converting methane into the building blocks of carbon-based life. This photochemistry is triggered by radiation from the sun, but similar chemical magic takes place in the presence of other types of radiation or from the energy in lightning.

Aside from temperature, some biologists believe that conditions on Titan today mirror prebiotic conditions on our own world some 4 billion years ago. Hydrocarbon solvents provide an environment that encourages the synthesis

---

[2] The known moons of the major planets, including those at Pluto, total 210 (this number does not count asteroid moons). Titan is second only, in size, to Jupiter's great ice moon Ganymede, which holds the record for largest moon in the solar family.

of organic chains like amino acids. And like the early Earth, before the cyano-bacteria oxygen revolution, Titan is devoid of gaseous oxygen. This lack of oxygen and free water actually preserves such reactions.

Intro to a planet-moon

If we ripped Titan from its orbit around Saturn and placed it onto its own circuit around the Sun, Titan would be a respectable planet in its own right. Larger than the planet Mercury and with a solid surface like the terrestrial planets, its atmosphere is second only to Venus in extent. Titanian air is denser than the Earth's, and Titan is the only world in our solar system besides Earth that holds nitrogen as the primary portion of the atmosphere. It is also the only other world that has liquid lakes and rivers on its surface, with active rainfall charging those "waterways." But it is cryogenically, bru-tally cold on Titan; its rain is not water, but rather liquid methane. Sunlight reacts with high-altitude methane to create complex organic chains that rain down through Titan's dense, orange fog. Titan's misty sky obscures the view of nearby Saturn, and even cloaks the Sun. Noon on Titan is a melan-choly affair.

Unlike the rocky terrestrials, this behemoth moon is a massive ball of ice with a rocky core. Frozen water—not rock—makes up Titan's landscapes. Its mountain summits are made of ice. Erosion has engraved deep canyons into the icy plains. Sandy beaches consist of ground up ice. All the familiar ele-ments of terrestrial vistas, boulders, rocks and gravel, are variations of Titan's frozen-water landscape. The giant moon's bone-chilling temperatures hover at an average of -297°F (-183°C). In these wintry conditions, water freezes to the consistency of granite (hence the mountains and boulders and canyons). Like the other planets and moons, Titan does have rock and metal in its core, but these materials—so familiar on our world—reside hundreds of kilometers beneath its frozen water-ice shell.

The features of this bizarre world bear some striking similarities to terres-trial planets. Its dense atmosphere resembles a subdued version of Venus, with superrotating "waves" of atmosphere moving across its face.[3] Its mountains, sinuous canyons, and volcanoes find analogs on Earth and Mars as well. Titan even has vast seas of sand dunes to rival Iran or the Sahara, and they are not inert, frozen piles of sand. Although movement of the dunes would probably

---

[3] Titan's equatorial winds are estimated to be just below 12m/sec (25 mph) ground speed. Its day lasts 16 Earth days.

have been too subtle to be picked up by the Cassini orbiter's radar, there is evidence that the dunes are modern. As Brigham Young University planetary scientist Jani Radebaugh puts it, "The dunes of Titan do not appear to be heavily eroded by all the methane rainfall we believe is happening, so in that respect they are still considered young and possibly active. " As if to underline the moon's alien nature, the sand in its dunes is likely made of hydrocarbon soot snowing from its exotic nitrogen-methane air. Future tourists need to take care: a spoonful of Titan sand, in the presence of oxygen, would be flammable! It is within the environs of this exotic moon that we find parallels with not only the terrestrial planets in general, but specifically with the prebiotic Earth.

Back then: life's expansion across the terrestrial globe

In the last chapter, we left our Earth with a thinning atmosphere and clearing skies. Little survives of that Archean world today. Life's fossil record has left us with a story that has gaps and tangents, puzzles and surprises. The farther back in the record that paleontologists go, the more incomplete the picture. There are clues, like the banded iron deposits. The Banded Iron Formations formed when layers of sea ice prevented the exchange of oxygen between the atmosphere and the ocean, allowing iron from underwater volcanic eruptions to build up in the seawater. But at the time life arose on Earth, little of life's chronicle remains. Over three and a half billion years ago, living things had already populated the seas of the young Earth. In the Canadian wilderness, in rocky outcrops of the Australian desert, and in the stony records of Greenland, we find the oldest surviving rocks on our planet. Within them hide the chemical and—perhaps—fossil traces of microbes. It seems biogenesis, the beginning of life, visited our world as far back as there have been rocks to tell about it. We have not been able to find rocks that clearly predate life, and it may be that none have survived our planet's erosion-mangled history to tell the tale.

Chapter 2 provided an overview of the four great eons of Earth history: the Hadean, Archean,

Proterozoic, and Phanerozoic. The earliest evidence of life appears in the Archean layers, but its planet-wide spread became most robust in the Phanerozoic. Leading up to this time period, life made its way into the shallows and depths of the oceans as the first continents provided shelves on which to grow in sunlight. The journey wasn't easy. Mass extinction-class impacts befell the planet from the beginning, and undoubtedly threatened life's tenuous hold early in Earth's history. In the Phanerozoic, the planet suffered a series of ice ages. Some of these coolings were so drastic that geologists have coined the phrase "Snowball Earth" to describe their extent (see Chapter 6 for more details). We will visit the Phanerozoic in detail in a moment, but to draw meaningful comparisons between Titan and the prebiotic Earth, we need run the tape much farther into the past, to do a quick survey of the advent of life on our own world.

How soon life?

One of the more baffling and wondrous things about life on Earth is how early it got started. 4.6 billion years ago, the Earth was a congealing sphere of slag. In stone formed less than 700 million years later, we find evidence of life's presence (see Appendix G: Early Life Timeline).

The evidence comes not from fossils (we'll get to them), but rather from analysis of the carbon isotopes within the rock. The isotope carbon-12 is more common that carbon-13. When organisms take in carbon, they can metabolize the carbon-12 version more easily than the heavier carbon-13. This means that living tissue contains a higher amount of carbon-12 in comparison to carbon-13, than the background isotopes of the rocks. Some of the oldest rocks on Earth, found in Greenland, contain this life-indicating carbon isotope ratio. The rocks are older than 3.85 billion years, because they are infused with igneous rock that dates to that age, so the rock must have been there even earlier. Similar signs of life have been found at other sites dating back to 3.8 billion years ago. Other isotopes back up the early date from elements including nitrogen, iron and sulfur.

If life was present in the Greenland rocks over 3.8 billion years ago, it is reasonable to assume that it was present across much of the Earth. But impact studies (Chapter 2) show that the Earth suffered major impacts after that time, rendering the surface molten once again. With this in mind, life may well have arisen within a few hundred million years of when the Earth's surface stabilized and cooled. That is a remarkably short time period. Life may have arisen rapidly here. Did it elsewhere?

We also find evidence of ancient microbial life in the fossil record. Discoveries of microscopic fossils are usually met with controversy. Many non-biological processes lead to cell-like spheres or ovals. Additionally, the most ancient rocks have been altered by geological forces, compressing and warping fine structures within. But in Warrawoona, in western Australia, paleontologists have discovered complex forms within 3.465 billion-year-old chert (fine-grained sedimentary rock). Electron micrograph images reveal linear, segmented forms, some with external sheaths. Further testing shows that the rock formed near a deep-sea hydrothermal vent. While controversy still rages as to whether the structures are organic (chemistry seems to argue so) or non-biological, the forms offer a reasonable case for biological origin a very long time ago.

One example of ancient fossil life is not debated: stromatolites (Figure 2.22). Stromatolites look like petrified lasagna, with striking layered structures. Today, we can see living stromatolites in places like Shark Bay, Australia. They form when microbes assemble into algal mats. The uppermost layer of microbes is photosynthetic, converting sunlight into energy. Beneath this layer, other microbes use the organic waste from the top tier as food. Sediment intermixes with the microbial blankets. Fossilized stromatolites are twins to the modern, living ones. But their sediments date back to about 3.5 billion years. Assuming that they worked the same in primordial times, photosynthesis must have been taking place within their upper layers. This advanced metabolic operation must have taken time to arise, so the stromatolites give us only a minimum age for life's origin. Organic life must have been advancing for some time before the rise of the stromatolites.

Biogenesis on Earth

Ben Clark was searching for life on other worlds when most astrobiologists were preparing to graduate from Kindergarten. With degrees in Physics and nuclear engineering, the biophysicist was on the science team for the landmark Viking Mars missions, where his invention first analyzed Martian soil. In addition to Titan biology studies, Clark has turned his attention to the organic chemistry in comets and asteroids. Clark sums up the prerequisites for life this way: "You need metabolism (the processes that captures energy chemically), and you need the RNA to serve as repository of the genetic information. And then you need a membrane to keep them semi-isolated. They can't be totally isolated because they have to be able to take in nutrients and export waste products. The membrane is considered a quite important step."

The first requirement for life—on Earth or anywhere else—is the right construction material. But where might it come from? As we saw above, the short answer is: everywhere! But while the universe is filled with biologically relevant components, we can be a little more specific.

For some time, researchers knew of three nucleobases—compounds linked to RNA and DNA—that resided within meteorites. Five critical nucleobases combine with sugars, phosphates, and other ingredients to construct genetic material in all known terrestrial life. But what about those missing two bases? Remember that meteorite that struck the driveway in Winchcombe? Like other space rocks, it contained amino acids, foundational to biological operations. But it and other meteorites analyzed in 2022 also contained cytosine and thymine, the missing two compounds critical to Earth life. Many astrobiologists feel these five bases, now all documented in material from the solar nebula's leftover asteroids, are important precursors to life. Most recently, uracil was detected in samples returned from asteroid Ryugu by JAXA's Hayabusa 2 spacecraft. Meteorites found on Earth have contained biogenic chemicals like uracil, but terrestrial contamination was often cited as a possibility. But Ryugu's samples, direct from space, are pristine and not suspect. The discovery team, from Hokkaido University in Sapporo, Japan, looks forward to carrying out similar tests on samples returned by NASA's OSIRIS-REx, which retrieved material from asteroid Bennu in 2020. As we are about to see, there may be other sources of prebiotic compounds closer to home.

All of this chemistry leads to an obvious question: how does a pile of chemicals become a living thing? With all of our discoveries in physics, cosmology, meteorology and geology, this, to many, is the greatest mystery of the ages. In the 1920s, two biochemists independently proposed a concept now named after them: the Oparin-Haldane[4] hypothesis. Both Jack Haldane and Alexandr Oparin conjectured that life arose in gradual steps from inorganic molecules like amino acids, combining over time into more and more complex systems. Neither offered an explanation of how this took place; they were merely attempting to describe initial processes that eventually led to living systems. Both scientists reasoned that Earth's biogenesis might have gotten going by a primordial mix of methane and solar radiation or charges from lightning. The early Earth, they reasoned, had these materials in abundance. So, to some extent, does Titan.

---

[4] Aleksandr Oparin was a Russian biochemist, and J.B.S. Haldane was an Indian/British geneticist and biologist.

Experimental confirmation of complex organics issuing from simple chemicals first arrived in the 1950s in the laboratory of Harold Urey and Stanley Miller. Other experiments soon followed, and many have been carried out since, using various initial conditions.

Biochemists have peered more deeply into life's mysteries, and in their late-night sessions in the laboratory or at the local pub, they've come up with clever ways of reconstructing life's beginnings. The first approach is to look at life from the bottom up. Biochemists examine the makeup of primordial chemistry and see how it would lead to more complex forms. This tactic relies heavily on laboratory simulations of prebiotic processes. Coming from the opposite end, the top to bottom method studies less and less complex organisms, ultimately looking for the most primitive and simple forms that life takes. Whether the scientist is a bottom-to-top, simple-to-complex kind of person, or a top-to-bottom take-the-high-road investigator, all operate within one of four different paradigms:

1. Metabolism first: this perspective assumes that metabolism was the first step toward the beginning of life. Metabolic operations are the chemical processes that organisms use to maintain life.
2. First things first: membranes! This perspective sees the most important first step as the formation of a membrane that can gather together needed organic material for living systems. It is widely thought that a boundary layer, separating the environment from delicate biological systems, is a prerequisite for life. Membranes can concentrate key compounds, speed up chemical reactions, and shelter the whole batch of life-related materials from the outside environment, which can be unfriendly to biology's operations. A family of molecules called amphiphiles form membranes, and amphiphiles have been found within meteorites. On Earth, membranes are usually built of lipids, which can create a boundary between interior organic material and exterior water. Some astrobiologists have suggested ways that such membranes, made of other materials like acrylonitiriles, might form in Titan's cryogenic methane environment.
3. Drips first. Coacervates are minerals that form spheres. These non-living forms tend to split when they become large enough to be unstable, just as a too-large soap bubble bursts. Coacervates are thought to have been precursors to cell-like structures.

4. Put on your genes first. This perspective sees self-replicating molecules (like RNA) as the starting gun of life's marathon. The scenario is bolstered by the fact that all life on Earth has a common genetic foundation. RNA is the instruction guide for assembling organic compounds into something that can pass on information and reproduce.

Biologists tend to gather into two cohorts when it comes to pinning odds on the origin of life. One group says that life will emerge at just about any site with Earth-like conditions. With a recipe of the right amounts of solar energy, radiation, organic soup and a few other minerals, voila! You've cooked up life. The debate is more complicated, of course. Did life get going in a comfortably warm primordial stew ponded on the surface of a young world? Or were life's chemistries fashioned in frigid clouds of interstellar space, bathed in the radiation of nearby stars? Perhaps life began at the sites of undersea volcanoes on the ocean floor.

The second faction of biologists suggests that life's genesis is incredibly rare, but once it gets going, it disperses throughout the universe, whether in spaceships or aboard meteorites. The implanting of life on a planet from an outside source is known as Panspermia, a concept first offered by British astronomer Fred Hoyle. It may be that the instigation of life can only occur within a narrow set of circumstances constrained by chemistry in the environment, star age, gravity, a planetary magnetosphere, the right geological processes (like volcanism, plate tectonics and erosion), along with just the right mix of minerals and atmosphere. But once life has a footing, it can serve as the seeds of life throughout its stellar neighborhood. Hoyle had good reason to suppose that organisms or their precursors could be transported through the void of interplanetary or interstellar space. Biologists have recorded cases where bacteria in permafrost, thousands of years old, were revived when warmed. Other bacterial spores have been resurrected after lying dormant for long periods of desiccation in dry environments. The bacterium Streptococcus mitis may have even survived for three years in the vacuum on the lunar surface aboard a Surveyor lander. Isn't it possible, then, that viable organic material could make the journey through interstellar space intact? After all, meteorites from Mars appear to house biologically related chemistries, and perhaps even fossils, that have made the interplanetary voyage from Mars to Earth.

Both of these biologist camps rely on a planetary surface as the progenitor source of biological material. Astrobiologist Caleb Scharf offers another possibility. Perhaps, Scharf and others propose, life might arise where the environment around a forming star system leads to a wide variety of chemistry and physical conditions. Prebiotic conditions may be a natural part of the cycling of materials from nebulae to star formation to planetary birth. Scharf asserts that, "…the conditions in a proto-planetary system are such that extensive organic chemistry must take place."[5] In other words, infant solar systems become chemical factories, pumping out biological material. In early solar system formation, billions of planetesimals drift through the cloud of dust and debris surrounding the infant star. Radioactive material is abundant, especially in the larger proto-planets. But the process of differentiation among the forming planets is often interrupted by the impact of planetesimals, starting the mineral mixing all over again. This processing transpires again and again, morphing into the same kind of organic material that today exists in many comets and asteroids. In this way, the space surrounding a new star—far from being a sterile void—may be an organic chemical paradise ready to blossom. But to explore life on planetary surfaces like the early Earth and Titan, we first turn to the one place where we know it actually happened.

Early ideas of life's beginnings

Put a slab of meat out on your back patio, and within a few days, you'll have life—little insects crawling out everywhere. At least, that's what appears to happen. The concept is called spontaneous generation, and it's a very old idea. Aristotle (384-322 BC) spoke of it, asserting that life arose from non-living matter as long as that matter contained what he called "pneuma," or "vital heat". The idea was the leading theory for the origin of life until the 19th century. Spontaneous generation proposes that the simplest forms of life are spontaneously generated by decaying matter (like rotten meat or fruit). Belgian physician Jan Baptist van Helmont even came up with recipes for living things. For example, if you want to make mice, dust a piece of cloth with wheat flour, wait for 21 days, and your project will yield mice!

Fast forward to the 17th century, when polymath Sir Thomas Browne and scientist/philosopher Robert Hooke questioned the concept. Hooke crafted the first drawings of microbes that he observed in a microscope of his own making. His 1665 revelations were followed by Antonie van Leeuwenhoek, whose sketches portrayed what were probably protozoans and some forms of bacteria. Van Leeuwenhoek experimented with sealed and exposed meats, concluding that spontaneous generation was not the answer to the Earth's biogenesis.

Spontaneous generation was not the only ancient guess at life's beginnings. In the fifth century BC, Greek philosopher Anaxagoras put forward the idea of panspermia. The philosopher alleged that the universe is filled with life, and that life came to Earth rather than originating here. Panspermia has evolved

(continued)

---

[5] *Extrasolar Planets and Astrobiology* by Caleb Scharf (University Science Books, 2008)

over the centuries, and presents itself in modern times in similar fashion: that life began somewhere else in the cosmos and was transported here on cosmic dust, meteors or comets. The idea gained traction with the discovery of possible chemical and fossil evidence of life in a Martian meteorite transported to Antarctica in an impact on the red planet. Though the Allen Hills meteorite/biology connection is still controversial, the possibility of interplanetary transport is an acknowledged one. Panspermia does not explain the genesis of biology, but rather puts it off of Earth and on to other unidentified sources in the cosmos. Critics argue that the concept is not, technically, a scientific one since it cannot be tested experimentally.

By the 1800s, Herbert Spencer and William Turner Thiselton-Dyer were writing about life arising from a chemical bath in a pond or lake. Charles Darwin took up the message, writing of a "warm little pond" in which a primordial soup of organics led to life. J.B.S. Haldane and Alexander Oparin both proposed that the first cells came from molecules that self-organized in a primordial stew of prebiotic organics. Variations on this theme continue today (see the box "Miller-Urey and other experiments" below).

Earth's biology "explosion"

Whether terrestrial biology came from above, issued from volcanic vents, or had its beginnings in primordial ponds is unknown, but once life did take hold on Earth, there was no stopping it. Living things emerged in the Archean and expanded through the Proterozoic, bringing us to the current Phanerozoic eon. The Phanerozoic eon comes in three acts: the Paleozoic, Mesozoic, and Cenozoic eras, the last of which we are performing in now. The Paleozoic—Act I—opened with the Cambrian explosion, which arose amidst escalating levels of oxygen in the atmosphere.

For nearly the first half of our planet's history—for more than two billion years—the water and air of Earth were devoid of free oxygen. No samples of atmosphere from that time have survived, so how do we know it lacked oxygen? The remarkable change in the atmosphere, wrought by trillions of aquatic microbes, left its mark in iron-rich rock formations. The rocks in question, referred to as the banded iron formation, are 2.5 billion years old. Stripes laminate the remarkable stone, layers of alternating chert[6] and iron, creating a candy-stripe pattern. The rocks originally made up the ancient sea floor, and this tells us something significant. The rocks on the sea floor today do not contain iron. To create the kind of deposits that became the layered rock from

---

[6] Chert is a gray-green to brown brittle sedimentary rock, consisting generally of quartz crystals.

ancient times, iron must have been drifting through ocean waters freely. This can only happen in an oxygen-free environment, because iron quickly reacts to oxygen, generating iron oxides. The Earth's modern seas have very low iron levels. Those orange candy stripes of rusty iron tell us that the metal settled through waters free of oxygen. Gases constantly migrate back and forth between seawater and atmosphere, so this, in turn, tells us that the air above was also essentially oxygen-poor.

Life thrived in oxygen-starved biomes, and it still exists in them today. Where most terrestrial and aquatic plants use sunlight to create food from carbon dioxide and water, some microbes instead use hydrogen sulfide or dissolved iron. These anaerobic ("oxygen-free") life forms provide us with at least a fuzzy portrait of the earliest life on Earth.

Along the margins of the oldest super-continent relics, red sandstone marks the regions of continental shelves, shallow waters where photosynthesis first took place. The sandstone formations match the geology we would expect with an abrupt rise in oxygen levels. Geologists estimate that this drastic increase in oxygen took place between 2.2 billion and 2.0 billion years ago.

Another past-oxygen gauge comes in the form of "fool's gold", iron pyrite. We saw how iron pyrite may condense on the mountain peaks of Venus. It does so because of temperature and pressure, but it remains there because of a lack of oxygen. Pyrite has a propensity to combine with oxygen, oxidizing into sulfates. It carries out this operation quickly, on a scale of decades or even years. On Earth, we don't see much fool's gold in the sands washed down from ancient continental rock that makes up our mountains. The reason is that the pyrite eroding from the high country combines with oxygen and disappears. This process gives us another important clue to Earth's past air: pyrite encased within 2.4 billion year old sandstone (the coastal rock of ancient seashores) washed down from high ground, down mountainside, along rivers, all the way to shorelines of the primordial continents, and in its extensive travels, it never had the chance to combine with free oxygen. It could only have been preserved in an atmosphere devoid of the gas we so greedily breathe. In rocks younger than 2.4 billion years, few such grains of fool's gold can be found.

These and other evidences confirm the stunning transformation that life brought to our atmosphere. It is called the Great Oxygenation Event (the GOE), and it was a game-changer. The GOE played out as an eons-long dance between geology and biology. As continents rose from the global seas (Figure 5.3), minerals like phosphorus eroded into the waters, changing the ocean chemistry in favor of life that used oxygen rather than other energy sources. Those same sediments buried the organic matter left behind by dying microbes, food that might have been used by non-oxygen-breathing microbes,

Figure 5.3 A billion years before the Great Oxygenation Event, most land masses consisted of volcanic islands or raised impact crater rims. Ocean water would have been a remarkable green from suspended iron, and photochemicals would have tinted the sky toward an orange hue. The continents began to rise from the waters just over 3 billion years ago. (from a painting by the author)

and the dance was complete. The tango between Earth's rocky continents and the life that brought oxygen into the air transformed the planet. It didn't happen all at once. Even before the GOE got going, before 2.4 billion years ago, chemicals trapped in primeval sedimentary rocks point to "puffs" of oxygen here and there. These transient oxygenation events must have been localized and fleeting. But eventually, the atmospheric tide turned, and the Earth's primordial, oxygen-poor atmosphere collapsed, replaced by the oxygen flowing from the tiniest of life forms.

All of this moving and shaking in the chemical realm had consequences. It led to a biological upheaval, an outburst of life forms more diverse than we can know (although each year brings more discoveries). The fire behind that biological explosion was cyanobacteria, sometimes called "blue-green algae". These unassuming bacteria are usually one-celled microorganisms that get their energy from photosynthesis. Cyanobacteria number among the oldest confirmed fossils, and they live on today as an important part of Earth's ecosystem. In fact, the chloroplast portions of plants, the part that carries out photosynthesis, is essentially a cyanobacteria living within the plant tissue.

The Great Oxygenation Event had humble beginnings. The landscape at the start of the Phanerozoic saw no galloping animals or banks of forest. The landscape was barren, save for ponds and lakes collared by rainbows of color. The sludge at water's edge glistened spectrally in the sunlight, colonies of microbes drifting in the water and sticking to the shoreline. The green, brown and yellow scum, some of it invading the land here and there, did its work slowly, relentlessly. Soon, it would make way for a new and different world.

The dark side of the GOE: mass extinctions

The GOE began when the Earth's first major glaciation, or "snowball event", covered the planet in ice. As that ice retreated some 2.4 billion years past, the iron-banded formations overlaid themselves above earlier sediments. Their mere presence is the signature—frozen in stone—of the blossoming of oxygen-producing organisms. With the gale of oxygen flooding the atmosphere at the hands of the GOE, the iron and manganese in the oceans precipitated out of the water, becoming manganese oxide and iron oxide. Microbes found it harder to access iron in the seawater around them, and they found themselves immersed in an environment filled with that nasty, toxic gas oxygen. The world was changing, and any living thing needed to change with it.

While there were winners in the aftermath of the world's freshened air, there were also losers. Entire groups of organisms simply could not survive the global changes in the atmosphere. The GOE paved the way for new life, but it also constituted one of the greatest mass extinctions in the Earth's history. There were others, some of which came before the planet's rocks were able to record such events. The earliest life forms had soft bodies and little structure, so the fossils they left behind are rare or non-existent. From planetary birth 4.6 billion years ago to 3.8 billion years ago, the heavy bombardment is thought to have periodically autoclaved the surface of the Earth from most life, turning some or most of the surface into molten slag. From 2.5 to 2.2 billion years ago, other kinds of extinctions befell our home world. We have seen the consequences of increased oxygen levels, which undoubtedly brought to an end many anaerobic microbe species. The GOE may have coincided with the first of a series of "snowball Earth" periods. From 750 to 600 million years ago, several (perhaps as many as four) of these events brought glaciers and extensive sea ice to the entire globe. Ice encased continents as glaciers flowed across wide swaths of new land. Widespread extinctions appear to have wiped out many microbial stromatolite communities, and many of the common microorganisms called acritarchs[7] came to an end. (For more on the snowball Earth, see Chapter 6). The paucity of fossil remains prevents us from knowing the details. But we can track mass extinctions in far more detail within the past 500 million years or so, since the Cambrian explosion (see Figure 5.4). Before the great flowering of life in the Cambrian, another wave of extinctions wiped out the first reef-building organisms, the highly successful archeocyathids. Concurrent with their disappearance, many groups and families of trilobites and primordial mollusks came to an end. Then, some 445 million years ago, the planet was dealt a one-two punch. Life was still relegated to the oceans, and an abrupt chill rerouted the oceans' global current patterns. Shifting continents played a starring role. The new undersea flows blocked the supply of warm-water foods like algae, causing mass starvation. Added to this, an intense period of glaciation froze the fresh water supply from the world's oceans, lowering sea levels radically. This great dying is called the Ordovician-Devonian Extinction, but it was only the preamble to the greatest catastrophe to visit the long march of terrestrial biology: the Great Dying, or the Permian Extinction.

It all happened 252 million years ago. In the span of less than a million years, our planet's atmosphere and waters changed. The radical changes may have

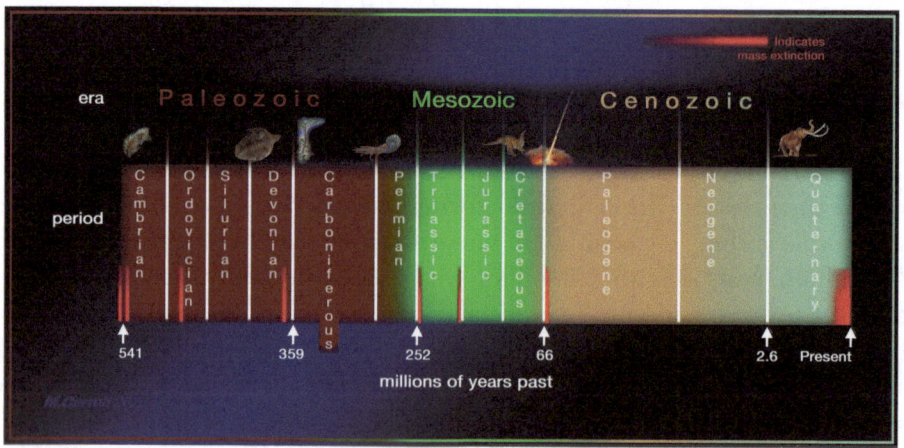

Figure 5.4 Ages of the Phanerozoic. The Phanerozoic Eon spans from 541 million years ago to the present, and is broken into three eras, the Paleozoic, Mesozoic and Cenozoic (the most recent). These, in turn, are divided into twelve periods. Several mass extinctions are marked by red lines. (diagram by the author)

---

[7] Many fossils of ancient single-celled microbes are simple in structure, but acritarchs tend to be larger, with complex spines and outer skins. The term is a generalization, and covers a host of poorly understood forms.

been caused by an asteroid impact, although there is, as yet, no evidence of such an event. At the time, increased volcanic eruptions tore across the landscape and sea floors, dumping greenhouse gases like carbon dioxide and water vapor into the air. Recent work[8] at Australia's University of New England suggests that massive volcanic activity in eastern Australia—in the New South Wales region— peaked some 253 million years ago, raising carbon dioxide levels and weakening the world's entire ecosystem. The continents continued to shift, too, contributing yet another variable to the mysterious dying. Less than a third of land species (plant and animal) survived, and 95% of all sea life perished. At the time, the synapsids—a cross between a wolf and an iguana—ruled the land as the dominant species of the world. None of them made it through.

Conventional paleontology wisdom had it that the world's oceans were nearly dead wastelands for millions of years after the Permian extinction. But in 2010, scientists examining a group of fossils in China called the Luoping biota concluded that the ocean's ecosystems bounced back within 10 million years. While this was remarkable, other teams digging through strata from such widely separated beds as sites in China and the western U.S. found that marine biology rebounded within just 3 million years of the mysterious Great Dying. Most recently, a team from the China University of Geosciences unearthed Triassic lobsters, predatory coelacanth fish, sponges and spiraled ammonites, all interspersed with volcanic ash that could be radiometrically dated to 250.8 to 250.7 million years ago. Their conclusion: life came back within a mere million years after the great Permian extinction. Life on planet Earth is tenacious.

Permian ambassadors

Perhaps the most iconic of life forms that died out toward the end of the Permian were the Pelycosauria (more correctly, primitive synapsids), a family of creatures that included the famous sail-backed Dimetrodon. Often mistaken for dinosaurs in the popular press, the carnivorous Dimetrodon was an apex predator, growing to a length of 3.5 meters. While the pelycosaurs became extinct, they seem to have been related to the therapsids, a group that superseded the pelycosaurs and eventually included mammals.

The ecosystem took another hit at the end of the Triassic Period. This mysterious cataclysm reduced biological diversity on the planet by half. An impact similar to the K-Pg dinosaur-killing impact (Chapter 2) may have been

---

[8] "Pulses in silicic arc magmatism initiate end-Permian climate instability and extinction" by Chapman, et al, *Nature Geoscience*, May 9, 2022

involved, and a suspicious crater in Quebec may be the resultant scar. The Manicouagan Crater is about half the size of the Chicxulub (K-Pg) crater in the Yucatan. It is dated to about 214 million years, just slightly older than the border between the Triassic and the Jurassic. But other environmental changes could have either contributed or been the primary culprit. Some paleontologists suggest that another change in the ocean currents or chemistry could have depleted oxygen levels in the shallows, cutting off the food chain at its knees. But these transformations should not have affected life on the land, which also took a major hit. In each of these mass extinctions, the road was paved for new kinds of life, and none was more important in this way than the Great Oxygenation Event.

Oxygen's new Earth

The rise of oxygen-using life forms during and after the GOE led to higher energy and productivity. There is power in oxygen: it yields more energy than metabolism that uses other fuels. The new kids on the block were the eukaryotes, critters whose genetic material (RNA and DNA) was protected within a nucleus. Eukaryotes had, and have, advantages over bacteria. Their internal membranes and molecular chains enable them to grow larger and take on a variety of shapes. They could grow into colonies and into complicated suites of cells making up creatures far more complex than single-celled bacteria. Thanks to these advances, we now have fish and frogs, reptiles and birds and mammals. You may personally know a eukaryote. In fact, you are one.

The ensuing parade of creatures—apparently unlike anything that has happened on any other terrestrial planet in our solar system—adds to the uniqueness of our world (Figure 5.4). As the Earth's skies flooded with oxygen, arthropods, mollusks, squid-like cephalopods, fish and exotic reef-building sponges multiplied. Worms slithered and trilobites skittered across the sea beds beneath the large predators like Anomalocarids. As carbon dioxide levels dropped to about 15 times what they are today, arthropods and other creatures inhabited the shorelines and plants invaded the landscape. Reptiles and amphibians visited the fresh coastlines and wandered inland. The Cambrian age came to a close some 490 million years ago. To put that timeline into perspective, the period leading from Earth's birth to the present is ten times as long as the period between the end of the Cambrian and today. Life's story is a long one.

The Cambrian was followed by the Ordovician period, an interval dominated by an energetic diversification of plankton, invertebrates and cephalopods. Several mass extinctions and an ice age closed out the Ordovician 67 million years after it started. Throughout the unfolding Phanerozoic, a total of twelve periods added to the parade of life that has marched across planet Earth. The kingdom of the trilobites came and went. Feathery tube worms and blind crabs huddled around the searing throats of sea mounts. Hadrosaurs and stegosaurs and tyrannosaurs had their day, and then followed the pterosaurs as they flew away. Giant mammals of the Quaternary ice ages—including Mammoths, camels and sabre-toothed cats—became dominant after the great dying at the end of the Mesozoic (in the ensuing Cenozoic era), and grasses appeared on the plains. The mastodons came and left before the elephants and deer and foxes took their places. The Phanerozoic eon continues today, with the appearance of us humans who try to understand it all. In fact, we are at the end of a roughly 9500-year interglacial period, where global temperatures have been incredibly stable. Oxygen-18 isotope levels can indicate global temperatures, and ice cores show us that the Earth's temperature leading up to this interglacial period varied by as much as 10°C on timescales of centuries. But over the past 9500 years, worldwide temperatures settled down considerably, and have varied by no more than ± 1.25°C. We should be entering another ice age about now, but as we will see in Chapter 9, evidence is mounting that our climate is no longer following its natural cycle.

Miller-Urey and Other Experiments

In the 1920s, two biologists came up with a major new hypothesis to explain how life began on Earth. Called the Oparin[9]-Haldane[10] hypothesis, it is widely accepted by astrobiologists, and forms a working foundation for modern astrobiology. Their hypothesis suggests that life can be understood through the lenses of physics and chemistry. If Oparin and Haldane were correct, they reasoned, scientists should be able to synthesize life-related materials by straightforward chemistry experiments in the laboratory.

Researchers began to craft laboratory tests that simulated conditions thought to exist on the primordial Earth. Experiments yielded amino acids and other key ingredients related to living things. The most famous experiment was the groundbreaking Miller-Urey test. Stanley Miller and his professor, Harold Urey, built a glass apparatus filled with gases to simulate conditions thought, at the time, to parallel those of the early Earth. They filled the first container with water to represent the ocean. This was heated to form water vapor. Tubes car-

(continued)

---

[9] Alexander Oparin (1894-1980) was a Soviet biochemist.
[10] J.B.S. Haldane (1892-1964) was a British geneticist, biologist and mathematician.

ried the vapor to a second flask, filled with methane and ammonia gases (known as a "reducing atmosphere"), simulating primordial atmosphere. They introduced electricity into their mix. The gas was cooled to condense the vapor into "rain" and then was recycled back into the first globe containing water. Within a week, the test generated a brownish organic sludge known as tholins. The experiment yielded over 20 amino acids found in DNA.

More recent work suggests that the early atmosphere was less reduced, containing gases from volcanism some 4 billion years ago. If true, the atmosphere was then dominated not by methane and ammonia, but rather by carbon dioxide, nitrogen, hydrogen sulfide and sulfur dioxide. Experiments using these gases rather than the reducing gases used by Miller and Urey have also yielded organic materials similar to those in the Miller-Urey trials. Current laboratory tests have resulted in even more material critical to life processes: amino acids, peptides, sugars, lipids and components of nucleic acids. But the transition from these substances to active life—known as biogenesis—is still a great mystery. Said Stanley Miller, "I would give anything to know what that step was."[11]

At least the place has nice atmosphere

Titan has as much atmosphere as a five-star restaurant. The air there is roughly one and one-half times as dense as that on Earth at sea level. When we look around the solar system at the atmospheres surrounding bodies with solid surfaces, Titan is second only to Venus. As we have seen, its atmospheric methane condenses into rainfall, filling rivers, lakes and seas (Figure 5.5). The complex organics in its skies lead to a "snow" of sooty hydrocarbons that bank into vast sand dunes and percolate in methane ponds and rivers. Ben Clark describes the Ferris wheel between Titan and its atmosphere. "The stuff that falls from the sky can be quite complex. It's been floating around up where the influence of the Sun makes it so that you can have photochemistry. The stuff stays up there pretty efficiently because it's made of such tiny particles, but still they're trying to fall back down."

Our overview of this alien world seems to emphasize its differences from the Earth, but it is precisely its hydrocarbon-rich environment that brings us to its strongest parallel with ancient Earth: the processes that operated on Earth when life began may well be occurring on Titan at this very moment.

The Cassini/Huygens robotic expedition found no evidence of Titan lightning, but there's plenty of solar energy forcing methane and carbon into creative forms that might have something to do with life. With all of its

---

[11] Personal interview for *Space Reflections*, 1982

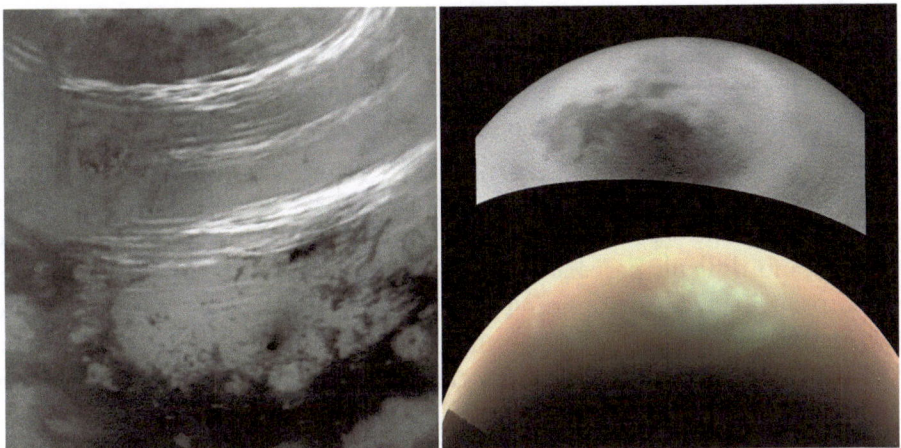

Figure 5.5 Left: methane clouds condense over the planet/moon's northern hemisphere. Right: The Cassini orbiter could see different atmospheric details using different filters. (images NASA/JPL/SSI)

carbon-based excitement, one would think that Titan would be crawling with life just like us. The problem, of course, is temperature. Terrestrial biological processes grind to a halt in temperatures like those on Titan. Chemical reactions of any kind (including biological ones) are sluggish. Could all those organic compounds—called tholins—dissolve in Titan's methane lakes, combining into life-friendly mixes?

But getting from non-living organic sludge to dolphins and people is a big, baffling leap. NASA Ames astrobiologist Christopher McKay points out that life appears to have gotten started very early in Earth's history, within a few hundred million years of the time when the planet's surface had cooled enough to solidify. "It happened quickly on Earth," McKay says, "but we don't have a clue of how you get it going. The Miller-Urey experiment made everybody think that was the solution to the problem. But it's not. The Miller-Urey experiment [demonstrates] how you get the hardware; where you get the bricks or, in computer terms, where you get the silicon. The real problem is *how do you write the program?*"

What McKay is getting at is this: the Miller-Urey experiment proved that creating life's building blocks is fairly easy. But biologists have come to realize that there are two important components to life's equation. These can be seen in terms of hardware and software. The hardware equates to amino acids, proteins, nucleic acids and other biochemicals. Miller-Urey and other studies showed how nature generates that hardware. But the second part of life, the software portion, is the hard part, McKay says. "Somebody's got to come up with the genetic code and program it in. At the time Miller did his experiment in the 1950s, we hadn't invented computers, so the notion of software wasn't in anybody's hat. Hardware was all they could think about. So they said, 'Aha! We have the hardware problem solved, so problem solved! The origin of life is within our grasp.'"

What those researchers didn't realize was that the aspect of biogenesis explained by their experiments was only the physical component. The more complex and difficult part had to wait for concepts from computer science, which brought an understanding of the difference between hardware—the raw materials of living systems—and software—the assembly instructions. Biologists, says McKay, had to realize that life, like a computer, has both. "Just assembling the hardware doesn't make your computer do anything interesting. You've got to have the software too. How do you create self-programming out of thin air? Miller showed self-assembling hardware." But Miller's experiments had nothing to do with the software side.

It's a puzzle that intrigues Ben Clark as well, and in the case of Titan, life will be tough. "Life as we know it just can't tolerate much below zero. It needs liquid water. Theoretically you would think that hydrocarbons might provide the solvent, but the problem is that nobody knows how to do that chemistry. It's simply different chemistry. We find organism down in gasoline storage tanks, but it turns out that they're living in a little bit of water that separates out. We haven't found any life that actually lives in this stuff. Maybe its there and we don't know how to detect it." And the software is another matter, he says. "DNA is the secret code that makes the correct sequence. It's like a binary code, but instead of having ones and zeroes, you have zero, one, two and three [representing DNA's four simple compounds: adenine, cytosine, guanine and thymine]. Chemically you have to have the right ingredients. Then you have to make a way to link them together to make these long chain polymers. It's all quite tricky."

Chris McKay hopes that somewhere out there, researchers will devise a similar experiment that will show us self-assembling software, a software analog of Miller's hardware experiment, "where you somehow put things together and hit a spark and it starts writing code, just like it made amino acids. It sounds ridiculous, but I bet before Miller's experiment it would have been just as ridiculous to say you put stuff in a jar and hit a spark and you make the elements of life."

The all-important software solution has been elusive. The reason may be, in part, because our conceptualization of the problem arose only recently. It has come about with our awareness of computer software. The computer has framed our ability to even ask the question, 'where does this software come from?'

As for the complex organics that the Miller-Urey team synthesized, there are plenty on Titan. Titan's lowlands are dusted with organic material, the product of interactions between Titan's atmosphere and sunlight in its upper reaches. The Sun's ultraviolet radiation severs the carbon-hydrogen bonds in methane and other hydrocarbons in a process called photodissociation (and more broadly, photochemistry).[12] The left-overs of these broken homes recombine to create more complex hydrocarbons. Nitrogen and methane stay behind, while the lighter hydrogen drifts into space. Since methane is constantly destroyed and transformed through photochemistry, it must be continually replenished, perhaps from Titan's interior through evaporation of lakes or recharging of the atmosphere by cryovolcanic eruptions.

Methane rains waft from the clouds, slicing out canyons and settling into great seas. Those seas mix with the organic material, generating a slurry of carbon-based goo akin to the chemistries of organic life. The giant moon possesses at least the hardware side of life's grand equation.

Titan and Earth: familiar features

While the first life on Earth may have been relegated to ancient ponds, coastal seas, or hydrothermal vents on the ocean floor, today it inhabits mountains and valleys, glaciers and hot springs, deserts and canyons, volcanic throats and nuclear reactors. All of the terrains graced by life have their analogs on Titan (except, perhaps, the nuclear reactors). Despite vast differences in temperature and materials, Titan, too, has its deserts, mountainous regions, canyons, seas, and perhaps even something akin to hot springs (although "hot" is a relative term). Remember that the atmosphere of Titan is primarily nitrogen and methane. Sound familiar? These two constituents were common to Earth's early atmosphere, and they were the chief components, along with hydrogen, in many of the later Miller-Urey experiments.[13] Aside from Earth, Titan is the only solid body in our solar system with a substantial nitrogen atmosphere. What kind of world is this sibling of ancient Earth?

---

[12] Although Titan's nitrogen is more abundant than methane, its triple-bonded structure is much stronger, so solar energy cannot break it as easily.

[13] More recent laboratory experiments have used other gases, including nitrogen/methane atmospheres thought to resemble Earth's earliest air, and Titan's atmosphere today.

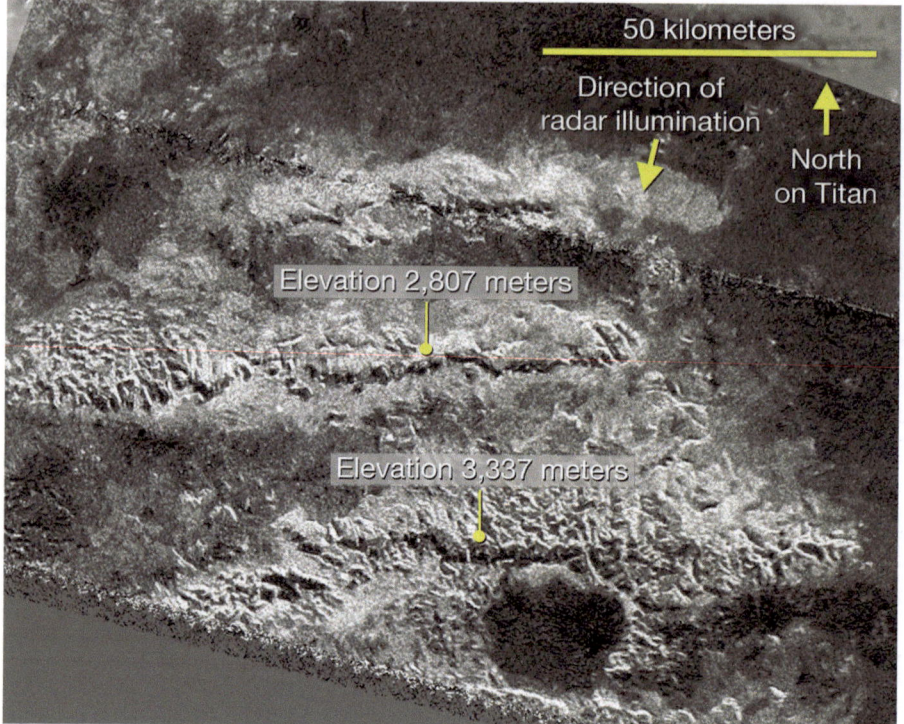

Figure 5.6 The mountains of Mithrim Montes are the highest yet discovered on Titan. This radar image, enhanced with a special smoothing technique, was taken by the Cassini Orbiter. Elevations are approximate. (Image credit: NASA/JPL-Caltech/ASI)

The High Country

To better understand parallels between Titan today and the pre-biological Earth, it helps to understand the lay of the land beneath that unique chemical blend in the sky, on the ground and in the lakes. Like the terrestrial planets, Titan's landscape rises in mountain chains, individual buttes and hilly regions. The highest ridges, a trio of peaks called the Mithrim Montes (Figure 5.6), reach into the foggy skies to heights of 10,000 feet (3050 meters). Among them, Titan's tallest summit towers an estimated 10,948 feet (3,337 meters) above the low-lying sand sea of Shangri-La.

Titan's highest pinnacles seem to huddle around equatorial regions.[14] Peaks of heights similar to the Mithrim Montes rise from another rugged province

---

[14] It's a little hard to tell, as Cassini was only able to map 45% of Titan's surface at the high resolution of its radar. Radar swaths were dictated by the geometry of each flyby, and that geometry changed as the spacecraft's trajectory was continually redirected to accommodate encounters with Saturn and other moons. Mid latitudes were especially missed by the orbital tour. The equator and poles were imaged most fully.

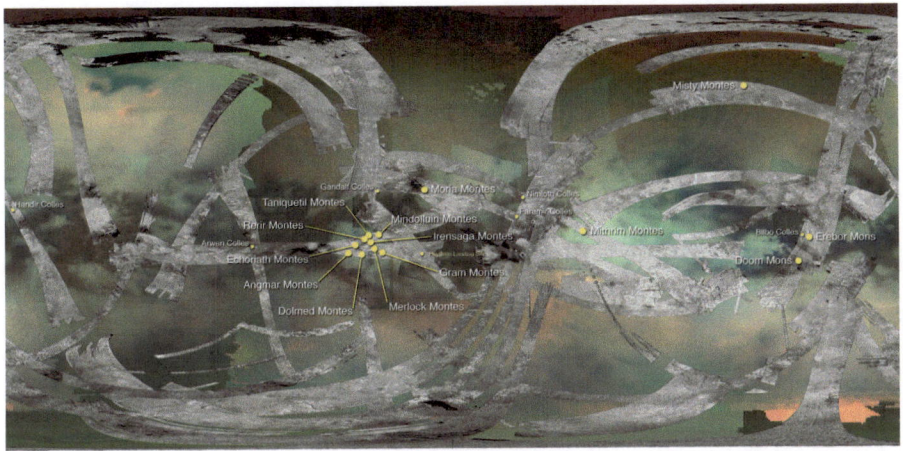

Figure 5.7   Mountain chains or solo peaks appear in many parts of Titan's globe. Here, they are labeled on a global map that includes the strips covered by Cassini's radar imaging system. By convention of the International Astronomical Union, mountains on Titan are named after the mountains from J.R.R. Tolkien's fictional Middle-earth. (Image credit: NASA/JPL-Caltech/University of Arizona/USGS)

called Xanadu. They also rear up in isolated "ridge belts", some of which are near the ESA Huygens probe's landing site (Figure 5.7).

Charting mountain chains is important to geologists: mountains on Earth denote the edge of active plates or faults. We have seen (Chapter 2) how the Andes, Himalayas, and Rocky Mountains mark the meeting or outcomes of such plate movements. The highest mountains are the youngest. (For example, the humble Appalachian Mountains of the eastern US, which top out at an altitude of about 6700 feet/2040 meters, were once similar in scale to the Rocky Mountains today.)

Planetary geologists are searching for evidence of such tectonic movement on Titan as well. There may be evidence of the ice crust fracturing and then drifting into new positions, but a comparison between terrestrial mountains and the mountains of Titan is difficult on two counts. First, the mountains are made of different materials. Because they are water ice, Titan's surface cannot hold massifs as great as the Himalayas or Alps. Planetary geologist Jani Radebaugh explains, "since water ice ultimately gets soft at the base, it will not be able to hold up super high mountains." The second weakness of comparing the two cases is a dramatic difference in erosion, Radebaugh says. "It could be that a big contributor [to Titan's flatter landscape] is the heavy erosion Titan is undergoing, and has been undergoing for a long time, which has brought the tall things down and the low things up."

We find a similar dynamic of sinking terrains on other large ice worlds. The closest relative to Titan—both structurally and in terms of size—is Jupiter's

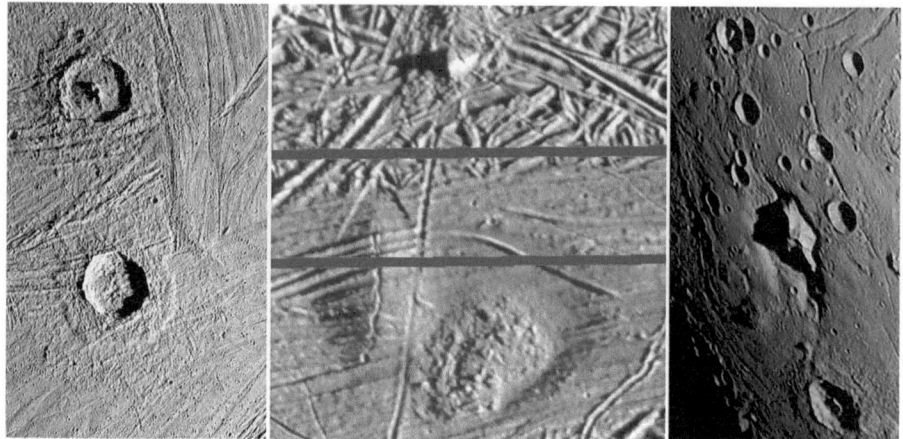

Figure 5.8 (left) The highest points on Jupiter's moon Ganymede tend to be on crater rims or central peaks. (center) The water-ice surface of Europa is depressed around two mounds of material in the lower part of the frame. (right) Charon's Kubrick Mons is a 3-4 kilometer-high crag surrounded by a deep moat. (left and center: Galileo images courtesy NASA/JPL-Caltech; right: New Horizons image courtesy NASA/JHUAPL/SwRI)

Ganymede, largest moon in the solar system. Despite its size, Ganymede's high ground seems comparatively subdued, with its uppermost locales on crater rims and central peaks. Few rises reach altitudes of one kilometer,[15] and the landscapes surrounding them tend to flow and flatten. Glaciers on Earth flow downhill, moving under their own weight. The warmer the ice, the more fluid the movement. Water ice on Pluto, Triton and other outer solar system members freezes to the consistency of rock, but on Ganymede, it is softer. Since the mountains there are made of water ice, they cannot stand for geologically long periods. Astrogeologist Paul Schenk of the Lunar and Planetary Institute comments on Ganymede's subdued terrain. "Everything is within 1 km of the local mean elevation. Slopes can be quite high, averaging 15 degrees, but up to 30-40 degrees on the freshest features. The highest relief, he says, is in the Gilgamesh region. There, "we may see 1.5 to 2 km high individual peaks." But as on other icy bodies, these are the exception, as the "bedrock" water ice cannot support higher elevations. Even on Pluto's ice moon Charon, where temperatures hover at a frigid -365°F, the giant mountain Kubrick Mons is surrounded by a deep moat, which may be a sign that the structure is sinking[16] (Figure 5.8).

---

[15] The highest ground on Ganymede is a great dome 550 km wide that rises at exactly the subjovian location—it points directly at Jupiter. The dome crest is 3 km high.

[16] Another nearby mountain, Clarke Montes, is at the center of a more subtle depression.

Volcanoes, a different kind of mountain

Planets and moons use a variety of techniques to build mountains. Sometime a planet's crust folds or buckles due to movement and stress in the crust. Others appear as the result of crustal blocks thrusting up from forces beneath. The larger impact craters leave a central peak, a dramatic mount rearing up from the crater floor. Still others ascend in the form of volcanoes, like the Earth's Mounts Fujiyama, Rainier and Vesuvius. These mountains bring molten rock from the subsurface, piling it up over time in either floods of liquid rock or explosions of gas, boulders and magma.

A different kind of volcano visits the frigid worlds of the outer solar system, worlds of rock and water-ice surfaces. Rising from these icy moons are cryovolcanoes, mountains erupting super-chilled "cryo-magmas" of ice water, ammonia and other liquids. Planetary geologists have tentatively identified several cryovolcanoes across the alien face of Titan (Figure 5.9). In the southern hemisphere of Saturn's great moon, one of the highest mountain ranges in the solar system rears up, culminating in the rugged Doom Mons. The 1450-meter-high massif, named after the dark mount in J.R.R. Tolkien's *Lord of the Rings* trilogy, shares features with many classic volcanoes. At its apex is a collapse feature called Sotra Patera. From this depression, flows of material have flooded the mountain's flanks and surrounding plains. The most extensive, Mohini Fluctus, has been partially buried by sand dunes. Sunken into the flanks west of Sotra Patera, a roughly 500-meter-deep indentation surrounds a circular pit that drops by another 400 meters in depth. Despite a constant rain of dark hydrocarbons, liquid methane and ethane from Titan's upper atmosphere, the summit of Doom Mons appears to be water ice, perhaps scoured clean by incessant winds.

Identifying a cryovolcano through the soup of Titan's fog is a difficult task, and not without a few false starts. NASA's Cassini Saturn orbiter carried a radar system that could peer through the clouds to image Titan's surface. Some resolutions came back at 320 m. Cassini's first flyby yielded a detailed radar strip some 12875 kilometers long. Titan was an alien world indeed, showing baffling formations that puzzled geologists. One of the landforms was a great, dark circle that bore some resemblance to the volcanic pancake domes of Venus. They christened the 180-km-wide formation Ganesa. With the available data, researchers assumed it to be a cryovolcano, but later imaging of the same terrain showed that rather than being a dome, the area was flat. Although Ganesa doesn't follow the same profile as a conventional domed volcano, it does have a central crater-like feature with flows issuing from it, so cryovolcanism is still considered likely by many. Another puzzling formation

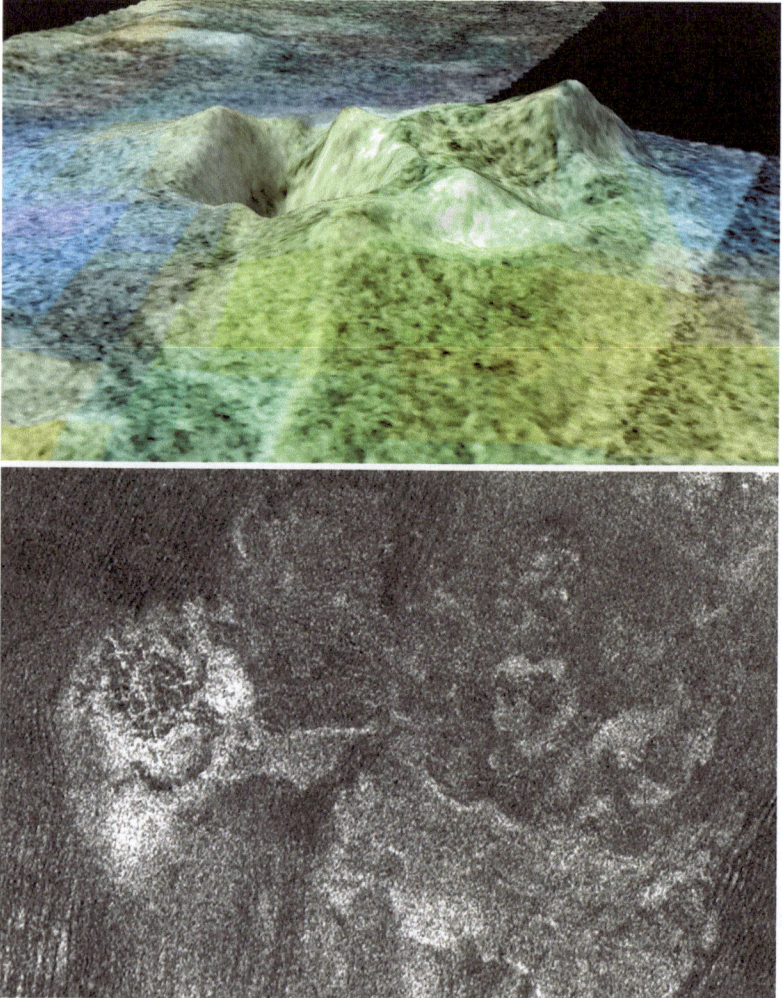

Figure 5.9 Top: Cassini radar image—exaggerated in height by a factor of ten for clarity—reveals Titan's Doom Mons rising some 1450 meters. A deep depression pierces the mountain near the summit. Flows meander down the flanks, and may be streams of cryolavas. False colors show different materials as they appeared to Cassini's Visual and Infrared Mapping Spectrometer (blue indicates fresh water ice). (NASA/JPL-Caltech/ ASI/USGS/University of Arizona, Tucson). Bottom: flows emanate from Doom Mon's presumed caldera, Sotra Patera, in this overhead radar view. (NASA/JPL-Caltech)

is called Tortola Facula, nicknamed the "snail" for its corkscrew shape. The feature first appeared at low resolution in Cassini's Visual and Infrared Mapping Spectrometer, but later radar imaging showed the area to be plains rather than a mountain.

Still, the presence of active cryovolcanism on Titan could explain a long-standing mystery about its air: where did all that methane come from?

Ganymede, larger by a bit, floats within the vacuum of space, with the odd molecule of oxygen bouncing around its surface here and there. Ganymede's higher temperatures make it more difficult for the giant moon to hold on to an atmosphere, but Titan's formidable atmosphere cannot be explained solely by temperature: it must be coming from somewhere.

Titan's core is substantial enough to retain large quantities of radiogenic materials that shed heat. Its interior acts like a baker's oven, splitting nitrogen and carbon from the complex organics that rain out of the sky or form within the moon. Nitrogen and carbon would tend to recombine into methane and nitrogen, making their way back up to recharge the atmosphere.

Although methane is abundant on Titan, sunlight and other gases tend to destroy (or transform) it quickly. Something must be recharging the atmosphere at a steady pace. A major suspect is cryovolcanism. So, it seems, some type of volcanic emanations are carrying on even today on this gloomy, cryogenic world. In fact, Cassini may have caught sight of a cryovolcanic eruption in action at a site called Hotei Arcus (Figure 5.10). Valleys and even flow

Figure 5.10 The arc of mountains making up the Hotei Arcus formation loom out of Titan's mist at right. In the foreground, flows of viscous cryolava ices form lobate hills. Beyond them, methane has carved out river beds, now dry, that issue from the rugged highlands. (art by the author)

features score this arc of mountains. In several successive passes, Cassini's VIMS showed a brightening between October 2005 and March of 2006. Some data suggests that the brightening coincides with increased ammonia levels. Ammonia is a candidate component of cryolavas. The work remains controversial, and studies continue, some of which indicate that the center of Hotei Arcus may be a dry lake bed. Several other sites sport crater-like forms with flows issuing from within. One of these is Tui Regio, a formation with a depression at its center like a caldera, or volcanic crater, with a thick, serpentine flow issuing from it, similar to silicate lava flows. But flows on Titan are enigmatic and tricky to interpret. Obviously, material has flowed across the surface, but the flow patterns are so diffuse that they could be thin or thick, and they could be volcanic or caused by methane rain flooding. Whatever they are, Titan's flood plains and river valleys add to its exotic, Earth-like demeanor.

Craters

Like Venus, Titan has no small impact craters (Figure 5.11). It's missing them for the same reason that Venus is: the dense atmosphere. A meteor or comet will create a crater roughly eleven times its diameter. Titan's Selk crater[17] spans a diameter of 70 kilometers, so the rock that made it was roughly 6 kilometers

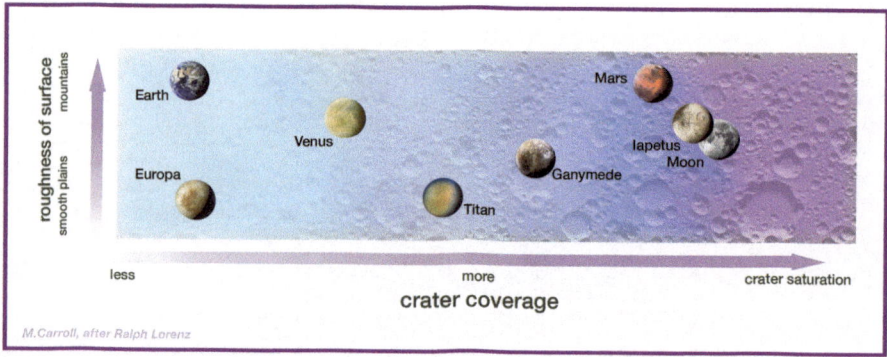

Figure 5.11 We find impact craters on every body of the solar system except Jupiter's volcanic moon Io. The surfaces of Earth and the ice moon Europa erase craters quickly, while worlds with little geologic activity and little or no atmosphere (such as Saturn's Iapetus or Earth's Moon) preserve craters. Crater counts can reveal the geologic age of a region. (art by the author, after Ralph Lorenz)

---

[17] Selk is scheduled to be explored by the upcoming Dragonfly mission, a helicopter mission to Titan.

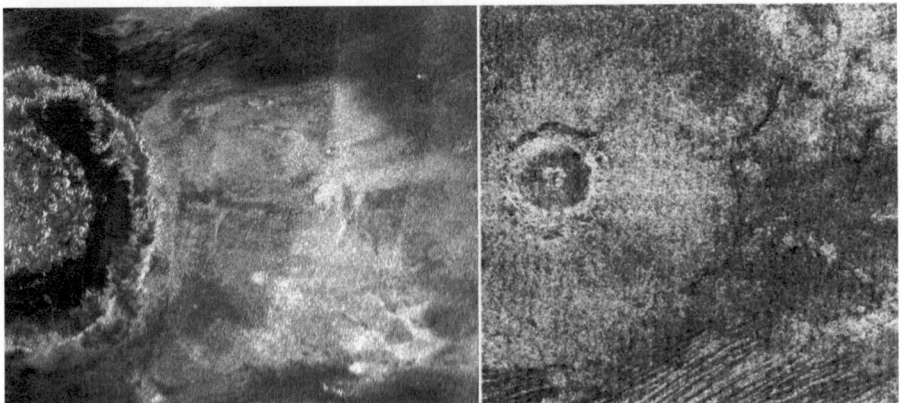

Figure 5.12 Two impact scars show the diversity of Titan's craters. At left, Menrva is the largest impact crater so far identified. The smaller crater Ksa, at right, displays a central peak and ejecta fanning out beyond. (images NASA/JPL-Caltech/SSI)

across (that is at the time of impact; the meteor began as a much larger body, but partially disintegrated as it burned through the atmosphere). But a small meteor that might make a small crater is slowed down so much by Titan's dense atmosphere that it will hardly make a dent in the surface. Many don't make it down, breaking up in the air. Consequently, Titan has few craters smaller than 10 km, as is the case for Venus .

Titan's craters are quite shallow with flat floors. Often, only the rims show, and often crater features are partially obscured by sand dunes. Ganymede, a twin in terms of general crust material and gravity, has a variety of deep craters more similar to those seen on other airless worlds like Mercury or Earth's Moon (Figure 5.12). Wind and rain erode Titan's craters. The best preserved of Titan's craters seen, named Sinlap, is 80 km across and 1.3 km deep, the deepest crater on the planet/moon. Many crater rims and ejecta show a spectrum indicating water ice. Its assortment of craters shows different stages of erosion. Menrva is the largest confirmed impact crater on Titan.

Seas of sand and rivers of methane

River valleys etch their way though craters, down mountain slopes, and across plains (Figure 5.13). In the north and south, they wind their way to the shorelines of methane lakes and seas. These dendritic (branching) networks extend for hundreds of kilometers, often fanning out into alluvial fans, triangular regions covered in cobblestones and rounded boulders. Alluvial formations on Earth form when fast-flowing rivers pour from highlands, fanning out as they

Figure 5.13 Extensive erosion from rainfall or flooding bisects Titan's landscape into a labyrinthine maze (left). Dendritic (branching) channels spread from a wandering canyon. (right) Networks of river channels lead to the rugged coastline of Kraken Mare in this oblique view, which shows the methane sea as black at lower right. (Cassini images NASA/JPL-Caltech/ASI)

reach flatter ground where they drop stones from higher altitudes. This type of river is characteristic not of gentle rainfall, but rather of flash flooding. It appears that Titan's rainfall is episodic. At the landing site of ESA's Huygens probe, the rounded ice stones appear to have been worn by the action of rolling in liquid flows, much like terrestrial river rock.

Vast sand seas blanket Titan's plains. Dunes ripple across flatlands, swirl around buttes, and bank up on the flanks of mountains. They fill valleys and cover craters. While Cassini's resolution was insufficient to see any movement from one pass to another, images revealed that the dunes are not heavily eroded, despite the fact that they are under nearly-constant wind and methane rain. The dunes must be fairly fresh.

Liquid, liquid everywhere, but not a drop to drink

Perhaps the most Earthlike processes on Titan are those that mimic Earth's water cycle. On Titan, that role is filled not by water, but by methane. Coleridge's "Ancient Mariner" would have felt the same thirsty frustration had he set sail on Titan. Aside from the occasional meteor impact, the only liquid water on the surface—if there is any—erupts from cryovolcanoes and remains for only a short time. The methane rivers, lakes and seas appear to come and go seasonally (though some seas are extensive enough to be permanent features of the Titanian marine landscapes, as we will soon see).

We have seen how rains scour the surface, and how they flow downhill, cutting valleys into the countryside (Figure 5.14). Their appearance is strikingly familiar, similar to flash-flood plains in the Earth's deserts. Bright rings surround some smaller lakes, essentially "bathtub rings" marking exposed shores where the lakes have retreated, probably by evaporation. Other methane "waterways" display bays and lagoons, peninsulas and complex chains of islands. The great seas like Kraken and Ligeia cover territory similar in scale to Earth's Black Sea. On some beaches, rivers spread into deltas as they dump methane into the seas. At other locales, rugged coastlines assume the dramatic appearance of the fjords in Scandinavia. Similar fingerlike coasts on Earth are the result of areas flooded by rising water levels, much like Lake Powell and Mead in the desert southwest of the US.

Many of the lakes exhibit rounded borders with steep sides, depressions sometimes hundreds of meters deep. They share the same characteristics as those of karstic lakes on Earth, called kettle lakes. Terrestrial kettle lakes can be born of collapsed caves, or they can be remnants of melted "icebergs" left behind by retreating glaciers.

On Earth, water evaporates, leaving behind contaminants. This pure water condenses as clouds and falls to the ground, where it interacts with briny saltwater in the oceans below. A curious parallel plays out between salt and freshwater on Earth, and methane and ethane on Titan. On Saturn's moon, as methane evaporates it leaves behind ethane with traces of butane and propane. Methane rains down into lakes and seas containing ethane, which corresponds to Earth's saltwater. Rainfall is more abundant at the poles, so fresh methane is flowing into the lakes farthest north, like Punga and Ligeia. Farther south lies Titan's largest sea, Kraken. Kraken's radar reflection indicates that it is made of something slightly different than the lakes farther north. Planetary

Figure 5.14 Monsoon-like storms of methane rain create a flash flood that churns through one of Titan's river channels in this artist's conception by Marilynn Flynn. The temporary falls spill over an eroded cliff, exposing layers of frozen hydrocarbon sediments and hard-as-rock water ice. The rushing liquid picks up silts and sandy materials, giving it a murky orange color. It is mid-summer at this high northern location in Titan's lake district. Saturn peers through an imagined break in the clouds with its north polar region completely illuminated. (Digital painting courtesy Marilynn Flynn ©2014)

scientist Ralph Lorenz[18] suggests that "fresh-water" methane is charging the northern lakes, flowing into the more southern Kraken, where methane evaporates and leaves behind ethane (Titan's "saltwater"). Lorenz compares the situation to that of the Black and Mediterranean seas. Fresh-water rivers flush salts from the Black Sea into the Mediterranean (Figure 5.15). "Like the Baltic Sea or Black Sea on Earth, both of which are fed by fresh rivers, the gradient in precipitation could drive …liquid from Ligeia into Kraken…The methane is able to evaporate at any point, but it will do so particularly at Kraken's lower latitude, where it is warmer. There the [ethane accumulates], rather like salt does in the Mediterranean Sea."[19]

---

[18] Lorenz is also an inventor who helped engineer the Huygens Titan probe.

[19] Lorenz explores the concept in his book Saturn's Moon Titan, Owners' Workshop Manual (Haynes 2020)

Figure 5.15 (left) Fresh-water rivers flush salts from the Black Sea (at top) into the Mediterranean (below), much as "fresh" methane rivers may force ethane from Ligeia Mare into Kraken Mare on Titan. For scale, Mayda Insula is about the size of the big island of Hawaii. The two images are not to the same scale. (left: Earth image courtesy NASA/Goddard Space Flight Center and ORBIMAGE; right: Titan map NASA/JPL-Caltech/ASI/USGS)

Titan methane lakes congregate in the high northern latitudes, with the exception of the large Ontario Lacus toward the south pole[20] (Ontario Lacus is surrounded by a huge hourglass-shaped ancient shoreline, indicating that the lake used to be much larger, and will be again). The middle real estate of Titan is dry, with very little evidence of past lakes. The lakes appear to be somewhat seasonal. Their propensity to gather in the northern hemisphere has a parallel on Mars, where a residual polar cap lasts through the year in the south, but not in the north (Figure 5.16). This imbalance comes from a straight-forward cause: the south is colder. The Martian orbit is eccentric: the red planet does not follow a perfect circle around the Sun. This results in a southern summer that is a little hotter than summer in the north, but much shorter, leaving a thicker glacier of carbon dioxide ice at the south pole. In 25,000 years the situation will be reversed, with a shorter, hotter northern summer. Carbon dioxide will, at that point, build a permanent cap in the north, and the southern one will vanish (for more on this, see Chapter 6).

---

[20] And a handful of small ponds

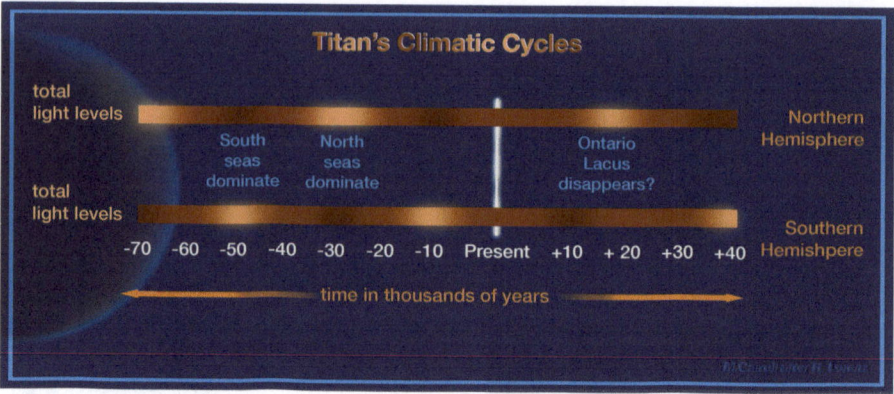

Figure 5.16 Due to Titan's wobbling spin axis and changing orbit over long periods, the amount of sunlight falling on each pole changes over time. The shift in climate drives methane from one pole to another, so that the seas come and go. Today, the northern hemisphere is Titan's "lake district", but thousands of years from now most of Titan's methane lakes will gather in the south, and the north will be as dry as the south is today. (diagram by the author, after work by Ralph Lorenz)

This long-term cycle is the result of the wobble of the Martian axis. Combined with the eccentric orbit, the effect is known as a Croll-Milankovich cycle.

Like Mars, Titan has uneven seasons. Saturn drags Titan along in its orbit around the Sun, and that orbit is eccentric in the way Mars' path is. Consequently, Titan's southern summer is shorter and warmer than the summer season in the north (are you having Martian deja vu?). The summer heat bakes volatiles like methane and ethane, and winds carry their vapors northward, where they condense in cooler climes. It's the perfect recipe for northern lakes and seas. Titan may well have its own long-term Croll-Milankovich cycles, as does Mars, because while the south is now much drier than the north lake district, it is peppered with dried-out lakebeds, ghosts of liquid methane pools long gone. Cassini watched as the seasons changed, but the south still remained fairly dry, and those phantom lakebeds stayed empty. One day, as Titan cycles through a ~40,000 year period, these may again fill as the north arrives at its cyclical millennia-long drought (Figure 5.17).

The ponds, rivers and seas of hydrocarbons, constantly filled by complex organics from the sky, constitute sites that are thought to have had analogs on pre-biotic Earth. Within these rich chemical cauldrons, circumstances may play out much like those that led to life on our own world. But Titan hides another place that might engender a second Eden: beneath its hauntingly familiar landscapes, within Titan's deep ice crust hides an ocean to rival any on Earth. It's a global ocean of salt water. From October 2005 to May of

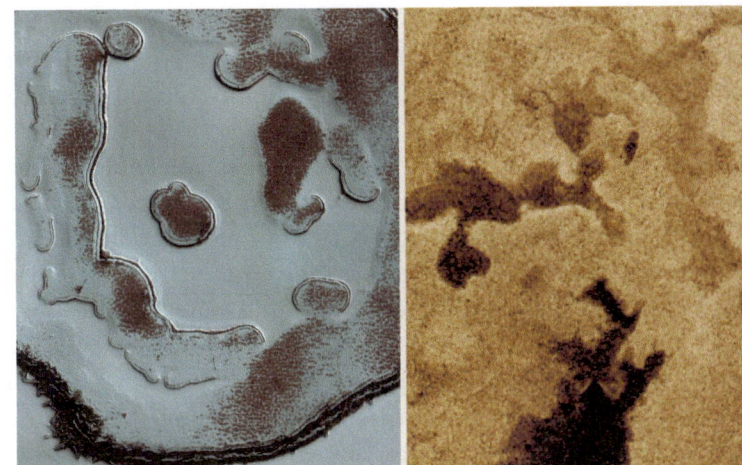

Figure 5.17 Left: The Martian southern polar cap takes on exotic forms as its carbon dioxide sublimates (evaporates from an ice directly into gas). Right: Titan's lakes show different stages of development. Darkest areas are filled with liquid methane/ethane. Lighter depressions are empty. Many show a classic "bathtub ring" at the margins. Eventually, the dry lakes may fill again. (left: courtesy University of Arizona/HiRISE/LPL; right: courtesy NASA/JPL/SSI)

2007, Cassini charted a shift in Titan's crust of some 32 kilometers, implying that the surface is floating on a subsurface global ocean or viscous ice. Added to this evidence, Titan generates the kind of low-energy radio waves that a buried ocean would. Additionally, the Cassini science team assembled gravity maps of the moon from Cassini's 127 targeted flybys, and these also fit with the existence of a subsurface water ocean.

Chemically, liquid water is a wondrous thing. It enables biochemical reactions and combinations far more readily than other compounds. The existence of water is a prerequisite for all life on Earth, but it is not *sufficient*. Water's mere presence does not equate to life. Earth life needs all those chemical goodies like carbon, hydrogen, nitrogen, oxygen, phosphorus and sulfur (CHNOPS), along with other building blocks used for cell membranes, mitochondria, etc. The bad news is that Titan's saltwater ocean is sandwiched between layers of water ice. But the good news is that it is not sequestered from the organics above, nor is it cut off from the minerals in the core below. Ice is permeable, and Titan's ice likely flows vertically as well as horizontally, mixing life-related organics into that hidden subsurface ocean.

And what of liquid water on that cryogenic surface? Titan's bitter cold slows down chemical processes, making life's functions difficult. No life on Earth can take temperatures anything like those on Saturn's mystery moon. Many astrobiologists believe that biogenesis must take place at higher temperatures than Titan's back yard. As it turns out, Titan periodically experiences higher temperatures, temperatures warm enough to preserve liquid water. These heat waves come in the form of meteor impacts.

While cryovolcanoes hold some promise for warm sites amenable to life, some investigators find greater potential in newly-formed craters, because they form at the hands of much higher energy. Meteor impacts produce larger quantities of melted ice, and that liquid takes a long time to freeze, on the order of centuries to millennia. Biologists consider liquid water a critical element in the creation of complex chemical reactions. In an impact, water is heated far more energetically than it is in prospective cryovolcanoes, which probably tend to erupt cryolavas at temperatures barely above the freezing point of water. In the case of an impact crater, higher energy and temperature results in more robust and rapid chemical reactions. Within these warmer environments, amino acids, proteins and other complex organics are fostered more readily than in the natural cryogenic conditions on Titan's surface. The idea of ponds as the nativities of life is not new. Even Darwin spoke in a written note to a colleague of "warm little ponds".

Into the deep weeds: Titan's solutions for life

Miller and Urey did not carry out their experiments under Titan's low temperatures. The moon's chill would take the blush off of most chemical reactions seen in those and later experiments. But all is not lost. Despite cryogenic temperatures (hovering around -150°C), Titan's environment generates materials that might be incorporated in exotic metabolic pathways, including acetylene, hydrogen and heavier hydrocarbons that drift from the sky. Organic reactions in hydrocarbons are nearly as flexible as those in water.

Biochemistry on the surface has limited choices; life will find fewer elements accessible on a world made of water ice, but there are some. For decades, NASA's strategy for finding life beyond Earth has been to "follow the water". In the case of Earth's biology, that water must be in liquid form (even if it is vapor) for life to thrive. Glacial microbes tend to congregate along the margins of the ice fields where water is melting; the expanses of the hard-frozen glacial surfaces are nearly sterile. In our terrestrial environment, water acts as a universal solvent. It lubricates biological structures, and it opens pathways for biochemical functions. But its role as a universal solvent for living systems relies on its liquidity and on its chemical interaction at Earthly temperatures and pressures. In the case of Titan, the liquid permeating the environment is methane. If biologists are to consider life that uses this as its solvent, all of our assumptions about interactions with the environment must be reassessed in a different light. Recent studies indicate that organic reactions in hydrocarbon liquids carry on with as much joie de vivre as they do in liquid water[21]. Perhaps the hydrocarbons so ubiquitous on Titan might be a biological analog to water in the Earth's biosphere.

As methane reacts to sunlight in Titan's upper atmosphere, photochemicals combine to create organic molecules. A carbon haze drifts downward, eventually blanketing the water-ice surface and mixing with the methane lakes, resulting in even more complex materials. Titan's lakes and seas become rich brews of the stuff of life, complex organic chains floating in methane baths. If carbon-based life is what you're after, Titan is the place to look.

Although the mineral-rich rocky layers of the moon are cut off from the surface by the ice crust, comets and meteors constantly deposit elements like iron, copper, nickel, sulfur, calcium and sodium (salts). But life in Titan's alien environment may not even need the kinds of elements useful to Earth's biology. Because of the cryogenic temperatures on the misty moon, water molecules might be used to fill the role that metals play for enzymes within Earthly biology. Water molecules could act as a catalyst for hydrogen-bonded structures in the same way that metals do for redox reactions (oxidation-reduction reactions critical to photosynthesis and respiration) in terrestrial biochemistry.

Terrestrial biology builds structures using proteins. As they interact with water, individual proteins fold into specific shapes necessary for life functions. In methane ponds and seas, no such proteins can exist, but substitutes may include hydrocarbon chains, aromatic ring structures, and carbon nanostructures like graphine. And with life thriving in a methane sea, the environment will change. On Earth, increased levels of oxygen, carbon dioxide, methane and nitrogen in the air are the direct result of biology. On Titan, the most promising gases for consumption by microbes or Mammoths are hydrogen, acetylene and ethane, all of which would affect the balance of gases in Titan's lower atmosphere. Acetylene offers some interesting interaction for potential Titan biology. Welding torches combine acetylene with oxygen to superheat metals. But on Titan (where there is no free oxygen in the air) acetylene can connect with hydrogen, with a similar result: it releases energy (though more slowly). The operation generates methane as a waste product. To the biologist, this means that acetylene is at least one possible "food source" for Titan organisms.

---

[21] Bains, W. Many chemistries could be used to build living systems. *Astrobiology* 2004, 4, 137–167

A recent NASA Ames research paper concluded, "The simple low temperature life forms and communities envisaged would have very low energy demands and would grow slowly. Life on Titan may be not much more than [chemically simple] reactions encased in azotomes.[22] However, if it had genetics…what a wonderful life it would be: a second genesis different enough from Earth life to suggest that our Universe is full of diverse and wondrous life forms."[23]

Life on Ice

The exotic, methane-dominated environment of Titan may find a corresponding one here at home, on the ocean floor. A kilometer beneath the surface of the Gulf of Mexico, in frigid temperatures under punishing pressures, a unique ecosystem thrives on ice ledges. As methane gas leaches from the ocean floor, it seeps into the chilled ocean water. There, it freezes into mounds up to three meters across. On the honeycombed façades of these great yellow mounds of ice, marine biologists have discovered entire colonies of rose-colored worms called *Hesiocaeca methanicola*. Each two-inch creature uses rows of oar-shaped fins to crawl across the methane ice surfaces. Biologists suspect that the creatures are grazing on bacteria that feast on the methane ice itself.

Creature quest

Knowing that life is possible on Titan is quite a different challenge from actually looking for it. But there are ways. Organisms bring change to their environment: they ingest material from their ecosystem and they pollute it (humans are very good at both of these!). On our own world, biological feasting and contamination have created a global signature of life that would be visible across planetary distances. Biotic waste products on Earth are primarily oxygen (as we saw in the GOE) and methane (just ask a cow). Oxygen holds the most convincing promise of life's byproduct, says astrobiologist Chris McKay. "The classic case is oxygen in our atmosphere. It's got to be produced by biology. Now, there are other ways to make oxygen, so it's not an absolute marker of biology, but it's a pretty good marker."

Life also leaves visual tells. Plants and animals flash assorted pigments to get the attention of others of their kind. Some décor comes in the visual range of humans, while other fashion statements are made in the ultraviolet or

---

[22] a membrane composed of nitrogen, carbon and hydrogen
[23] McKay, C.P. *"Titan as the Abode of Life". Life* **2016**, *6*, 8

infrared, invisible to us but striking to, say, a feisty octopus or bumblebee. We may be able to detect such pigments and colorations telescopically or via spacecraft.

There's a new twist to the search for Titan life: terrestrial biology imparts a characteristic spin—or *chirality*—to its light, a phenomenon noted by French chemist Louis Pasteur. In 1849, while Pasteur was filtering out tartaric acid from wine leaves, he noticed that the acid from the living matter bent the polarized light passing through it. That was interesting, but what really got his attention was the light coming from tartaric acid that had been chemically synthesized in the lab. This artificial acid had no such spin, even though its composition was the same. Pasteur showed that the acid's crystals form mirror images, one that polarized its light clockwise and the other counterclockwise. He realized that the molecules were asymmetric, resembling each other in the way that left and right hands do. More importantly, Pasteur realized that the organic molecules consisted only of one type of polarization. Non-organic material showed no such polarization. This chiral phenomenon is sometimes referred to as "handedness". It occurs at the molecular level in all of Earth's biogenic matter. What is more, it is visible using certain types of imaging. Astronomers are now considering the polarization of reflected light, and its twist caused by chirality, as a possible marker of biological activity.

Extreme entities

It's been nearly half a century since marine biologists discovered volcanism along the Galapagos Rift zone on the Pacific Ocean's sea floor. For years, heated undersea plumes had baffled researchers. Their nature and source had remained a stubborn mystery until an expedition the robotic submersible Alvin. On a deep-sea dive off the Galapagos Islands, the craft returned images of sulfurous stone chimneys rising from the ocean floor. Researchers came to realize that the number of volcanoes on the globe's sea floors must outnumber the roughly 600 active ones on the surface. Undersea hydrothermal vents break through along the mid-ocean ridges where the Earth's mantle comes to within a few hundred meters of the surface. Seawater percolates through the crust, heading downward. When it eventually makes contact with the 1200°C magma, the water heats up to 540°C. The surrounding water pressure prevents it from boiling. The heated fluid makes its way back up through fractures in the rock, leaching minerals along the way. When it finally flows up into the ocean, it is laced with a complex mineral broth. The rich water streams from these vents, building delicate structures of spires and columns. Some of them tower dozens of feet above the sea floor. The vents themselves rise up in stony pinnacles, blobs and delicate curtains. Some chimneys grow twenty feet high in just a year. One chimney, called "Godzilla", had grown as high as a 15-story building when it collapsed, but it is building again even now.

(continued)

With the detection of the undersea volcanoes came the discovery of entirely new biomes around them. Marine biologists discovered that within that eternal darkness, life thrives in a food chain anchored to the minerals flowing from erupting vents.

The sea-floor vents of sites like the Galapagos and the Juan de Fuca ridge are so deep that the water pressure around them could crush a bus like a soda can. But in this high pressure and frigid darkness, where the water is nearly freezing and the liquid squirting from the volcanoes is nearly boiling, giant tube worms nearly three meters long undulate in the flowing water, bobbing like lilies of the field. Reddish feather gills stick out of long white body tubes, lending the appearance of a gigantic lipstick. Half-meter-long clams bask in the bubbling brew, accompanied by exotic one-eyed shrimp (their single eye is on their back). Blind crabs and Mohawk-gilled Pompeii worms thrive in water as hot as 80°C, the hottest temperatures yet recorded for living creatures.

SETI astronomer Jill Tarter comments that, "microbes are extraordinarily fit for environments that humans are not suited to at all. There's a lot more potentially habitable real estate out there than we might have imagined. We're finding creatures capable of doing all kinds of ingenious things to make a living."

Life's challenges and promise

Titan throws some major challenges at any life that might try to take hold there. First, compared to the diverse materials available in the terrestrial environment, Titan is a frozen desert wasteland. Its life-making materials are extremely limited. Second, low temperatures make it difficult for chemicals to dissolve in liquids and to combine into more complex forms. Third, liquid water's polar nature actually empowers life to carry on biological processes. Finally, Earth's organisms use proteins to build themselves and carry on daily activities, while the available molecules for such a role on Titan are hydrocarbons, which are more limited.

But not to despair! Titan also offers a few olive branches to organic life. First, "food" constantly falls from the sky like manna, generated by the photochemistry above. Second, because of the polarized nature of water, it tends to attack and break down organic molecules, but methane is far more benign. It leaves biomolecules to their own business. Third, unlike the Earth, Titan lacks the same destructive solar radiation at its surface that we find on terra firma. Finally, cold temperatures may actually preserve biological materials (as they do on Earth).

These facts lead some to argue that a separate biogenesis may have already taken place beneath the orange haze of this ice world. To Ben Clark, the margins of Titan's lakes seem a good place to start. Ontario Lacus and other liquid methane bodies appear to have "paleoshorelines", rings of brighter material left behind by retreating lake margins (Figure 5.18). "We think the wet-dry

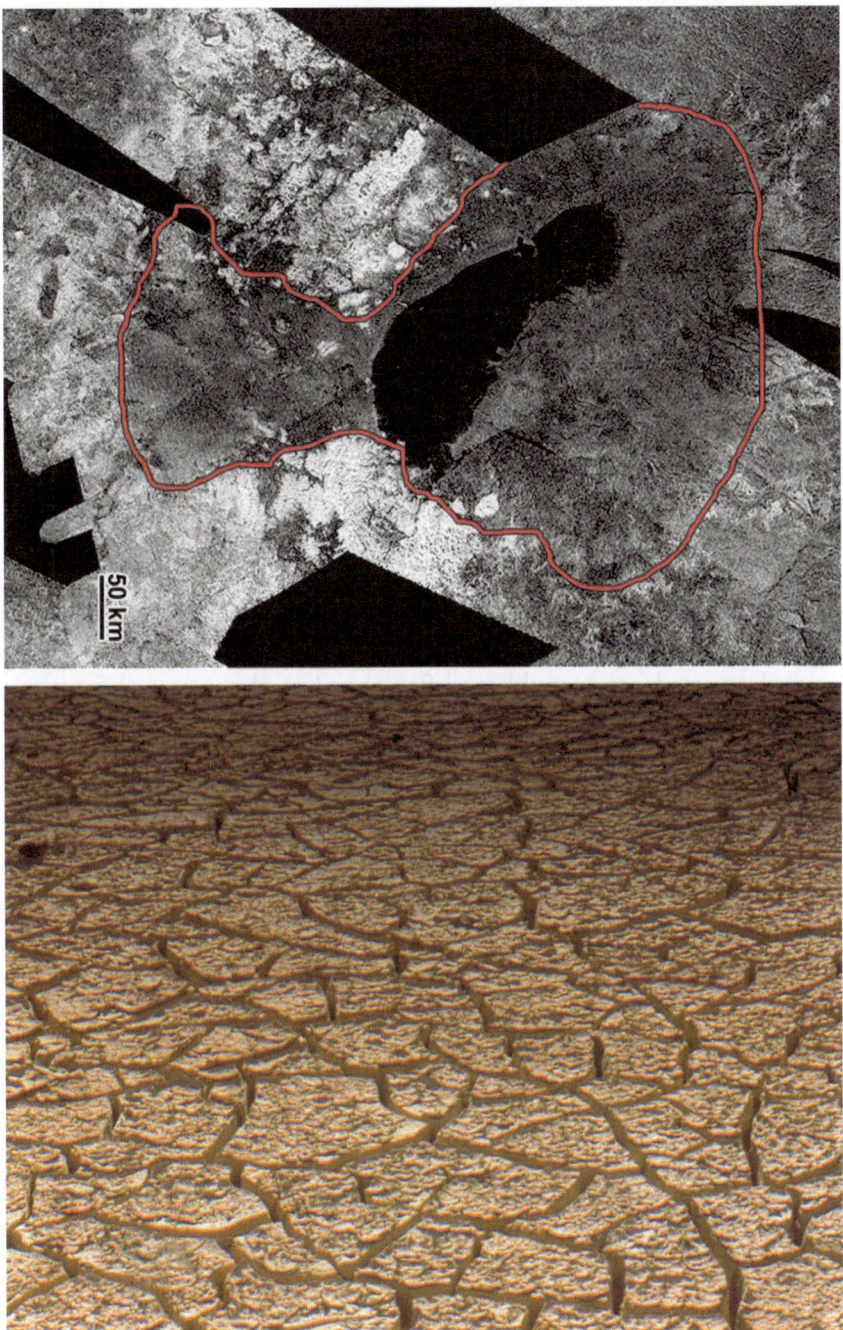

Figure 5.18 Top: the putative paleoshoreline of Ontario Lacus is outlined in red, showing the probable extent of the lake in the past. The margins of lakes, with their shifting conditions and wet/dry cycles, are seen as potential nurseries for primitive life. Bottom: Cracked and bowed surfaces, like this terrain on the floor of Death Valley's Ubehebe caldera, may be prime real estate for new microbes. (Top: courtesy Ralph Lorenz; bottom photo by the author)

cycles is the way you make proteins. And for that matter RNA. It forces the chemistry the way you want it to go. These bathtub rings (around Titan lakes and ponds) are exactly what you want. These days I'm high on mud cracks. The thing about mud cracks is they form when the little lake or pond gets larger and smaller again, and you get this repeated drying of the surrounding mud, especially when it has clay in it. As soon as you modify some of the ordinary basaltic minerals to clays, the clays expand and contract a huge amount." The resulting cracks are arranged in a variety of different orientations with respect to the Sun. On early planets, strong UV radiation from the Sun irradiated the surface. The cracks that Clark is so fond of may provide the just-right environment for life's genesis. Ultraviolet illumination can get into some of the cracks and not others, providing a variety of environments. Some of the methods of forming chemicals like proteins involve getting ultraviolet radiation into the solutions to activate organic chemistry. "The islands between the cracks on top become bowl-shaped. It's like having a bunch of little test tubes out there with different environments. Eventually you put all these things together, and out pops something like Covid 19!" Clark quips. In the regions where methane seas ebb and flow, mud cracks may well form within claylike margins. The same conditions visit Mars, where dry lakebeds leave behind cracked clay surfaces. "You get these mud crack patterns that are oriented in lots of directions compared to the incoming sunlight." Some of the ultraviolet radiation forms chemicals like proteins. Too much destroys organics, but in the tiny labyrinths of cracked ground, some of the radiation levels will be in the sweet spot where organics are formed, but sheltered enough from UV to be preserved.

If, in fact, an independent biogenesis has taken place on the misty moon, Chris McKay envisions a simple ecological biome there. In a recent paper he states, "Given these limitations, it may be that if there is life in the liquids on Titan's surface it may be simple, heterotrophic, slow to metabolize, and slow to adapt with limited genetic and metabolic complexity. The simple molecules needed for metabolism may be widespread in the environment and in the methane/ethane liquids, but the complex organics needed for structural or genetic systems may be hard to obtain or synthesize. The communities formed may be ecologically simple—perhaps analogous to the microbial ecosystems found in extreme cold and dry environments on Earth."[24]

---

[24] "Titan as the Abode of Life" by Christopher McKay (MDPI 3 February 2016)

Any Titan life forms will contrast in fundamental ways with carbon-based Earth biology. In one sense, this is good news. The search for a second, independent biogenesis on Titan differs from the search for life on Mars in that the environment of Mars may be in danger of contamination from Earth microbes, because of its similarities to present-day terrestrial conditions as well as its apparent resemblance, in ancient times, to the warmer, wetter Earth.[25] Mars also presents the danger of back contamination for our biome on Earth. The prospects of water-based Earth microbes contaminating Titan, where any surface life will likely be based on the methane cycle, are essentially nil. The same is true for returned samples from Titan to Earth. Biology on the cryogenic moon will be so different from ours that cross-contamination is not an issue.

Titan, then, is a chemically preserved portrait of pre-biotic Earth in many ways. As we peer into its soupy orange atmosphere at its methane rains and organic-encrusted shores, we may be watching a rerun of processes that took place at the dawn of life on our own planet. Says NASA's Chris McKay. "It would be really cool to find something happening on Titan, because that would tell us that life is not just common, but that it can be bizarre. It's not all cookie-cutter stamped-out-of-water-based life like us, but there are things that are really different. We should expect to be surprised." If Titan is any indicator, we will be. But there are more surprises to come. (Figure 5.19)

Looking Back: A Brief History of…Titan

As we see from Chapter 1, the planets coalesce from a great, churning disk of gas, the solar nebula. But in a miniaturized version of the process, each planet forms its own disk of material, rotating in the same direction as the orbit of the planet around its sun, and circling the planet in essentially the same plane as its equator. These smaller disks lead to the systems of moons that we see orbiting the planets today. Of course, every rule is made to be broken: some moons come from elsewhere and are later captured by their parent planets, such as Triton at Neptune. Our clue to Triton's vagabond nature comes in the direction of its orbit—which is opposite to that of Neptune and the other moons—and its highly inclined orbit, which brings it far above and below Neptune's equator. Evidence points to a Titan formation in the conventional way: the moon appears to have been birthed within the cloud surrounding young Saturn, so it is likely that Titan formed as part of Saturn's protoplanetary cloud. But the Saturn system shows

(continued)

---

[25] Some theorists have even proposed that if Mars hosted a second genesis in its early days, current conditions would be largely hostile to any surviving microbes, and that humans may opt to intervene in Martian conditions to enable ancient Martian life forms to thrive again.

some odd traits. Titan's orbit is not perfectly round (it is "eccentric", as are some art directors). Its circuit around Saturn should have settled into a near-perfect circle as it lost energy due to tidal forces within its ice and surface seas, so the path it follows is a bit enigmatic.

Titan's remarkable atmosphere probably came early. The huge moon's interior may have pumped methane into its skies during three developmental periods. A large portion of Titan's construction materials came in the form of comets, which brought water and other ices, along with a plethora of organic material. In its infancy, as Titan gathered itself together within Saturn's own cloud, its rocky core amassed beneath a deep water mantle. Its core is estimated to be 3380 kilometers in diameter. A water-ice crust topped the mantle, trapping a subsurface ocean inside. During the first several-hundred-million years after accretion, heat from the moon's formation—along with the warmth of radiogenic elements in the core—combined to melt through the crust here and there, occasionally releasing methane.

The second major methane release probably took place two billion years ago, when Titan's silicate core became hot enough to convect. This geological surge of heat again melted the ice crust, causing more outgassing. Ammonia, salts and methanol mixed with the water ice would have helped to serve as a natural antifreeze, and may have triggered widespread cryovolcanism.

As Titan's geological violence subsided, a mix of methane and water ice would have formed a lattice, called a clathrate. This clathrate crust would gradually thicken above a layer of pure ice. Convection would have begun within that outer crust itself, freeing the methane trapped within as geyser plumes or gas leaking out of the ground.

The picture we get is of a geologically young and active world rich in water, methane and other volatiles, forming in the cold outer solar system. As Titan differentiates, rock and metal settle to the center, while water moves to the surface, forming an ice crust. Later, various processes—including volcanism and the breakdown of gases due to the Sun's radiation—lead to the nitrogen/methane atmosphere familiar today.

Within Titan's frozen mantle, pressures and tidal heating fashion a global ocean of water, locked beneath the ice crust above. Methane seas form on the surface, establishing a methane cycle buttressed by cryovolcanism. Meteors bring organic material and water to the surface and leave their mark beneath the thick atmosphere. But some mysterious event triggers a facelift of the planet-moon. Geologists studying Titan's land features tell us that something appears to have reset the geological clock, much as something caused a global resurfacing at Venus. Titan's face has the countenance of a planet only 500 million years old, a scant tenth of the age of the solar system. We do not yet know what that resurfacing event is, but from this convoluted history emerges the unique organic-rich world we see today.

Figure 5.19   The calm before the storm: artist Marilynn Flynn envisions a squall line advancing toward a river channel with cracked features along its borders. The seasonal river flows from the northern lake Jingpo Lacus. Its "crusty" banks may be ideal for the production of pre-biotic chemistry. Some Titan lakes appear to have outflow channels where hydrocarbon streams carry sediments, ending in alluvial fans. When the dry season arrives, stagnant shallows along the streambed may slowly dry out, leaving a crust of sediments and evaporated organics on the surface. (Painting courtesy and ©Marilynn Flynn)

(Find a brief history of Earth in Chapter 2, and one for Mercury at the end of Chapter 3. A brief history of Venus is found at the end of Chapter 4, and for a brief history of Mars, see the end of Chapter 6.)

6

Earth=Mars: a snowball world

It's been a billion years since the Earth emerged from the solar nebula. Sheets of ice blanket the rocky landscape, with flurries of snow drifting from an overcast sky. Lakes, rivers and entire seas hide beneath crusts of frozen water, hued in deep blues and greens. The scene is the Earth beneath a faint young Sun, but it could just as easily be the Mars of an earlier era.

Mars is a world locked in a profound deep freeze, an ice age billions of years in the making. Once the planet settled down into a stable world, it had periods of much warmer conditions than it feels today. The highest air temperatures on modern Mars top out around 20°C in its tropics, but temperatures that high remain close to the ground, where sunlight heats the surface and the surface, in turn, heats the air adjacent to it. An astronaut standing in summer daylight on the Martian surface would enjoy room-temperature shoes, but her helmet would work hard to shelter her head from temperatures dropping below the freezing point of water. And that's the situation during Martian highs. The polar regions drop to 140°C below zero in the winter, so cold that the air mutates into carbon dioxide snow and falls to the ground as "dry ice." The atmosphere on the red planet is as rarified as the air on Earth at a height of 30,500 meters. Any surface water would explosively boil into vapor, leaving behind only traces of water ice. Even the water ice can't last long: exposed to the low pressure, it will evaporate in years or centuries.

While diminutive Mars is too small to have held on to substantial atmosphere (due to weak gravity and a fading protective magnetosphere), it seems to have had the added problem of a lack of tectonics. But considering the

Figure 6.1 The "snowball Earth": 700 million years ago, our planet may have been covered from pole to pole by vast ice sheets. Here, coastlines of the supercontinent Rodinia are just visible through the planet's global ice covering. A band of open water and clear land along the equator begins the process of global melting, perhaps due to increased volcanic activity. The Moon's bright rayed crater Copernicus is very young, and Tycho has not yet appeared. (art by the author)

magnetic patterns in the Earth's sea floor—remember Mr. Wegener?—planetary scientists wondered whether any such record might exist on Mars. They got their chance to find out in 1997, when the Mars Global Surveyor became the first orbiter to map the Martian magnetic field imprisoned within the rusty landscape.[1] Within this crustal map, some researchers suggest that localized mirror images—similar to those on the sea floor of the Atlantic—exist in several areas across the globe (Figure 6.2). If so, Mars had Earth-like tectonics during its earliest epochs. That field is long gone. Its remnants remain within Mars' stony skin, but not everywhere. Several large impact basins disrupt the global pattern of magnetic fields, so the death of the Martian dynamo can be traced back to those events, yielding an estimate of more than 3.8 billion years ago. A second, faint burst of magnetic activity from the core may have taken place in the Hesperian period, which began 3.7 billion years ago.

But the story of Martian plates may not be so straight-forward, some research scientists contend. They assert that more recent maps don't show that pattern as clearly. The contours in later magnetic mapping do not follow the geometrical patterns, such as mirrored traces and arcs of great circles, that were established by the theory of plate tectonics for mid-ocean ridges. Some Mars experts conclude that there is no credible evidence for early plate tectonics on Mars.

The northern hemisphere has been resurfaced by lavas and, perhaps, long-standing water. The magnetic anomalies in the southern regions may indicate a buildup of the crust in stages, perhaps similar to that found in the lunar highlands.

Moving plates or not, the jumbled record of magnetic imprints within the Martian rocks tells a tale of dying magnetic fields. As we have seen on other worlds, a magnetosphere is important for a healthy atmosphere. Without one, a planet's airy blanket is open to erosion by the Sun's furious winds of energetic particles. Magnetic fields work in tandem with gravity and temperature (higher temperatures mean more movement of molecules in the atmosphere; too much movement, and those molecules ping off into space) to preserve an atmosphere. At some point in the past, Mars' core lost its protective ability.

The core of the matter

Ask any confectioner about a box of assorted chocolates: it's what's at the core that counts. The outsides all look pretty much the same. You have dark chocolate and milk chocolate, and some of the pieces have decorative swirls or blobs of marzipan on top. But the real story is beneath the chocolate shell: from maple crème to raspberry truffle, surprises abound. No planet has a chewy nougat

---

[1] The plucky orbiter continued to produce high quality science until its last communication in November of 2006, tallying an impressive near-decade-long mission of exploration.

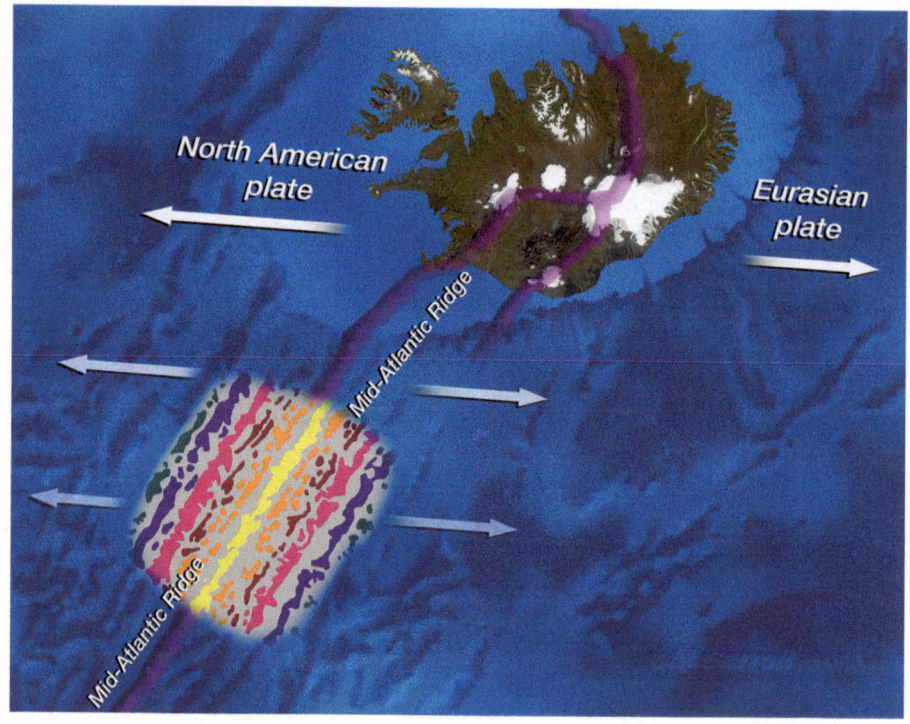

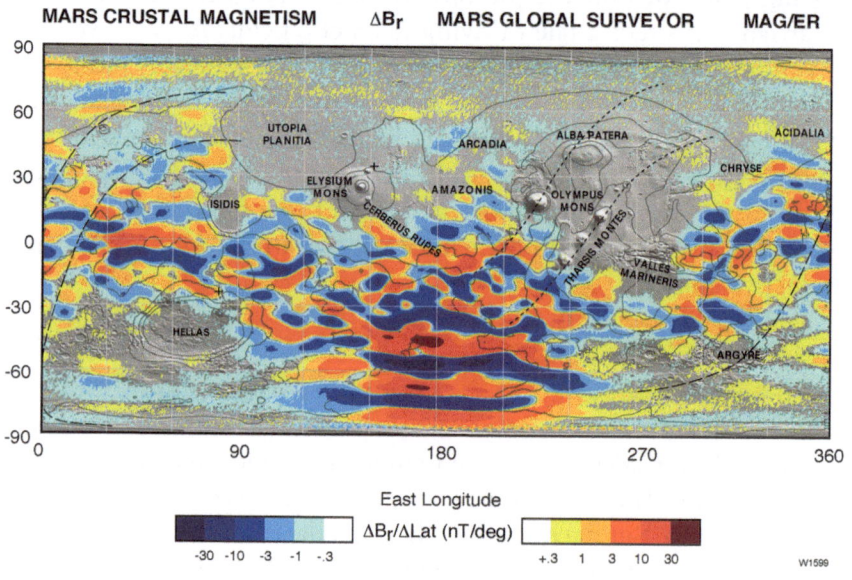

Figure 6.2 Telltale Martian plate tectonics? Maps of the Mid-Atlantic Ridge show a mirror image of magnetic direction locked into the rocks as plates spread and the Earth's magnetic field changes. Do Martian magnetized rocks show similar echoes of moving plates in ancient times, before plate tectonics—and an active magnetic field—shut down? (top: art by author; bottom: MGS magnetometer data courtesy Connerney, J.P. et al, NASA Goddard)

core,[2] but the interiors of worlds form a critical part of their essence. Where planet Earth is concerned, the soaring mountains, fruited plains, abyssal canyons and running rivers are all window dressing; it's what's inside that counts (although all that window dressing is an expression of what lies beneath).

Chapter 2 gave us a view to the process of differentiation, and differentiation is the crux of core formation. All the flotsam and jetsam of the solar nebula, the asteroids and comets and interstellar dust and ices that came together to make an infant planet, collapsed under their own weight, so that planetary interiors heated up. Metals melted and dribbled toward the center, while a dross of lighter material rose to form the crust. All four terrestrial planets have heavy cores surrounded by a lighter mantle, capped by a crust. Mercury, Venus, the Earth and Mars all went through the initiation of fire that is differentiation.

Earth's Core Values

It might give us a secure feeling to think of our planet's core as a stable, reliable center supporting the rest of the planet in predictable ways. But the center of our planet may be a bit more capricious. The Earth's solid iron core floats within a molten outer core. Some earlier models indicated that the core turned at the same rate as the mantle above it, dragged along within the molten iron "sea" around it. Other data pointed to a slightly faster rotation of the core, overtaking the layers above by about 1° every day (the Earth spins 360° each 24 hour period). In fact, the Earth's inner, solid core may have reversed relative to the rest of the world, completely pausing in its rotation compared to the mantle and crust above.[3]

The Earth's heart may not be finished with its aberrant behavior. The inner core's rotation may be continuing to slow, so that eventually it will reverse direction compared to the spin of the planet's outer layers. The changes in rotation may be part of a 70-year cycle that affects not only the length of day but also the magnetosphere.

The core had been rotating more quickly than the layers exterior to it, but in 2009 the inner core nearly came to a grinding halt. Scientists monitor seismic waves triggered by earthquakes. These waves move through the core, but recent readings show that over the past 30 years, the waves have taken different time periods to move through the Earth's center. In 2009, those differences vanished, implying that the core had begun turning in sync with the rest of the planet. Later, the differences appeared again in the opposite direction, suggesting that the core is now turning in the opposite direction to the mantle and crust. Further studies of records from Alaskan earthquakes seem to show another reversal of heart in the early 1970s.

Alternatively, some investigators propose that the inner core is stationary, turning with the rest of the planet. They assert that the differences in seismic waves come not from the spin rate of the core, but rather from changes in its shape.

---

[2] That we know of, but we're still exploring.

[3] For more, see *Nature Geoscience*, "Multi-decadal variation of the Earth's inner-core rotation", by Yang and Song, Dec 5, 2022

Half a century ago, researchers had pieced together a preliminary picture of Martian geologic history, and that picture painted the small red planet as differentiating gradually over a billion years. The thinking was that it would take that long for radiogenic heat to build up to the point where differentiation could occur. But in the 1980s, scientists began to sketch in a different portrait. The Viking landers sampled Martian air directly, and the isotopes they found enabled meteor specialists to determine that an entire class of rare meteorites on Earth had come from Mars. These precious chunks of Mars' upper layers contained lower levels of siderophilic—or "iron- associating"—elements than predicted. What this meant was that iron made its way down to the planet's core so quickly that a surprising amount of it escaped from the crust. Instead of a billion years, the process of differentiation probably took less than 50 million years. Mars matured early.

The Earth and Venus are large enough to hold on to primordial heat, and they have large inventories of radiogenic materials (like Uranium, thorium and potassium) to warm their interiors. This drives some of the active geology that we see etched upon their faces, and in the case of Earth, a molten core combined with our fast spin of 24 hours generates an energetic, protective magnetosphere.[4] Mercury, much smaller, cooled more quickly. As it did, it shrank and wrinkled, leaving behind its craggy rupes. Like Venus, it rotates in a leisurely fashion, and its smaller core can only produce an anemic magnetosphere (although compared to the size of the planet, Mercury's core is enormous).

Mars falls somewhere in between the Earth/Venus duo and Mercury (Figure 6.3). Its core spans about 3590 kilometers, nearly half the planet's diameter. This is much larger in relation to the planet than the Earth's is. But Mars is less meaty than the Earth, and its less dense nature gave its magnetosphere a shorter life span. Any global Martian magnetic fields—likely beginning as a smaller version of those here on Earth—died away, allowing the solar wind to strip its atmosphere. They remain today as shadows of their former selves, weak localized magnetism within the rock.

---

[4]Venus may well have a molten core, but its slow spin appears to preclude the generation of a magnetosphere.

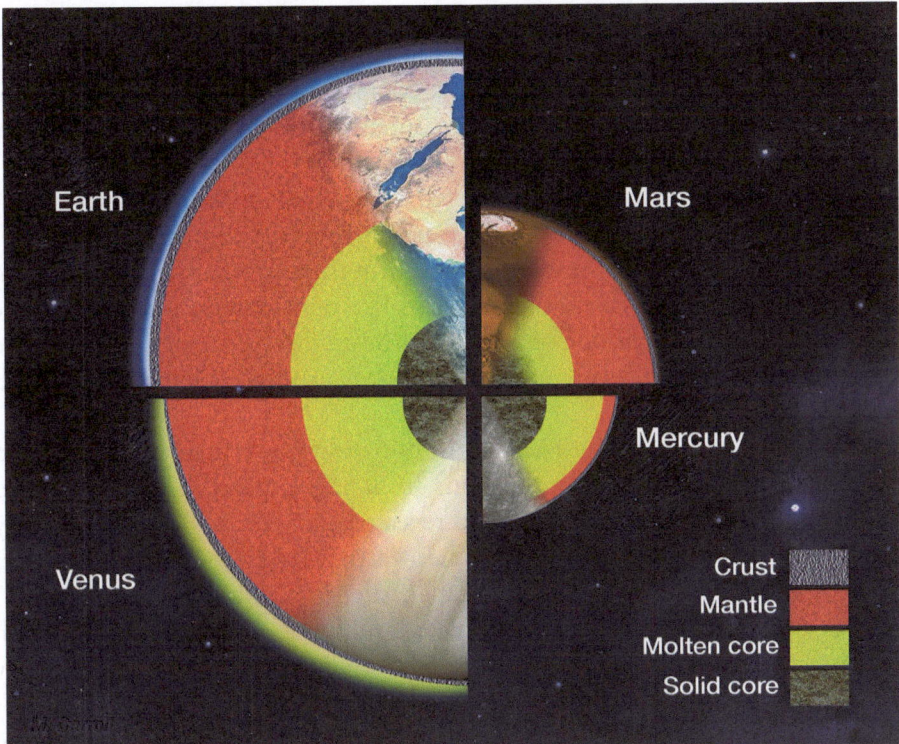

Figure 6.3   Cores of the terrestrial planets compared. Clockwise from upper left: Earth, Mars, Mercury and Venus. Note the large solid iron core of Mercury. Inner cores of Venus and Earth are thought to be similar in size, while the size of Mars' solid core is not yet well constrained. (art by the author)

How active are the Martian guts today? We have gained "insight" into the issue with the mission of a very simple Mars lander, a craft carrying a single, important experiment. The InSight spacecraft (for Interior Exploration using Seismic Investigations, Geodesy and Heat Transport) made landfall on Mars on November 26 of 2018. Aboard nestled a carefully crafted seismometer. Seismometers had been carried to Mars aboard the Viking landers, but Viking 1's failed to deploy, and Viking 2's instrument was overcome by vibration from winds for most of its operation. The location of Viking 2's seismometer on the lander's flight deck meant that it was better at detecting the lander's own operations (scooping soil, sifting samples, etc.) than it was at sensing Marsquakes. What investigators needed was a seismometer on the ground, and they got it with InSight.

In a sense, a seismometer can make a CT scan of the interior of a planet by mapping waves as they travel through solid and molten rock. Natural earthquakes generate the waves (or in this case Mars quakes). Data also comes from meteor impacts, which send a pulse coursing through the nether regions below ground.

Beginning in 2019, InSight detected six distinct marsquakes that consisted of a three-part pattern. The pattern fit several computer models predicting how waves would travel through differing internal structures, from solid rock to complex layers of material. The best match revealed that the Marsquake waves had to be bouncing off a boundary between a solid mantle and a liquid core. That boundary lies fifteen hundred kilometers beneath the Martian plains.

In October 2022, InSight detected the two most powerful impacts recorded on Mars in real time. The concussions from the impacts sent shock waves through the Martian crust. That data revealed that some Marsquakes likely come from magma flowing beneath the Martian landscape, despite the fact that volcanoes on Mars have likely fallen silent. Combining data from 20 quakes, a Swiss team has determined that liquid magma still gurgles beneath the surface of a region called Cerebus Fossae (Figure 6.4). The low frequency of the events is similar to that generated by magma flowing in the Earth's underground. A large magma chamber could explain the tremors coming from the region, and volcanic dust at Cerebus indicates active eruptions within the past 50,000 years.

InSight unveiled a planetary interior different from the one scientists expected. The Martian mantle is not deep enough—and its pressures not high enough—to create a separate lower mantle. On Earth, heated, dense rock in the lower mantle helps to insulate the planet's core. Because Mars lacks such a layer, its core likely cooled off more quickly. This rapid cooling may have helped heat migrate through the core during the earliest epochs of Martian history, generating currents that led to an early magnetosphere. Traces of that magnetosphere may be the patterns mapped in the southern crustal rocks by Mars Global Surveyor.

It would seem that after a geologically brief time, no tectonics sustained volcanoes to recirculate the atmosphere, and no magnetosphere shielded the gossamer gases from solar wind. At the cold outer edge of our solar system's habitable zone, Mars had no chance of retaining the kind of active surface biome that Earth possesses.

Figure 6.4 Cerberus Fossae is a region that may still be volcanically active today. Here, dark volcanic dust appears to issue from a 13-km long fissure. (NASA/JPL/MSSS/The Murray Lab)

And yet…the Martian lowlands have been engraved by a calligraphy of wandering rivers and braided flash flood plains. Rivers north of the equator all appear to have drained into a vast low plain girdling the northern hemisphere, with terraces and curved ridges as probable high-water marks of an ancient paleo-shoreline. Similar beach-like features encircle the great northern plains. The Mars Odyssey orbiter mapped the Martian terrain with lasers, and found an almost-unique profile stepping down from the equatorial and mid-latitudes to the northern lows. Only one other place in the solar system has a similar profile: the continental shelf along the Earth's oceans. Some researchers refer to this purported ancient sea as "Oceanus Borealis".

Some narratives show ancient Mars as a desert world visited periodically by temperate moments, transient warm spells that came and went like collegiate fads. Other researchers say that data favors a lot of water over long periods of time. Hints to the past climate may come from several sites imaged from orbit, areas which evidence the action of powerful tsunamis. Ancient tidal waves left their mark not once, but twice within several million years. Investigators at Cornell University[5] propose that two mega-impacts triggered the events as meteors landed in a shallow northern sea (one impact candidate is the crater Pohl). It was the Martian version of Earth's Chicxulub. In fact, the size of the 110-kilometer crater indicates that the impact was similar in extent to the Earth's Mesozoic-ending asteroid catastrophe.

The shorelines under study date to about 3.4 billion years ago (Figure 6.5). Liquid water from the first impact receded, leaving backwash channels as the tidal wave drained back toward the ocean to the north. The soggy event was no place for watersports. The impact would have mixed the seawater with mud and boulders, creating a deadly flow of debris.

After the first impact, the Martian climate became colder, and the ocean boundary pulled back farther, building a second shoreline. The rushing waves from the second event left a different signature: lobe-shaped mounds that froze in place. Because the ice was deposited in position for a long period without returning to the sea, it is likely that the water in the tidal wave was already partially frozen. The well-defined boundaries of the waves' farthest extent, along with the flow features preserved after the impact, demonstrate that the oceans of Mars were cold, briny deeps, environments thought to be conducive to life.

The coastlines where those proposed tidal waves crashed ashore are part of a larger formation that completely surrounds the northern hemisphere of Mars. The northern ocean mentioned earlier would have covered nearly a third of the entire planet. A study carried out in 2010 by the Laboratory for Atmospheric and Space Physics (Boulder, Colorado) found that seventeen river deltas form at an altitude consistent with an ancient coast. The deltas are part of a widespread network of now-dry river valleys that flow from the higher Martian equator toward the low northern plains (the area that would have been the ancient sea floor).

Features left by ancient water can often resemble lava flows and other volcanic features. But data from the European Space Agency's MARSIS radar system aboard the Mars Express orbiter indicated that the region consisted of low-density sedimentary layers or large subsurface deposits of water ice. Volcanic plains would have yielded a very different result.

---

[5] The Cornell team was led by Alberto Fairen from Madrid's Center for Astrobiology. Related work was carried out by a team led by Alexis P. Rodriguez of the Planetary Science Institute in Tucson, AZ.

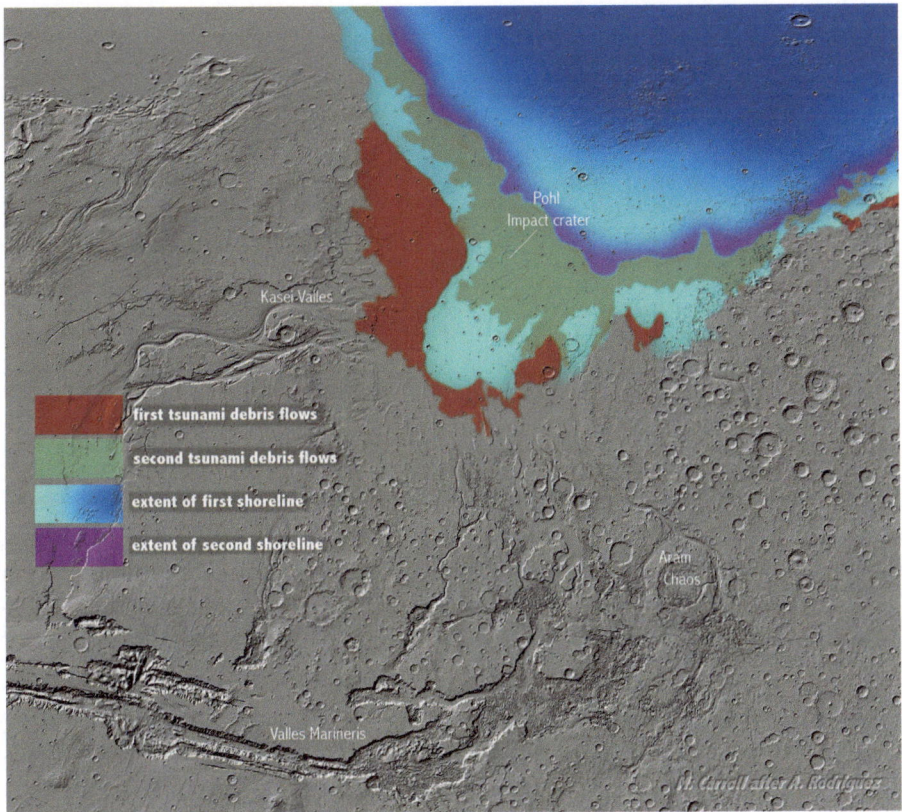

Figure 6.5 Evidence of two possibly impact-related tsunamis is mapped here. The first, occurring 3.4 billion years ago, is indicated in shades of blue. Its debris appears in red. The second impact took place after the ocean had receded northward (purple outline) as the Martian climate cooled. Its remnant debris flows are shown in green/tan. (author image after work by Rodriquez et. al. ESA/DLR/FU Berlin via creativecommons.org/licenses/by-sa3.0/by/igo; background shaded relief Mars map by NASA/JPL/MOLA science team)

Some prehistoric Mars rivers have been carved by flash floods, dramatic single events that probably had little long-term effect. But the valley networks toward the northern plains of Mars point to a more stable water cycle involving precipitation, runoff and condensation of ponds and lakes. Other areas on Mars point to long-standing bodies of water. Some features appear to be the remnants of beaches or lakeshores, where long-lived wave action has eroded the landscape in forms unique to lakes or seas (Figure 6.6). As JPL's Timothy Parker puts it, "Standing water forms a surface that intersects

Figure 6.6 On the plains of Acidalia, the Deuteronilus Mensae region hosts evidence of glacial activity, as well as features resembling long-term wave erosion similar to that expected from an ocean coastline. This dramatic oblique view was generated from data via the European Space Agency's Mars Express orbiter. (ESA/DLR/FU Berlin)

topography at a fixed elevation around the margin of a depression. Simply stated, modern lakes and ocean shorelines on Earth are level and possess attributes modulated by storms and tides." Skeptics of the Mars ocean theories point out that many of the proposed Mars shorelines are broken or vary with height. But Parker has studied ancient lake shores on Earth, like Lake Bonneville in Utah. He finds that "abandoned" or dried-up shorelines on Earth are rarely preserved at the same level, but rather have been warped or tilted by later tectonic forces. The same would be true, he says, for paleo-shorelines of primordial Martian seas.

The great impact basins of Hellas and Argyre also bear some inklings of perennial past water. "U"-shaped hanging valleys excavate the hollows between the mountain rings surrounding these titanic impact arenas, valleys typical of glacial erosion (rivers carve valleys with a "V" profile rather than a "U"-shaped one). On several slopes, rocks have flowed downhill in long trains, leaving boundaries similar to the lateral moraines at the borders of terrestrial glaciers. We find rock glaciers—glaciers buried in rock and sand—on the Earth. It may be that some of these apparent rock glaciers on Mars are still active, filled with flowing glacial ice protected by sand and gravel from the thin Martian

Figure 6.7 Mars Reconnaissance Orbiter photo of classic glacial formations located within the Cassius Triangle. At lower center, moraine-like linear features end in a rounded "tongue" common to many terrestrial glaciers. (NASA/JPL-Caltech/Arizona State University)

air. Glaciers require centuries or millennia of snow deposition, compacted over time into solid rivers of slow-flowing ice (Figure 6.7).

Mars gives up other clues to big past changes, and some of those changes are tied to modern phenomenon. For decades, researchers broadly agreed that Martian history involved water flowing across its surface. From small gullies to vast flood plains, the evidence was conclusive. But by the close of the 1990s, most concluded that any Mars waterworks had shut down billions of years ago. But closer looks by orbiters like Mars Global Surveyor changed our picture: intermittent streams may be flooding down crater rims and mountainsides even today. MGS located dry gullies carved into crater walls that appeared to fill with bright material. It was baffling. How could a deluge rumble down a slope when temperatures are well below the freezing point and pressures are so low that liquid water should explosively boil away?

Several major explanations have floated to the surface. The first concept is that hot springs or other heat sources warm underground aquifers to the point where they melt and sporadically bubble up to the surface. This idea has problems. It doesn't explain the kinds of places where the gullies are seen, places apparently unrelated to volcanic activity. And to make things worse, the gullies issue from solitary mountains and even on the flanks of sand dunes (Figure 6.8).

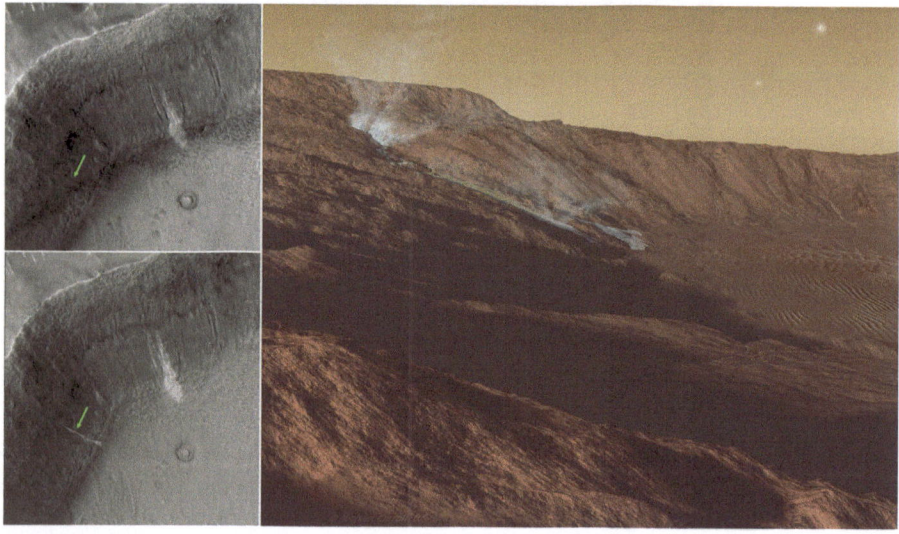

Figure 6.8 Left top and bottom: Mars Global Surveyor orbital images of a gully filling with bright material, perhaps water ice (green arrows). Gullies like this tend to form on crater rims or hills facing poleward, where sunlight may warm the ground to temperatures above freezing. (NASA/JPL/Malin Space Science Systems) Right: artist's concept of a Martian "gully-washer". (art by the author)

But what about at times of extreme obliquity, others ask. A second camp of researchers proposes that the flows formed back when Mars' axis was steeply tilted. Their idea is that air pressure at those times would have been great enough for liquid water to flow, as temperatures rose above the freezing point. In the summers, the sunward-facing slopes would warm enough for the ices beneath the surface to melt. The gullies do, in fact, seem to prefer the sun-facing sides of mountains and craters (and even sand dunes).

Yet a third proposal comes from observations in the Earth's arctic regions, where a protective cap of ice cracks, allowing water and loose rock inside to escape downhill. The debris slopes end in fans, much like those seen on Mars. But what would be the heat source? The mystery gullies of Mars serve to demonstrate that Mars and its alien geology are often baffling.

Farther north and south, the remarkable layered sediments at the Martian poles also bear testament to changes in the environment. Like the rings of a tree, annual cycles of dust storms and freezing events lay down alternating layers of dust and ice. Winds erode the surface, eating troughs into the polar cap's flanks and exposing an elegant spiral of laminar patterns radiating out from the north pole. The deposits alternate in light and dark tiers. They extend a thousand kilometers across. Radar studies from ESA's Mars Express and NASA's Mars Reconnaissance Orbiter reveal that the ice is up to 3000 meters thick. The patterns in images taken hundreds of kilometers apart repeat in matching arrangements, implying that the formations have a global, climatic cause and origin. Because the polar cap is smaller and more transient in the south, the southern ice archives are not as pronounced as those in the northern arctic. But in the north, we witness the full extent of dramatic global dust storms as they have played with winter ice and frost over the relentless march of seasons. Just how much history does the ice chronicle? Assuming that the northern polar cap built up gradually after the last extreme in the Martian obliquity (when the axis was tilted to its greatest extent), the ice record may represent five million years of time.

Antarctica's checkered past

The Earth's history has been recorded in the sands of the sea floor, in the annual deposits of coral reefs, in the rings of trees and in the layers of ice put down over millions of years. Snows in summer differ from those in the winter season. Under its own weight, the snow becomes ice. As it does, it traps particles, acids and gases from the air. Trapped air bubbles preserve conditions present when the ice formed, such as overall climate, local temperatures, volcanic eruptions, greenhouse gas levels, and even solar activity.

Across the face of Antarctica and Greenland, scientists drill cores through thousands of feet of ice, pulling out tubes that constitute a time capsule of frozen water and ancient material. The US, Russia, Japan, Denmark, Germany and France have all perfected deep core drilling. The longest ice core dug so far represents 800,000 years of Antarctic history.[6] It shows that carbon dioxide levels have risen and fallen with changes in climate, from 180 parts per million during ice ages to 280 parts per million through interglacial (warmer) periods. Today's levels are approaching 390 parts per million.

Sections of a 3.5-kilometer-long core from eastern Antarctica displays material from 330,000 to 450,000 years ago, a period that includes a full ice age and the warmer interglacial periods before and after it. Each sample point was separated by only 300 years, giving researchers a "high resolution" timeline for the samples. The samples showed eight distinct episodes when levels of carbon dioxide spiked, creating a pulse of heat lasting a century or more. While volcanoes can cause such spikes, researchers see no evidence of erupted dust in the core. Instead, they assert that the changes were a result of shifts in the global ocean currents. The Gulf Stream periodically weakened, bringing less warm water to the North Atlantic. In a chain of events, the ocean's surface temperatures changed, triggering shifts in the weather of the tropics. These changes, in turn, shrank wetlands. Vast areas of tropical plants died and decomposed, sending carbon dioxide into the atmosphere to warm the climate.

Rocks tasted, drilled, scraped and selfied by our rovers and landers sometimes have the profile of rounded cobbles, stones rolled over great distances by rushing water coursing through riverbeds (Figure 6.9). And their chemistries seem to indicate warmer, wetter periods. Some sites appear to be the remains of hot springs. Others may mark the edges of acidic ponds. Martian volcanoes might have brought thickening air and greenhouse warming at various times, but it is clear that the Mars of today is a different world than it was in the past. Percival Lowell and other observers in antiquity postulated that Mars was a desert world, withering away in a planet-wide drought. They supposed that intelligent beings—if there were any—might construct a global web-like

---

[6] The most ancient ice core from Greenland dates back 130,000 years.

Figure 6.9   NASA's Curiosity rover charted several locations that exhibit evidence of an ancient riverbank, like this one. This "sedimentary conglomerate" arose when cobbles were cemented together, likely at the edge of a flowing stream. The rounded shapes and sizes of the pebbles indicate this type of erosion. (NASA/JPL-Caltech/MSSS)

network of canals and rivers to bring water from the icy poles to the desiccated, red deserts. After all, some mysterious change took place on the planet as the seasons came and went: the poles shrank as they melted, and as they did, a wave of darkening spread across the face of Mars, as if the dark regions grew more defined with the growth of vegetation. The dark markings on Mars, intrinsically grey, take on a distinct green hue in many telescopes as their optics cast a blue tinge to the image.

At the end of the day, those dark markings turned out to be volcanic dust blown by seasonal winds, not long artificial waterways dug into the ancient plains of a dying world. But Mars really must have been a place of rivers, ponds and even oceans in the distant past. It's now an ice- wrapped rock. What happened?

We have seen that Mars lost its atmosphere over time because of a one-two punch: its low gravity could not hold on to substantial amounts of gas, and without an active magnetosphere, the Sun's radiation had the nasty tendency to strip away air molecules and cast them into space. Add to that the shutting-down of any tectonics that might recycle atmosphere from the rocks (through volcanism), and it wasn't looking good for a Martian paradise.

Scientists knew the planet lost its atmosphere early from inference and indirect evidence, but a Mars orbiter opened entirely fresh lines of evidence. NASA/Goddard Space Flight Center's MAVEN (for Mars Atmosphere and Volatile Evolution) settled into orbit in the fall of 2014. Its mission was—and is[7]—to chart the atmospheric loss at Mars. The planet has lost about 66% of its air over time, but the details of when and how were a bit fuzzy. MAVEN's reconnaissance cleared some things up. By comparing levels of two isotopes[8] of argon gas (argon[36] and argon[38]), researchers hoped to find out just how much carbon dioxide stole away into space over the Martian lifetime, and who the culprit was behind the theft. The lighter of the two versions of argon could escape into space more easily than the heavier isotope, and atmospheric physicists knew the beginning ratio ahead of time, so all they needed to do was add up what was left. A team of researchers at the University of Colorado at Boulder measured abundances of the heavier argon at the top of the atmo-sphere, and compared it to what had settled to the lower layers of air, charted earlier by the Curiosity Rover. Argon has the advantage of being a noble gas, which means it does not react with the Martian rocks as other gases (particu-larly carbon dioxide and oxygen) would. The only thing that could strip away the argon was the solar wind. Once the team had figured out the total amount of argon[36] lost to space, they could determine how much carbon dioxide and other gases were lost over time. The cargo of information beamed home by MAVEN indicates that the hydrological cycle on Mars was robust and active during its first few hundred million years, complete with rainfall and standing water. According to MAVEN results, the atmosphere declined gradually.

---

[7] As of this writing, MAVEN is still operational and continuing its reconnaissance.

[8] Isotopes are atoms of the same element with different atomic masses.

Astrobiologists find this to be an important point. They assert that if Mars had developed an active biosphere, the gradual departure of the Martian air may have given any microbial life[9] a chance to retreat into the recesses of the planet, where it may still reside today. The Martian magnetosphere appears to have died at about the billion-year mark, and Mars began to hemorrhage its atmosphere on a grand scale. Today, Mars is losing about 100 grams (3 ½ oz.) of air every second (MAVEN scientists compared it to losing one hamburger's worth each second). It seems a small amount, but it adds up to nearly 20,000 pounds of atmosphere each day. The weight of lost gas is equivalent to 1625 average SUV's every year. At current rates, Mars will be airless in two billion years.

It appears, then, that Mars had an edenic era, a time of warm days and cool nights, of raging floods and undulating rivers, and perhaps even gentle summer rains. But the time came to an end. The water evaporated or froze beneath the surface. The land became a dry, dusty desert as volcanoes quieted, the Martian rocks reacted with the atmosphere and "rusted", and Mars marched from summer rains to wintry snows of carbon dioxide.

Martian Reality

craters

Although Mars has no canals or ruins of ancient civilizations, the real Mars is a remarkable place in its own right. Like many other solid bodies in the solar system, from the terrestrial planets to asteroids to ice moons of the outer system, Mars displays a record of meteor, asteroid and comet bombardment. But unlike the Moon or Mercury, Mars has a dramatically lopsided crater covering. In the northern hemisphere, vast regions seem to have no craters at all, while in the south, the most ancient and well preserved impact scars lend a ruggedness to the planet. This contrast between hemispheres of a planet or moon is called global dichotomy. We find it on Earth's Moon, where vast dark lava plains cover much of the Earth-facing hemisphere, but are largely missing on the far side. In the case of the red planet, the dichotomy is between the north—with rare and heavily eroded craters—and the south, where craters are not only frequent but also are better preserved. The difference in crater density gives geologists a powerful tool. Since, in general,[10] all the planets and

---

[9] Or Martian mammoths, for that matter.

[10] Some data suggests that the rate and nature of meteor impacts in the outer solar system differed from that of the inner terrestrials.

moons have been subjected to the same bombardment of asteroids and comets, the number of craters shows the relative age of a surface. The heaviest rain of stone and ice probably tailed off about 3.8 billion years ago, as we noted in Chapter 2. Craters that form today trickle in at a slow rate, and most are small. The wandering giant planetesimals have all been gobbled up into planets and moons (as happened with Theia), or cast completely out of the solar system. Moons with little or no active geology have so many craters that any new impacts wipe out as many craters as they create. This state is called crater saturation, and we find it on the most ancient surfaces. For example, the Earth has very few craters due to its geologically young surface. Weather, volcanism and tectonics erode them away. Saturn's airless and geologically quiescent moon Mimas is covered in craters, many on top of one another. It is physically impossible to add more craters to it without wiping out others, so its total number remains the same. It is saturated.

Another time scale that craters offer has to do with the total number on any given surface. Surfaces with many craters are geologically older (less changed) than surfaces with few craters. Geologists can estimate the relative ages of terrains by comparing their crater counts. Crater counts afford an accurate measure of a given surface. On Mercury, for example, flood plains of lava are lightly cratered, but the surfaces they flow across may be nearly crater-saturated.

Surfaces give up clues to their ages by the preservation ("freshness") of their features, along with how those features relate to each other. If a channel cuts through a crater, the channel is younger. If, on the other hand, a crater has obliterated part of a valley, the crater came after the formation of the valley. Craters often overlap each other, providing another way of determining the order in which they arose.

Due to its history and makeup, Mars displays some unique craters unlike those on other worlds. Because the planet had (and my have) frozen or liquid water under the surface, strange formations accompany many impact features. The ejecta surrounding rampart (or "splosh") craters looks like a slurry of melted chocolate. When the impactor landed, it melted subsurface ice into a wave of liquid mud.

Close relatives of rampart craters are known as pedestal craters. Pedestal craters and the fan of ejecta around them stand above surrounding plains. The crater-forming impact hardens the material that creates the crater and its ring of disgorged material, so that the surrounding (softer) ground erodes away, leaving the crater raised up on its own pedestal.

Many crater rims on Mars have been breached by water floods (though we find similar forms on Mercury or the Moon where lava flows, rather than water, break through crater rims). Evidence in the lay of the land points to long-lived rivers or flash-floods rupturing the rims of craters and spreading through the interior in the shape of deltas. Jezero Crater, site of the NASA Perseverance Rover's explorations, is such a crater. Jezero has been around for 3.8 to 3.9 billion years. It's 40 kilometers across, and was once filled with water. A delta rich in clay fans out across its floor. From patterns within and outside of the crater, it is clear that water filled Jezero to the brim at one time. NASA engineers dispatched another Mars rover, Curiosity, to the Gale Crater, which was also the victim of past flooding. Other dramatic flood scenes lie within the craters Eberswalde, Holden, and Morella. The latter crater held a lake equal to the combined volume of Lakes Erie and Ontario on Earth.

Another type of uniquely Martian crater is called *exhumed*. In an old crater field, dust and ice blankets the surface, burying the craters and softening the details of the countryside. But as climatic changes come to the planet, underground ices loosen the surface material as they sublimate. Wind and temperature cycles erode the carpet of dirt and rock, gradually revealing the long-buried crater.

We tend to think of the planets and moons today as stable, quiet places. But orbiters have continually observed Mars since 1997, and they have found numerous newly-formed craters. Over that time, orbiters have witnessed the appearance of over a thousand freshly-blasted craters. Some leave dark halos of material around them, while others excavate fresh ice from underground. Similarly modern cratering events have been spotted from lunar orbit as meteoroids continue to pummel the Moon. Engineers continue to draw plans for human lunar or Mars exploration, and the frequency of craters gives mission planners pause. The fresh impacts demonstrate how dynamic our solar system still is.

Volcanoes

The largest volcanoes in the solar system rise from a plateau called the Tharsis Bulge. The leader of the pack, Olympus Mons, is a shield volcano nearly three times the height of Mount Everest, even taller than Hawaii's massive Mauna Loa (as measured from its base on the ocean floor.) At its summit spreads a great caldera 80 kilometers across, exhibiting layers of solidified lava pools from multiple eruptions over its long lifetime. The volcanoes of Tharsis may have erupted for more than a billion years. Terrestrial volcanoes like Vesuvius or Mauna Loa eventually shut off because of the movement of the plates on which they rest. The magma stream gets pinched as the rocks move laterally. The entire Hawaiian Island chain issued from one hot spot on the Pacific

Ocean floor. As the Pacific Plate drifted across the hotspot, volcanoes continually broke through, building islands and then falling silent as the plate continued to move. The most ancient of Hawaii's islands have eroded to the point where they are submerged. The "big island" of Hawaii is the newest to break the surface, and the most active volcanically. An even newer island, Loihi, is building on the sea floor to the southeast (it is currently a thousand meters below the surface, and is estimated to make its first above-water appearance between 10,000 and 100,000 years from now). But on Mars, no such plate movements transpired to shut off the volcanoes, so they continued to build into behemoth mountains. The Martian massifs are so huge that they help to govern the weather. Wind flows around them, and water-ice clouds form on their flanks. The trains of white mist can be seen in even moderate telescopes on Earth. In fact, before spacecraft revealed its true nature, the cloud-collared Olympus Mons was called Nix Olympica, the "Snows of Olympus".

In yet another hint of conditions in the Martian past, some volcanic shapes resemble mobergs, flat-topped volcanoes that erupted under glaciers. Their forms certainly imply vast areas of standing glaciers in the past. The Tharsis and other regions also exhibit some flow features that may be ice-related, but lava flows take on very similar forms.

Caves

On the flanks of some volcanic mountains in the Tharsis region and in the volcanic province Elysium, skylights in the roofs of underground tunnels punch holes through the surface, creating entrances to caves (Figure 6.10). A lava tube forms when an underground river of molten rock cools and drains away, leaving a rounded tunnel beneath the surface. Sometimes these structures collapse, but at other times, they remain intact except for an occasional localized cave-in of the roof. These events leave roughly circular skylights. Lava tubes have also been found on the Moon (none have been found on Mercury, where volcanism leaves behind lava flows of a different form). The Martian and lunar conduits are three times the size of those on Earth.

Mars has many other features related to past volcanism. In addition to Tharsis, another great volcanic province rises on the other side of the planet. Half way around the globe, Elysium hosts several large volcanoes. Other volcano-related features pop up all over the planet, including fractured crust (from magma rising underneath), cinder cones, and lava-flooded plains. In the stories of other planets, we have seen how volcanoes can affect atmosphere and climate. The same is true for Mars. Between the ages of ancient, wet Mars

Figure 6.10 Left: This open pit lies within an area etched by linear troughs, the collapsed remains of lava tubes. Many appear as dark, featureless holes, but if the sun slants at the right angle—as in this image—or if rock from the failed roof builds a pile of debris under the opening, detail within the cave will be visible from the prying eyes of orbiters. (NASA/JPL/ASU) Right: Astrobiologists would very much like to explore caves as possible Mars habitats for indigenous life, as envisioned here by space artist Pat Rawlings. (Painting courtesy and copyright Pat Rawlings)

and the current dry desert world, volcanic eruptions may have warmed the climate locally or globally, depending on their severity.

Dust and dust devils

Martian volcanoes are tied to two other important features: canyons and atmospheric dust. As we'll see below, the volcanic province of Tharsis pulled magma from the adjacent plains, splitting the surface and creating the largest canyons on the planet. But the volcanoes also generate the darker Martian fines and loft them into the sky. Most Mars dust is bright orange to tan, but the volcanic activity of the past (and possibly present) generates darker dust. Fine, dark volcanic dust is partially responsible for our myths, literature, poetry and movies about Martians and their canals. In the 1950s, Ray Bradbury described two Martian inhabitants who lived by a fossil sea: "Once they had liked…swimming in the canals in the seasons when the wine trees filled them with green liquors…" Edgar Rice Burroughs had his Princess of Mars, and H.G. Wells cast his Martians as villains in a cautionary tale of hostile beings from a dying, canal-covered world. The volcanic dust that astronomers mistook for linear features darkens and fades seasonally. Suspended in the thin air, it tints the sky a tawny pink until sunset, when the fines scatter sunlight, creating an eerie blue twilight. Annually, dust storms visit the desert world, and often they become global tempests, cloaking the planet in floating powder for months at a time (Martian dust is the consistency of talcum powder).[11]

Recall that the Martian surface is dramatically warmer than the air just above it. This temperature differential sucks warm air upward toward the cooler, lower pressure regions above. A column of air forms and begins to spin, pulling dust with it from the surface. Dust devils form on Earth using the same dynamic, but on Mars they grow to hundreds of meters across and many kilometers high. They appear most often in early morning hours during mid-summer in both hemispheres, when the solar heating is at its strongest. Their wind speeds can reach 100 km/hr. Orbiters have observed and clocked them from orbit, and several rovers have glimpsed and photographed them. The Pathfinder lander's weather station actually monitored several as they passed directly over the craft. Despite brisk winds, the air is so thin that even the largest of the dust devils would not cause any damage. In fact, on the solar powered Spirit and Opportunity rovers and the Phoenix lander, dust devil and wind events actually cleared solar panels, enabling the robots to carry on for extended missions.

For the artist in us, these miniature versions of tornadoes leave an elegant filigree across dune and plain, as they clear dark trails against the dusty Martian surface. Looking like a Tim Burton design, the curls and zigzags take on whimsical forms, designs scoured into the desert Mars scapes. They are transient masterpieces, erased by the next local dust storm. That is part of their charm, just as its fleeting nature brings beauty to each sunset.

Canyons

The volcanoes have created equally dramatic collateral effects across the russet landscape. The canyon system of Valles Marineris, one of the largest geological features in the entire solar system, fractured and spread as the Tharsis volcanoes pulled magma from the understory of the planet. Canyons and fractures are also associated with other volcanically-related regions. As we've seen, floods of water or lava have carved other canyons.

---

[11] The ubiquitous Martian dust is responsible for killing many solar-powered landers and rovers, and may have been the downfall of the Soviet Mars 3 lander, the first probe to survive all the way to the surface.

Snowfall's fingerprints

Blizzards of dry ice fall upon the polar caps—the Phoenix lander's LIDAR (Light Detection And Ranging) laser system detected precipitation—and water ice condenses on the sand and rocks at some latitudes. The desiccated landscape collapses to form pits and valleys, eroding into exotic forms like "brain" and "fingerprint" terrains. Brain terrain is also called brain coral terrain, due to its resemblance to the aquatic life form. It arises on debris fills within craters, valleys and alluvial fans. The convoluted surface likely overlays subsurface ice.

The bizarre fingerprint terrain is relegated to the southern polar cap (it is not found anywhere else on the planet). Curvilinear, slender depressions form as carbon dioxide ice sublimates, collapsing the surface.

Spidery geysers

Yet another unique feature on the Mars geology smorgasbord is a mystifying set of transient formations called "spiders" (Figure 6.11). The dark smudges

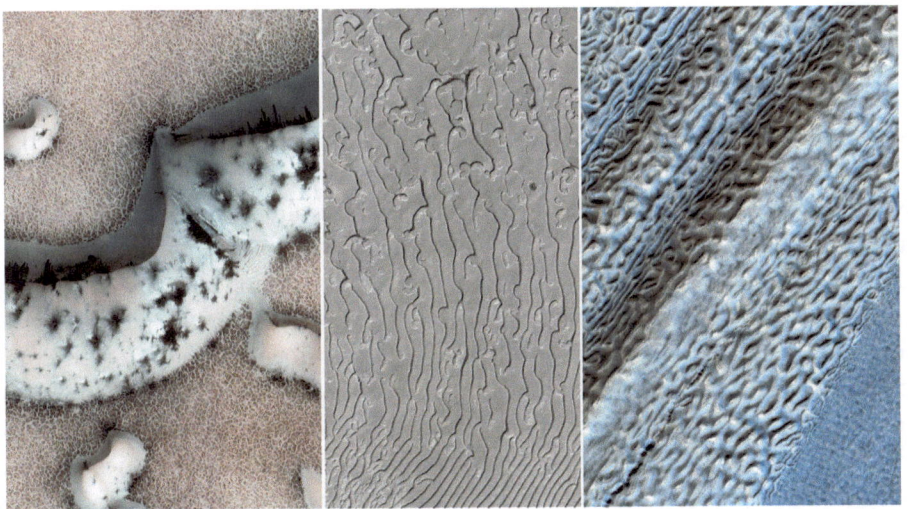

Figure 6.11 Left: Martian spiders—the stuff made of sci-fi "B" movies—appear on the face of a frosted dune. As carbon dioxide ice sublimates, freed gases explode into the air, casting dust in a spray reminiscent of a spider's legs. Note the polygonally patterned ground beneath the dunes, a pattern common to ice-infused tundra regions on Earth. Center: Fingerprint terrain found in the southern Martian arctic. Right: exotic brain terrain, here in false color. (all HiRise images courtesy NASA/JPL/University of Arizona)

spread across the frozen wastelands near the Martian south pole, and they appear seasonally. While carbon dioxide is the chief component of Mars air, winter temperatures are so low that some of the atmosphere condenses on the ground as frost. This sets up a carbon dioxide cycle between atmosphere and polar caps. With the spring thaw, the surface warms. Observers theorized that the carbon dioxide frost vaporizes, forming jets of gas that spew dust into the air. This substrate of material is darker than the bright surface dust, and overlays it in gray/brown rooster-tails of material.

Laboratory studies confirm that frozen carbon dioxide is to blame. Dry ice beneath the surface heats up as winter draws to a close. It does not melt, but sublimates directly from ice to gas. A carbon dioxide gas pocket builds up, eventually breaking through the surface and geysering dark material onto the surface in a spidery, radial form. Theory was confirmed in the laboratory when a team of scientists at Ireland's Trinity College simulated Martian conditions of pressure and temperature. They placed a block of dry ice, drilled with several holes, into the Mars simulation chamber and watched what happened as temperatures rose. The resulting vapors mimicked the arachnid-like shapes seen on Mars.

Dunes

A familiar, quite Earth-like natural marvel abounds on the red planet: sand dunes. A collar of dark dunes encircles the north polar cap, and dunes march across plains, filling canyons and covering crater rims. A team of scientists tracked some 500 dunes in images taken over the course of several years by Mars Reconnaissance Orbiter. They found that the alien environment of Mars affects the piles of sand in unique ways. Like the Earth, Mars has winds that drive and sculpt the sand. On Earth, dune movement can be retarded by plants or by ground water near the surface. But because the air pressure is so low, the Martian dunes behave differently. They move in slow motion compared to their Earthly counterparts, but a few break the speed limit. In several locations (including Syrtis Major, the Hellespontus Mountains west of the Hellas impact basin, and across regions near the north pole), dunes pick up speed. The mystery movement may be due to abrupt changes in topography and in temperature, both of which are exaggerated in these regions.

Polygonal ground

While the polar caps are the most obvious expression of frozen water, subsurface ice has also sculpted the landscape. The Viking orbiters of the 1970s

Figure 6.12 Top: Polygonal ground similar to that occurring in the Martian arctic can be seen near Kotzebue, Alaska. (photo by the author) Bottom left: Polygonal patterned ground on Mars indicates the presence of ice near the surface. While some polygons are found in ancient terrain where ice has long since disappeared, the patterns at right are generated by active ice layers just underground. Bright ice fills depressions surrounding raised dirt mounds containing frozen water. (NASA/JPL-Caltech/University of Arizona). Bottom right: rocket plumes from the Phoenix lander exhumed pure water ice just beneath a layer of surface soil. (NASA/JPL-Caltech/University of Arizona)

hinted at patterned ground in several areas, but most noticeably between 60° and 80° north latitude. The textures of the surface mimicked polygonal ground common to tundra in the arctic regions of Earth (Figure 6.12). As higher resolution images came in from increasingly capable orbiters, the patterns came into focus as virtually identical to terrestrial counterparts.

The difference: many of those on Mars are ten times the scale.[12] Still, the features are unmistakable.

Polygonal ground forms when ice-infused dirt expands and contracts as the ice increases and melts or, in the case of Mars, sublimates. The polygonal ground sites were some of the first clues that Mars surfaces have contained ice in the past, and geologists suspected that some of that ice remained to the present. But the polygons take time to develop, and many seem to document earlier epochs in the planet's past. Polygonal features exist in ancient regions, documenting past climate conditions very different from those today.

Not all of the patterned ground is ancient, however. When the Phoenix lander set down in the Martian arctic, its rocket nozzles blew surface dust from a layer of nearly pure water ice, showing that water remains near the surface in Mars' polar regions. Spectrometers aboard the Mars Odyssey indicate that some 70% of the surface north of 60° north latitude is marinated in water ice to a depth of at least a meter. The Phoenix lander carried an arm equipped to dig into the soil, and the tool uncovered deposits of frozen water near the surface. In images taken hours or days apart, white ice deposits within trenches can be seen dissolving in the thin Martian air, a testament to just how much water may be sequestered under the surface of the desert world.

Chaos regions

Some areas on the red planet seem to have subsided and collapsed, likely due to the retreat of subsurface ice or liquid water. One such ragged spot, Aram Chaos, is the subtle remnant of a primeval crater that has broken into jumbled blocks and hummocks. In the case of Aram, the crater likely hosted a long-standing lake earlier in Martian history. Over the course of a few million years, wind-blown debris settling into the lake solidified, becoming sedimentary rock. As planetary climate change progressed, a layer of permafrost built up. Something triggered a catastrophic melting of this ice (perhaps volcanism or a warming period). The ground collapsed, and water seeped to the surface to form a new lake. A narrow passage on the eastern side of the crater is all that's left of the floods that drained Aram's water away. Many such chaotic regions may have formed in similar ways.

---

[12] One explanation for the difference in scale may be that Martian seasons are twice as long as those on Earth.

Sandy stains

Other evidence of current Martian water activity might be found in several sites where dark, branching stains appear to move down slopes. These forms differ from the gullies we saw earlier, as they are darker than the surfaces upon which they spread. One such site is Newton Crater. There, spring and summer temperatures usher in the strange activity within the Martian regolith. The dark markings span from a half to five meters/yards wide, and they appear on steep slopes at several locations in the southern hemisphere. The features, called lineae, darken as they grow, and then fade again with colder weather in the autumn. They congregate on sun-warmed, equator-facing surfaces where temperatures rise well above -30°C, at times topping out at 27°C. Briny salt-water may explain the features.[13] A second possibility is that fragments of carbon dioxide ice may glide down the slopes, digging out the grooves as they coast on a thin film of sublimating gas. Yet another possible cause could be avalanches of dry dust, similar to those frequently seen on the face of Earth's sand dunes.

New data returned by China's Zhurong[14] rover found evidence of geologically recent water on dunes at its landing site in Utopia Planitia. Surface crust, cracks, granulation, polygonal ridges and what's called "strip trace" features all point to saline water gathering when frost or snow falls on the salty dune surfaces. The water may have accrued at the low-latitude site as water vapor from the poles moved toward the equator during one of Mars' extreme obliquity phases.

Today, Mars stands as a uniquely beautiful world in its own right. But we are still left wondering: how did we get to the Mars of today, a world so different from the comparatively temperate, watery planet of past ages (Figure 6.13)? The astonishing transformation of the red planet was forced by an assortment of contributing factors, but one factor may have trumped them all. At Titan, we saw how long-term climate changes result from long period cycles in a planet's spin and orbit. Could such cycles be responsible for Mars' fall into the deep freeze? Let's explore the cosmic conundrum.

---

[13] McEwen et al. Aug 5, 2011 *Science*
[14] Science Advances, April 28, 2023.

Figure 6.13 The Viking missions provided much proof of a warmer, wetter past Mars. If the Martian climate was once much more life-friendly, might it one day return to such a state? And if microbial or more advanced life lies dormant beneath its rock and ices, might the Martian biome be resurrected by the change in climate? Inspired by the data, an artist visualized a biome awakened from a long hibernation, with a Martian bird using the ancient Viking Lander for nesting material. (painting by the author, ca. 1977)

Creating climate change on Mars

"Red sky at morning, sailors take warning." The old adage is true: a vibrant sunrise may indicate the passing of a high-pressure system,[15] heralding tempests to come. The skies of Earth hold many clues to upcoming weather, but climate is another issue altogether.

For years the popular press equated climate change with global warming. While it is true that our climate is changing with an overall rise in global temperatures, climate models predict a complex set of changes beyond those of higher mercury in the thermometer. In fact, climate change entails shifting in the *patterns* of the world's weather from temperature to wind direction, from precipitation to storm frequency. Ironically, a global increase in temperature

---

[15] Unless, of course, you live on Mars, where every morning sky is red.

may mean more snowfall or extreme temperature swings—both up and down—in certain parts of the world. Climate is the long-term trend in our environment, while weather is the day-to-day behavior of temperature, rainfall, etc. As the saying goes, "Climate is what you expect. Weather is what you get."

Arguments arise as to the degree and causes of various aspects of climate change, and we'll visit some of these in our last chapter. But comparative planetology has much to teach us on the subject. We have seen how seasons on Titan govern methane rainfall or drought on the huge moon's exotic environment. Mars is another case that will give us insights into our own world. The red planet's river valleys, chemistry and beach-front property all speak to a drastically different climate than the cold desert we see today. Mars appears to be in an extended ice age that has lasted for the past five million years. The Martian axis puts a new spin on the subject.

Part of the reason for all this drama is something called precession. The axial tilt of the Earth today is 23.5° (compared to Mars' 25.2°). This tilt—called obliquity—engenders seasons and governs our meteorology. But axial spins are slippery things. Mars has a similar axial tilt to the Earth's, but within 50,000 years, Mars will be tipped over nearly on its side, by as much as 60°. A child's spinning top on a table rocks back and forth with a *precessing* motion. Planetary axes precess, too. As a world spins around once each day, any inhabitant would not notice the movement. But over millennia, the planet will slowly work its way over as its spin axis wanders. This shifting tilt has consequences. The most extreme example of a tipped axis in our solar system is Uranus, which is currently leaning over by 98°. This means that twice each year, its pole is pointed directly at the Sun, with the other pole pointing away from it (as if the planet is rolling around the Sun). This causes strange days with the Sun appearing to spiral around one spot in the sky. When the planet works its way half way around the Sun, the equator becomes illuminated in a more conventional day and night cycle before the planet once again returns to its extreme day and night.

The Earth's Moon is heavy enough that it stabilizes the precession in our spin. The result is that changes in our climate and seasons are more gradual and will never be as extreme as those conditions on Uranus or Mars.[16] Changes still come. The Earth's ice ages have been linked to the relatively subtle precession of our axis. But the Martian axial tilt has ranged from 0° to 60° over the

---

[16] The tilt of Uranus is likely not due to Croll-Milankovich cycles or precession, but rather may be the result of a major impact that tipped the planet over permanently.

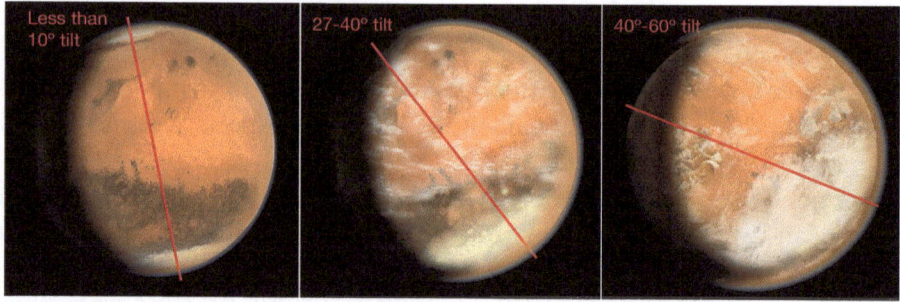

Figure 6.14   The Martian spin axis wobbles back and forth over a 124,000–year period as it precesses. Left: at obliquity of less than 10°, a permanent carbon dioxide ice cap (beneath a water-ice one) sequesters much of the Martian air. Center: the permanent carbon dioxide cap vaporizes, replaced by a larger seasonal one. Permanent water ice caps remain. The air thickens, and clouds and dust storms become extensive. Right: At Mars' most extreme obliquity, vast ice caps come and go seasonally, and the permanent ice caps are gone. Glaciers form near the equator beneath a dense, dusty atmosphere. (Mars base modified from image taken by ESA's Rosetta spacecraft; source: http://www.esa.int/spaceinimages/Images/2007/02/True-colour_image_of_Mars_seen_by_OSIRIS, art by the author, after Forget, Costard and Lognonne)

past 100,000 years, triggering more radical ice ages and alternately pumping up and depleting the entire atmosphere. Mars is cold enough that carbon dioxide—which makes up essentially all of its air—freezes into dry ice. As winter comes to the north or south (and especially in the south where temperatures fluctuate more), the Martian atmosphere actually gets thinner as carbon dioxide freezes to the polar caps.

Added to the precession effect of the axis is the egg-shaped Martian circuit around the Sun. Its eccentric orbit moderates temperature swings in the north, and exaggerates them in the south. But it doesn't stay that way: each 171,000 years, the extremes of the orbit's oval change with the effects of nearby Jupiter and the gravitational sway of other planets. Mars' course pulsates from nearly round to oval, so that at some times it is far from the Sun at winter solstice (accentuating low temperatures), and at others it is nearest at that time (moderating the winter seasons). The orbital cycle eventually ends up with reversed extremes of temperature from what we see today. Where wintry temperatures rule, south becomes north and north becomes south (Figure 6.14).

Three stages of Martian climate

Stage 1

The combined cycling of Mars' out-of-round orbit, along with the long-term swaying of the Martian axis, is called the Croll-Milankovich cycle (introduced in our Titan section). When the Martian obliquity (tilt) is small, seasonal changes are more subdued.[17] Less sunlight falls upon both poles, as the Sun is always low in the sky from the viewpoint of the top and bottom of the planet. Since temperatures in both the north and south are lower, more carbon dioxide freezes there, trapped within the polar ice caps and in the soil beneath, causing the global air pressure to drop. The thin Martian atmosphere would clear of dust, as weather would calm. The color of the Martian sky would shift toward a more earthly blue tint, perhaps turning to a metallic grey by whatever dust remains in the air.

Stage 2

When the red planet tilts farther over (with an obliquity of more than 27°), the seasons are more marked. As the poles warm during each of their summers, carbon dioxide is freed from its polar prisons. The amount of gas in these reservoirs is not certain, but some scientists estimate that the Martian atmosphere may increase by up to twice the current density. That is still low by Earth standards, but enough to dramatically increase the circulation of winds across the planet. Seasonal dust storms would warm the air, and water vapor clouds would form frequently, strengthening Mars' anemic water cycle.

Stage 3

At the extreme of the cycles, the Martian environment would vaporize most of the water at the poles. That moisture would migrate to the warmer equatorial regions, where it would precipitate as something quite familiar on Earth: snowfall. The snow would form glaciers in the high country. As mentioned above, we have, in fact, seen evidence of such glaciers, notably on the flanks of the great Tharsis volcanoes (including Olympus Mons) and in the mountains adjacent to the eastern rim of Hellas Basin.

Today, Mars finds itself as freeze-dried as an arctic raisin. Its obliquity lies between the first and second stages mentioned above. But the planet may be pulling out of its current cold spell. Studies of the southern polar cap reveal interconnected pits where the ice has melted away. Images taken years apart show a steady expansion of the melted areas. Mars may be undergoing a global warming epoch of its own.

---

[17] Seasons are caused by the tilt of a planet. For example, Jupiter's obliquity is only 3°, so it essentially has no seasons.

If wobbly Mars had a little more air pressure and a bit more liquid water (as it did in the past), the planet would have become a permanent "snowball" world sheathed in ice from pole to pole. As it turns out, there is a planet that has gone though such a metamorphosis. You are sitting on it.

The Earth's environment has endured dramatic changes throughout the planet's history. One of the strangest transformations, which took place on more than one occasion, saw the freezing over of the planet's oceans as glacial ice encased the continents, painting a sterile, frigid landscape much like that of Mars (Figure 6.1). It's called the "Snowball Earth", and, as mentioned in the last chapter, it was bad news for any heat-loving creatures. More than a few life forms were lost to the snowball events.

The planet's historical record seems to have taken a break at the same time. History is a loose-leaf notebook. A few pages have fallen out or faded. Some of its pages have been shuffled through tectonics. And an entire critical chapter seems to be missing.

First hints of "snowball" Earths

Our first hints that something was not quite Kosher with our narrative of the Earth's history came from something that wasn't there. A billion somethings, in fact. A billion years is a lot of rock layers to go AWOL. But that's about what seems to be missing. Some 5 kilometers of vertical rock has vanished at a place in the Earth's vertically layered sedimentary history. Geologists call it the "Great Unconformity." (Figure 6.15) The gap representing the missing billion years can be found in sediments across the globe. At the top of the layers leading to where the rocky record is missing, the stone chronicles a time some 700 million years ago, when life on Earth was single-celled and simple.[18] The most ancient—and deepest—layers of sediments end at an unusual level containing tillites, angular rock fragments that have been transported and then left behind by glaciers. Glaciers have left other clues in the scrape-patterns and grooves that document glacial dragging of boulders over solid rock. It appears that worldwide glaciers scoured the Earth on more than one frosty occasion, wiping out layers of the Earth's stony record. These events are collectively referred to as the "Snowball Earth."

Another line of evidence comes from Canada's Yukon Territory, where glaciers left their mark in 716.5 million-year-old rocks. Telltale signs included

---

[18] Long before spreadsheets and dentist appointments; many people long for the simple life.

Figure 6.15 The Grand Canyon of the Colorado River displays the great "unconformity" near its base. There, layers of rock stand nearly vertically beneath the horizontal strata of later epochs. The transition marks a region of missing history in the rocks. Inset: the green line delineates the transition between the two assemblages of rock. (photo courtesy Marilynn Flynn)

glacial deposits, striated clasts (rocks assembled from other, older rocks), ice rafted debris (carried from highlands by icebergs) and soft sediments presumably deformed by the weight of overlaid ice. The kicker is that at the time of the Snowball Earth, these rocks made up a landscape at sea level in the tropics, less than ten degrees from the equator. Analysts were able to lock down the location of the site by determining the composition and magnetism within the samples. They conclude that glaciation took place near the equator at the time, confirming that at the time, the Earth was a planetary cold storage locker.

We have seen how the Martian orbit and axial tilt can affect its climate. Could snowball events have been triggered by a disturbance in the Earth's orbit? Could the entire planet have tipped over for a period of time, bringing long days and nights to the poles, as does the planet Uranus? It takes a lot of energy to change a planet's orbit, and as we've seen in Chapter 2, our Moon stabilizes our axial tilt. Careful study of stromatolites, those colonies of cyanobacteria that we visited earlier, contradict the idea of extreme wandering orbits or tilting spin-axes. Cyanobacteria are very sensitive to day/night cycles. They lay down layers of sand and grit, forming something like tree rings. This banding shows us that the length of days and the path of the Earth's orbit did not substantially change throughout snowball glaciations.

The first such glacial erasure in the layers occurs at about 2.5 billion years ago, but a more defined double-punch of cold snaps arose between 800 million and 650 million years ago. This dual chill coincides with the breakup of the vast supercontinent Rodinia. It covers the two geologic periods known as the Sturtian (which spanned fro 717 to about 660 million years ago) and the Marinoan[19] (from 650 to 632 million years ago). For a listing of ice ages, see Appendix H.

"The reason [the Earth] gets into things like that is that once you have a series of cold seasons, you start freezing important things," explains biophysicist Ben Clark. "You freeze all the lakes and part of the ocean, let's say. Then when you get to warmer times, the Earth is not absorbing the sunlight as well because the ice reflecting the Sun's rays back into space. If you don't have greenhouse gases to intercept that, everything just gets colder and colder. This feedback system just keeps going, and things get colder and colder."

One theory about the onset of snowball epochs entails extensive volcanism coating the continents that straddled the equator. While volcanoes belch out warming greenhouse gases, they also send out floods of rock that soak up carbon dioxide as they erode. High temperatures near the equator would encourage the process, sopping up greenhouse gases and dropping the temperatures.

The Snowball periods would have rendered the Earth's surface into a very alien environment. The oceans may have been buried in sea ice 500-1500 meters deep. The weathering of the continents—critical for feeding minerals into the oceans and pulling carbon dioxide from the air—came to a grinding halt. Across the innermost regions of the continents, wind and evaporation cleared some of the icy terrain, exposing sterile, lifeless rock. Winds would have carried dust eroded from that bleak landscape out to sea, where it would tint many of the ice surfaces in tans and golden browns. From a distance, the Earth would have offered a brown and white face, devoid of any green and having very little open ocean to display the blues that we hold so dear. All of Earth's life may have retreated to hydrothermal sea floor vents and to a narrow band of open water that may have remained—or cleared early—around the equator.

---

[19] Also called the Varanger glaciation.

wobbles and iron banding

The snowball Earth periods must have starved the seas of oxygen. How did life survive when the world's oceans were cut off from the oxygen in the atmosphere? How did early life endure? With a global ice seal , the oceans would have had dwindling amounts of oxygen for its inhabitants.

The layers of banded iron—deposited over a period of roughly four million years—each document a time of low oxygen levels, but these layers are interspersed with layers of silica. New work indicates that these silica traces are evidence of "pulses" of oxygen.[20] While searching for explanations of how oxygen might have entered the ocean throughout the snowball periods, researchers measured the magnetism in the banded iron deposits. They found that variations in the magnetic field matched the timing of shifts in the Earth's orbit (those Milankovitch cycles). They suggest that Milankovitch cycles were at least partially responsible for the ebb and flow of ice sheets in concert with changes in solar radiation as the Earth's orbit brought it closer and farther from the Sun. The advance and retreat of glaciers periodically opened up the sea ice, letting oxygen mix with seawater.

The sneaky getaway: escaping Mars' fate

As ice accumulated during glaciations, it reflected sunlight away from the planet, causing even deeper cooling in a deadly cycle that could have led to a permanent, paralyzing chill of our world. But somehow, the Earth escaped its deep freeze. "Those conditions [of the Snowball Earth] never reappeared, and that's really all about plate tectonics," says Royal Holloway University quaternary scientist Scott Elias. "It has to do with the relocation of the continents and changes to currents in the ocean and atmosphere."

Harvard's Daniel Schrag agrees. He writes[21], "…the concentration of continental area in the tropics was a critical boundary condition necessary for the onset of glaciation, both because the existence of substantial continental area at high latitudes may prevent atmospheric carbon dioxide from getting too low and because a tropical concentration of continental area may lead to more efficient burial of organic carbon…"

As continents collide and merge into supercontinents, their collisions raise mountain ranges that shift weather patterns. Depending on their arrangement, they can cause a lowering of global temperatures, and they can cause an

---

[20] "Orbital forcing of ice sheets during snowball Earth," by Ross N. Mitchell, et al, *Nature Communications* July 7, 2021

[21] *A neoproterozoic snowball earth* by Hoffman, Kaufman and Schrag; 28 August 1998, *Science*

increase in various gases in the atmosphere due to shifting ocean currents or changes in the rate of weathering of minerals. Supercontinents cover large swathes of the mantle beneath, increasing volcanic activity by concentrating heat from beneath the crust. Volcanoes also spring from the boundaries of plates, and as those plates move, the volcanoes dump gas into the atmosphere and cover the ground with fresh minerals in basalt lava flows. As a supercontinent breaks up (all good things must come to an end), the changing map of land and sea can lead to another warming event. This happened at the opening of the Mesozoic era, when the super-continent Pangaea fractured, spreading and isolating vast populations of dinosaurs and other life. At the time, models indicate that global temperatures ranged from 14°C to 25°C. Northern hemisphere temperatures gradually increased in the Triassic, and then fell through the Cretaceous. In the south, the trend was roughly the opposite. Throughout the Triassic and Jurassic, ice was a rarity on Earth, but as Pangaea broke apart and sea levels rose, temperature swings became more pronounced.

At the time of Earth's great glaciations, no continents covered either pole. The continents spread across the equator and mid-latitudes. This arrangement is rare—in fact, unique—in Earth's history. The planet has not seen such an array since.

Today, most of our ecosystem's heat is stored within the tropical oceans, but at the time leading up to snowball events, the tropics were mostly landlocked. As shallow seas retreated from equatorial continents, energy-storing oceans drained away from the equator, leaving instead regions of reflective land surfaces. The continents couldn't store heat, but rather reflected it away, cooling the planet. But as climate cooled and the land went from tropic to arctic, ice cut off the surface processes like photosynthesis and weathering of silicate rock, two processes that remove carbon dioxide from the air. The growth of ice fields and glaciers ultimately led to higher concentrations of carbon dioxide, causing their downfall at the hands of greenhouse warming. The location of continents also regulates the flow of ocean and air currents. The continents on modern Earth are arranged in such a way that vast undersea "rivers" bring cold water to warm areas, and carry warm water to chilled regions. One such great river is the Gulf Stream (Figure 6.16). This great current originates in the Caribbean and Florida's Gulf Coast region. Heated water moves north along North America's eastern seaboard, then marches northwest across the Atlantic Ocean, warming northern Europe. The warm waters of the stream affect more than just the ocean: they also set up important air currents as air rises above the heated waters.

The scientific method yearns for independent confirmation, and we get confirmation of the Snowball Earth from biological signatures in the rock.

Figure 6.16 A NASA/GSFC visualization shows sea surface temperatures within the Gulf Stream, a warm underwater stream stretching across the Atlantic Ocean. The Gulf Stream is responsible for clement conditions in northern Europe. (courtesy NASA/Goddard Space Flight Center Scientific Visualization Studio)

Anomalies within isotopes found in rock layers from the correct time show a drastic drop in biological activity of marine life. This is consistent with a globe-spanning blanket of ice.

The complexities of planet Earth come through as we study the ice age cycles. 800,000 years ago, the Earth's ice age cycle underwent a dramatic shift, called the Mid-Pleistocene transition (MPT). At this time, ice age cycles lengthened from a 41,000-year to about 100,000 year length. The change brought increased variations in carbon dioxide levels, and as ice coverage ebbed and flowed, so, too, did sea levels.

The MPT results from an extraordinary convergence of variables. Climate researchers first thought the trigger for the MPT was a shift between two separate Croll-Milankovitch cycles, Earth's obliquity (tilt in its spin axis) and periodic changes in its orbit. Both of these factors play into shifts in our planet's surface temperatures, but the situation is more complex, and is as yet not clearly understood. We know that before the Mid-Pleistocene Transition, global surface temperatures were warmer than after the MPT. Over a dozen explanations have been put forward to explain why ice age cycles within the last 800,000 years have strayed significantly from a stable, predictable 100,000 year cycle.

Explanations include interplanetary dust clouds that periodically block out sunlight, decrease in atmospheric carbon dioxide, asteroid collision in the South China Sea, the uplift of the Tibetan Plateau and Himalaya mountain chain, and others. The overall effect of the MPT was to lengthen the period between glaciations. The longer phases of cleared land allowed for vast migrations across northern Asia and North America, distributing wildlife populations in ways that would not have been possible with the vast roadblocks of ice (for a time, the majority of Canadian real estate sat beneath a thousand meters of solid ice).

The Earth's history-in-the-rocks shows that after millions of ice-filled years, glaciers rapidly retreated toward the poles and high country, and the ice blankets swaddling the planet's continents and seas fractured and dissolved. Ben Clark picks up our story again: "One of the mechanisms is that you have a burst of extra strong volcanism. There are episodes on Earth where there were huge eruptions that came out of nowhere. What happens is you release a lot of greenhouse gases, and also you put a lot of volcanic dust in the atmosphere, and the dust absorbs sunlight like crazy. So all of a sudden you heat the Earth, the ice goes away, and you reverse the snowball. Everything gets much warmer. Later, if there's no volcanism, the dust settles out and you get another episode." In this way, volcanic activity could have played a role in both the cooling and warming of Earth's climate.

But another factor may have been at work in the Earth's recovery from its series of global glaciations. Research published in 2016 indicates[22] that during the breakup of Rodinia, sea floor volcanoes erupted along submarine ridges that formed along the fractured coastlines. These underwater eruptions generated abundant volcanic glasses called hyaloclastites which altered into palagonite (a compound resulting from the interaction of water and volcanic glass). The changes in minerals altered the chemistry of the ocean water during the episodes of glaciations, saturating the water with calcium, magnesium, and phosphorus. The release of these elements contributed to a rise in atmospheric oxygen following each snowball period.

During a welcome thawing period of between 4 and 30 million years, at the end of the second (Marinoan) glacial epoch, the melting ice exposed hundreds of meters of carbonate rock. This freed carbon dioxide, minerals like iron, and nutrients. Living things came out of hiding in a big way. Greenhouse gases warmed the air once again. Sunlight poured through the clearing skies. Photosynthesis was the order of the day. The oxygen flowed, the oceans became a bright green, and life forms spread out in amazing diversity. The stage was set for the bizarre, multitudinous actors of new terrestrial life, including animals. The theatrical production was called the Cambrian explosion, and it was biology's answer to the Big Bang.

---

[22] "Snowball ocean chemistry driven by extensive ridge volcanism during Rodinia breakup" by T.M. Gernon, et al, *Nature Geoscience* January 18, 2016

Drifting continents and megafloods

To get an idea of the profound effect continental drift has on climate, we need only look to the Mediterranean Sea, and we need only look back in time by 5.6 million years or so. At that time, tectonic forces sent northern Africa crashing into the Eurasian plate, raising a land barrier through the Straits of Gibraltar. The rising land cut off the Mediterranean from the Atlantic. The bridge changed patterns of various populations, making migration from north to south possible and mixing species that had never come in contact before. The natural dam caused Mediterranean sea levels to fall. Evaporating waters left behind great salt deposits. But within 300,000 years, the natural dam burst, and the waters of the Atlantic thundered in, resulting in the most extensive flood in history. As much as 90% of the catastrophic flood was over in less than two years, filling the Mediterranean basin, its level rising a remarkable 10 meters per day at times. The resurgence of the waters severed the land bridge once again, isolating populations from north to south.

Deep-sea drilling cores reveal that prior to the flood, areas of today's seabed had been dry woodlands dotted with briny lakes. The event is known as the Messinian salinity crisis. It is named after Messina on the island of Sicily, which contains one of the regions were salt from the ancient lakes can still be found. But the extent of the cataclysm has left scars as far away as northern Libya and southern Spain. The changes in water level, salinity, and even air currents in the region likely combined to create an entirely new local climate.

Snowball Earth: a summary

For a multitude of complex reasons, including the arrangements of continents, volcanism, ocean currents, the obliquity of Earth's axis, and dynamics of its orbit around the Sun, our planet enters an extended period of low temperatures. The world's oceans begin to freeze. As snow and ice cover an increasing fraction of the Earth's exterior, the white surfaces reflect more solar energy away from the ground, further cooling the environment. The carbon cycle in the ocean is largely cut off from the atmosphere by sea ice. Volcanic eruptions continue to pour carbon dioxide into the air, but the seas cannot absorb and process any of it. Rocks on the continents are also cut off, so carbon dioxide levels build, perhaps as high as 350 times what they are today. Increased levels of carbon dioxide create greenhouse warming, breaking the snowball Earth cycle. But taking into account the historical patterns of world temperatures, many geologists and climatologists consider the Earth to be just now emerging from a series of glaciations called the Quaternary Ice Age Cycle.

Looking Back: A Brief History of…Mars

During the formative years of the solar system, as the large planets migrated inward and then back out again, Mars was in the thick of it. In fact, Mars is the victim of a sheltered childhood. Jupiter began in the center of what is now the asteroid belt, and as it spiraled in toward the Sun, it disrupted all the material around it, shielding Mars from infalling debris. In effect, Jupiter prevented Mars from becoming something the size of the Earth.

Martian history is complicated. Planetary geologists divide it into four periods. The earliest, the Pre-Noachian period, began at the formation of the planet ~4.5 billion years ago, and lasted half a billion years. As it came to a close, a huge asteroid plowed into the red planet, digging out the great Hellas Basin. Several other giant basins formed at the time as well.[23] It was at this transitional time—between the Pre-Noachian and the Noachian periods—that something remarkable happened to the northern hemisphere. Whether it was a giant impact, increased volcanism, or something else, the events left Mars with a smooth province in the north and ancient, cratered terrain in the south.

The Noachian period continued to suffer from the Late Heavy Bombardment, spanning approximately from 4.1 to 3.7 billion years ago. During this era, Mars was wetter than today, probably with active rainfall, flowing rivers, lakes, and even seas filling the northern lowlands.

The volcanoes of Mars got to shine in the next epoch, called the Hesperian. Great lava plains and many volcanoes are associated with this period. Olympus Mons began to form atop the Tharsis Bulge, helping to create the grand canyon of Mars, Valles Marineris. A profound event also took place during the Hesperian: the loss of the Martian atmosphere. Something triggered a catastrophic climate change on the planet. Much of the water migrated underground. Oxygen was locked in the rocks, giving Mars its famous rusty tint. Lighter gases escaped to space.

The last phase of Martian history is the Amazonian, which began 3 billion years ago and continues today. Major volcanic eruptions tailed out, and glaciers replaced the flowing water of the past. Mars suffered its own "snowball Earth" period. The planet finally transformed into a parched and frigid world of ice, sandstorms and blue sunsets, a strange reflection of the Earth in its own snowball periods.

(For a brief history of the Earth, see Chapter 2. For Mercury, see the end of Chapter 3. A brief history of Venus is at the end of Chapter 4. Titan's brief history is found at the end of Chapter 5)

---

[23] These include the Isidis and Argyre basins.

7

Surrogate Earths: making homes away from home

Four centuries ago, Johannes Kepler was thinking about the planets. Unlike most of his contemporaries, he envisioned those wandering points of light in the sky as real worlds, places where humans might one day trod. He wondered, *"But who shall dwell in these Worlds if they be inhabited? … Are we or they Lords of the World?"* Half a century later, Christian Huygens had come to appreciate just how immense the cosmos is. He said, "How vast these orbs must be, and how inconsiderable this Earth, the theater upon which all our mighty designs, all our navigations, and all our wars are transacted…a very fit consideration, and matter of reflection, for those kings and princes who sacrifice the lives of so many people, only to flatter their ambition in being masters of some pitiful corner of this small spot."

We have seen that this "spot", small as it is, is a unique and special planet. And yet we wonder about other places we might inhabit. What nearby worlds could we "dwell" upon, anyway? Would there be good reason to? And how would we get there? Huygens could only explore the distant worlds through the imperfect lens of a refracting telescope.[1] We've made a few advances since, and gotten to know the neighborhood to a degree never imagined by Kepler and Huygens (Figure 7.1).

---

[1] Nevertheless, he discovered the true nature of the rings of Saturn, and was first to spot the moon Titan.

Figure 7.1 Before a complete planetary overhaul, terraforming visionaries foresee an intermediate step of tenting low-lying areas of Mars for agriculture and settlement. The "Worldhouse" concept would encase the entire planet in a globe-spanning pressure tent. Here, portions of Melas Chasma have been partially-tented, enabling agricultural and wilderness areas. More conventional domes in the background house local jungles. (Art by the author; Background: THEMIS data from Mars Odyssey courtesy NASA/JPL/Arizona State University)

Making the case for an Earth 2

The solar system is arrayed with planets both rocky and gaseous. The rocky terrestrials, most Earthlike in composition and structure, are an obvious choice for a home away from home. Just next door is the fifth largest moon in the solar system, Luna, our Moon. The rocket engineer Krafft Ehricke famously began some of his lectures by saying, "If God had wanted us to go into space, he would have put a planet next to us." At which point, with great flair, Dr. Ehricke would unveil a slide of the Moon. He referred to it as "Earth's seventh continent."

Krafft Ehricke was not alone in his thinking. For many aerospace strategists, sociologists, engineers and visionaries, the nearby worlds provide a natural extension of Earth's wilderness, a new frontier to explore and expand out into. The Moon has a lot going for it. Unlike any of the other terrestrials, it's only three to five days away. It will take humans six to eight months to get to Mars. Trips to Venus and Mercury are shorter, but the destinations present problems, as we will see. The Moon's gravity—just 1/6 that of the Earth's—makes for a relatively undemanding orbit and landing. Additionally, the airless Moon does have one precious resource: water. Water is handy for space travelers. It can be used for drinking. It can be split up into hydrogen and oxygen, with the hydrogen used as fuel and the oxygen for other obvious purposes (including oxygen bars).

But Mars advocates point to the red planet's comparatively vast resources. In addition to readily available water, assets include elements that we can cull from the minerals in the rocks and gases in the atmosphere to manufacture a host of materials. Temperatures on Mars are more subdued than those on the Moon, where daytime highs of 120°C (over the boiling point of water) drop to frigid nighttime lows of -130°C. The Martian climate sees equatorial highs reaching 20°C, with winter night lows of -153°C in the polar regions. The rarified Martian air still constitutes an advantage over the hard vacuum of the Moon, enabling infrastructure to be built of lighter-weight materials.

Mars enthusiasts proclaim that we should not be content with a few astronauts huddled in underground bunkers. They see the end game as a fully independent, self-sustaining civilization on another world. To that end, many mission scenarios take the long view, advocating incremental assembly of increasingly complex infrastructure between the Earth and the planets, moons and asteroids (for details, see "The Mars Direct approach," below). Aerospace engineer and inventor Robert Zubrin sees human expansion to Mars and beyond as part of what makes us human (as explorers) and as a "grand opportunity to diversify and expand our creative ability as a species" he says. "The challenge of new frontiers will drive the development of many new, highly inventive branches of human civilizations that will make many contributions advancing human life everywhere." Zubrin's vision sees Mars bustling with innovative city states, whose Martian ingenuity will have a parallel role in driving future progress to that which the Yankee Ingenuity born of the challenges of the American frontier has done historically. "Together to Mars, then together with Mars," he says, we will create a "free, open, and magnificent future."

Zubrin's paradigm of space exploration is obviously a positive one. He sees humanity at the doorstep of a new era of growth and prosperity. The Mars visionary asserts that humanity is off to a good start: we have migrated out of our caves, moved across the African savannah and now inhabit the far corners of our planet. Now, Zubrin says, it is time to build a new kind of civilization, one that lives, works and plays among the planets as an extension of our home world.

Like Robert Zubrin, SpaceX's Elon Musk sees human expansion into space as a matter of survival. Musk's motivations may be darker. He has alluded to a future Earth starved of resources and overpopulated, a planet in need of a pressure valve to release all that planetary angst. To Musk and others, settlement of the planets will provide back-up sites for humans to live as their planet becomes inhospitable due to pollution, extinction-level impact, disease or other unforeseen disasters. Musk, too, draws a link between Mars settlement and the colonization model.

Several analysts have raised the complaint that futurists like Zubrin and Musk popularize the mindset of empire and conquest. They suggest that the language of imperialism carries negative connotations and historical associations. But political sensitivities aside, many view the cosmos as the next frontier for humanity, a place that opens possibilities for a bright future beyond a new horizon.

Scoping out the neighborhood: once and future missions

Before we can safely venture beyond the Earth/Moon system, we need to know the lay of the land. We've already surveyed our solar family, and we have exciting plans for future exploration. We will need them; going off to Mars or the moons of Jupiter half-cocked will be a recipe for calamity at best, and doom at worst.

After centuries of struggle at the telescope, scientists made a breakthrough with the first successful planetary mission, the Mariner 2 Venus flyby (Chapter 4). Aside from the Earth's Moon, Venus was the easiest target; it was the closest planet, and flying to Venus had the advantage of falling in toward the Sun, requiring less fuel than going the other way, toward Mars and the outer system. The parade of Veneras, Pioneers and Mariners joined the exploration of the planet by the European Space Agency's Venus Express. Japan mounted its Akatsuki mission, resurrected in 2015 after a years-long delay. But after those missions, Venus-specific missions ceased[2]. That hiatus is about to come to an end with a new flotilla of planned spacecraft (see below).

Mars has been an exploration priority since the earliest days of space ventures. But Mars is a tough destination. So many spacecraft were lost that engineers began to suspect—with tongue in cheek to varying degrees—that in the great void between Earth and Mars lived the Great Galactic Ghoul,[3] waiting to devour any hapless space probe that dared sail through interplanetary territory. At the time of the first success, Mariner 4, the score stood at Mars: 6, Earth: 1. The treacherous journey to Mars would continue to gather its victims, with far more failures than successes. By the time of the first successful orbiter, NASA's Mariner 9, Mars had swallowed 22 attempts, with only 8 successes. Various space agencies of the world have improved on the ratio of successes to failures.

The solar system beyond Mars remained the purview of telescopes until the NASA/Ames/TRW's Pioneer 10 and 11 missions. Pioneer 10 coasted by Jupiter in 1973, after having survived the first crossing of the asteroid belt. Pioneer 11 added Saturn to the mix, encountering the ringed giant in 1979 after a 1974 Jupiter flyby. The small Pioneers were nearing the orbit of Uranus[4] by the time a follow-on mission, by the twin Voyager spacecraft, was mounted. Voyager 1 carried out a sophisticated reconnaissance of Jupiter and Saturn, including a detailed study of Titan. Voyager 2 was targeted to encounter Jupiter, Saturn, Uranus and Neptune using a rare planetary alignment known as the "Grand Tour". Both spacecraft opened up the outer system to our understanding, revealing such important concepts as cryovolcanism, tidal heating, and a host of other foundational processes. Now, it was time for the orbiters.

The Galileo Jupiter orbiter/probe and the Cassini/Huygens Saturn orbiter/Titan lander transformed our understanding of planetary science. Galileo carried a Jupiter atmospheric probe, and Cassini carried ESA's Huygens Titan atmospheric probe and lander. These were followed by sophisticated asteroid and comet explorers, as well as the Pluto New Horizons Mission. For specifics of successful planetary missions, see Appendix D.

---

[2] There were many flybys that obtained "drive-by" encounter science, including Galileo, Cassini, MESSENGER, Parker Solar Probe, IKAROS, Bepi-Columbo, and eight planned flybys of ESA's SolO solar probe (which are in progress).

[3] The term was likely first coined by JPL's John Casani, a talented engineer who went on to manage the landmark Voyager, Galileo and Cassini planetary missions.

[4] Neither of the Pioneers followed trajectories that would bring them close to either Uranus or Neptune.

Assuming our civilization begins to advance out across the planets and moons of the solar system, where will we go? What kinds of worlds will be best suited to afford us surrogate homes? Space travel visionaries have put together some general ideas and strategies from our past and on-going missions, but we clearly don't know enough yet to pack up and relocate. Still, plans are on the table and nearing the launch pad for future missions that will pave the way.

Desolate Mercury burns beneath a blistering Sun, rich in material wealth but a poor place for a home away from home. While Venus might not be the best candidate for future real estate investments, it may provide opportunities for floating structures, and as one of the most Earthlike worlds, it is important for future studies. First on the docket is VERITAS, the Venus Emissivity, Radio science, InSAR, Topography And Spectroscopy mission. The NASA orbiter is equipped with radar that can return images of the surface at far greater detail than the best we have to date, the radar images from the Magellan mission. A launch date is tentatively set for 2031. VERITAS will be arriving later than two other probes. The European Space Agency is set to dispatch EnVision to the hothouse world in the early 2030s. The orbiter is tasked with studying the interaction between the Venusian atmosphere and surface. Its four-year reconnaissance will be complemented by NASA's DAVINCI deep atmospheric probe, slated for arrival in June of 2031. DAVINCI (Deep Atmosphere Venus Investigation of Noble gases, Chemistry and Imaging) is a disk-shaped titanium probe a meter across. NASA calls it a "flying analytical chemistry laboratory." A parachute will initially slow its hour-long descent, but like Soviet Veneras, the chute will drop off as the craft makes it down into the dense atmosphere, where air friction will be enough to soften its landing. During descent, the probe will transmit aerial vistas of Venusian mountains in the Alpha Regio area, while the carrier craft flies by and charts clouds and terrain from above. While DAVINCI is designed to operate above the ground, engineers hope the probe will survive a 25 mph (12 m/s) landing to send data from the surface as well. India's relatively new planetary program plans to launch the Shukrayaan Venus orbiter at the end of 2024. The craft will map the Venusian landscape with an assortment of instruments including radar, infrared cameras, and a suite of instruments from several countries. The mission is in the science experiment selection phase.

Mars is often toted as the most likely to host the first permanent settlements outside of Earth. The red planet is a world of abundant resources, as we will see. Space agencies across our globe have focused on Mars since the earliest days of the space program, and continue to do so with plans for even more complex missions.

The NASA Perseverance Rover, currently operational in Mars' Jezero Crater, is caching samples for eventual return to laboratories on Earth. Jezero was chosen because of the apparent flood plains blanketing its surface. Perseverance has discovered that some of the floods were volcanic in nature, but other areas contain clays and minerals pointing to a standing lake within the crater for an extended period.

Returning samples from Mars has been one of the holy grails of planetary science. With samples in hand, researchers will be able to determine detailed time lines of various events on Mars using more sophisticated equipment than can currently be flown in space. But getting samples back to Earth is expensive and challenging. NASA and ESA are developing a multi-craft Mars Sample Return Mission designed to interface with the samples collected by Perseverance. Under the current scenario, Perseverance would continue to collect and store up to 43 samples of soil, rock and atmosphere in pencil-sized titanium tubes. The rover is depositing samples at a back-up depot in Jezero Crater called Three Forks, a flat area clear of large rocks (a cache of the first ten was officially emplaced by the end of January 2023). The Rover carries an identical primary sample set on board. Plans call for a Sample Retrieval Lander and Mars ascent Vehicle to land near the Perseverance, which would transport the samples to the lander. The combined lander and ascent vehicle would be equipped with a sample transfer arm designed to pick up the sample tubes. Two Ingenuity-type Mars helicopters, equipped to retrieve samples from the cache at Three Forks, would be standing by should the rover be unable to deliver them. Perseverance is depositing the tubes at the Three Forks depot in a zigzag pattern, leaving a working area around each sample for retrieval by the helicopter option.

The two-stage Mars Ascent Vehicle, powered by solid rocket motors, would carry the samples to an Earth Return Vehicle awaiting them in Mars orbit. According to the baseline mission, Martian samples would arrive back on Earth in 2033.

ESA's Rosalind Franklin rover promises to be the most advanced life detection spacecraft to ever visit Mars. It can drill samples from two-meter depths, and has advanced onboard laboratories to search for biomolecules and evidence of past biology. The rover was to be delivered by a Russian lander, but the crisis in Ukraine necessitated delays and redesign. With the redesign of the lander delivery system, the Rosalind Franklin is now scheduled for launch no earlier than 2028.

Japan plans to use the expertise it has gained from its two successful asteroid sample return missions[5] to retrieve samples from Mars' largest moon
Phobos. The orbiter will also study Deimos and Mars itself. Launch is scheduled for the mid 2020s. These combined Mars missions will contribute to our
understanding of past and current conditions, and possibilities for human
habitation.

<u>Voyages beyond the asteroids</u>

Beyond Mars and the main asteroid belt, the outer planets and moons open a
frontier rich in resources. The ocean moons may be prime candidates for
understanding the oceans of our own world, and the mountains, craters and
valleys of the ice moons will afford prime real estate for future outposts or
settlements. With the launch of its ambitious JUICE (Jupiter Icy moons
Explorer) mission, the European Space Agency spearheads the exploration of
Jupiter's ocean moons Europa, Ganymede and Callisto (Figure 7.2). Beginning
in July of 2031, the six ton solar-powered JUICE will carry out 35 flybys of
Europa and Ganymede —moons thought to hide deep internal oceans, and
Callisto.[6] JUICE will also perform studies of Jupiter itself. After the 35 initial
moon encounters, JUICE will settle into orbit around Ganymede for in-
depth studies of the solar system's largest moon. JUICE must first get into
orbit around Jupiter. It does so by using the gravity of Ganymede during a
close flyby on its approach to the system. Once captured by Jovian gravity,
JUICE will embark on its multiple moon flybys, with its first Europa encounter scheduled for a year after arrival. The spacecraft will establish a 500-km
circular orbit around Ganymede in December of 2034, becoming the first
spacecraft to orbit a moon other than Earth's Moon. JUICE is a richly international collaboration among the 23 member nations of ESA, the Japan
Aerospace Exploration Agency (JAXA), NASA and the Israel Space Agency. A
major goal of JUICE is to determine the habitability of the ocean within
Ganymede.

In the autumn of 2024, NASA joins the European Jovian exploration when
it launches its Europa Clipper mission atop a SpaceX Falcon Heavy booster.
Like JUICE, Europa Clipper is solar powered; its panels extend farther than

---

[5] Japan's successful asteroid sample returns were with the Haybusa at asteroid Itokawa and the Hayabusa
2 at asteroid Ryugu.

[6] Callisto may also harbor an internal ocean, but its structure is not well understood. Water within
Callisto may be spread through a jumbled matrix of rock and ice.

Figure 7.2   Early art concept showing the JUICE and Europa Clipper missions working in tandem to study the radiation environment, surfaces and interiors of the Galilean satellites. (art by the author, courtesy NASA/JPL-Caltech)

the length of a basketball court. Beginning in 2030, the spacecraft will follow an orbital tour around Jupiter that takes it on a series of 50 close encounters with Europa (much as Cassini orbited Saturn to study Titan). Over the course of three years, Clipper will study the moon's surface and interior with high-resolution cameras, spectrometers, and ground penetrating radar. Project scientists hope to determine the extent of Europa's ocean, as well as the composition of its crust, some of which may lend insight into what's beneath that frozen surface. Clipper must spend most of its time outside of Europa's orbit, distancing itself from Jupiter's deadly radiation. Its Europa flybys will take it deep into Jupiter's magnetosphere, but the meetings will be quick, allowing the craft to retreat to safer havens after each flyby.

Europa and the other Galilean satellites are not the only ice moons that NASA has its sights on. The Johns Hopkins Applied Physics Lab is fabricating its Dragonfly quadrocopter for a 2027 launch to the foggy Saturnian moon Titan. An international team includes many NASA and private engineering centers in the US and abroad. The list includes NASA/Goddard Space Flight Center, NASA/Ames, NASA/Langley, Lockheed Martin, Penn State University, Malin Space Science Systems, Honeybee Robotics, and the Jet Propulsion Laboratory. International partners include the Laboratoire Atmosphères, Milieux, Observations Spatiales (LATMOS), the German Aerospace Center (DLR), the Japan Aerospace Exploration Agency (JAXA), and the Centre National d'études Spatiales (CNES). Dragonfly is a hybrid lander and drone crafted to maneuver in the dense atmosphere of Titan. It will study the air and surface composition, flying to many disparate locations during its planned two-year, several-hundred-kilometer mission. The nuclear powered hovercraft[7] will determine the habitability of the environment, and attempt to establish just how far prebiotic chemistry has progressed on the methane-soaked moon. The probe is even equipped to search for the "tells", or chemical traces, that might result from hydrocarbon-based life there.

All of these missions will contribute to our understanding of the boots-on-the-ground conditions of other worlds. These are the type of studies required for any establishment of a home base on another world. In this kind of space travel, with a permanent presence in mind, the adage "you can't take it with you" takes on new meaning. To set up housekeeping millions or billions of kilometers from home, we must learn to use the local resources.

Living off the land

As we weigh out future options across the solar system, Venus might seem a prudent place to skip. Living on the surface of Venus is well beyond our current technology. Although aquanauts have survived in high pressures of the deep sea, they have not endured anything like the surface pressures of our sister world. The time when humans will witness, first hand, the craggy lava fields, soaring mountains, and deep canyons of the planet is not in our near

---

[7] The light at the surface of Titan is too dim to use solar power for this high-powered craft.

future. But humans may well view the planet from a different viewpoint, watching the lightning-laced Venusian clouds billowing in the furious winds, sulfuric acid virga dropping into the fog below, and the metallic blue of a sky 55 kilometers up. There, the air remains just a bit warmer than room temperatures during the day, and the pressure is equivalent to that at Earth's sea level. At the visible cloud tops, winds reach 210 mph (340 km/hr), but that breakneck speed is relative to the ground. Aboard a floating colony or outpost, winds would be manageable as the station drifts along on the air currents. Because the Venusian atmosphere circles the entire planet every four days (a phenomenon called superrotation), cloud inhabitants would experience a much shorter—and comfortable—day/night cycle.

A long-term inhabited facility above the clouds of Venus will find some benefits from the environment. Super-pressure balloon probes have been proposed for both Venus and Mars, similar to high-altitude weather balloons used for meteorology on our own planet. On Venus, a highly pressurized balloon is not necessary. Instead, an envelope can be pressurized to match the pressure outside of it. Such a balloon will be buoyant using a mix of nitrogen and oxygen, similar to that found on Earth. Within Venus' carbon dioxide atmosphere, these benign gases will actually be able to lift 60% of what a helium-filled balloon above Earth can. A 400 yard/meter-radius balloon—the size of a small sports arena—can lift 350,000 tons, a mass equivalent to twice the weight of the Statue of Liberty. Balloons with equal pressure to their exterior have another advantage: if they are punctured or tear, the gas will leak out at a leisurely pace. No catastrophic "pop"!

Despite the hellacious conditions just a few kilometers below, life in the clouds would be benign (Figure 7.3). Any human Venus colonists living in the sky would require only light garments to protect them from the acid-infused air. An oxygen mask would take the place of a bulky pressure suit. Advocates for Venus aerial condos also point out that Venus is rich in resources necessary for life: carbon, hydrogen, oxygen, nitrogen and sulfur. Its dense atmosphere provides protection from cosmic radiation, in contrast to environments on Mars, the Moon, or asteroids, other targets for human exploration and settlement. Venusian gravity is 90% that of Earth, so the long-term physical problems that may arise on Mars from bone calcium loss are not an issue there. Abundant solar energy at high altitudes, and wind energy in Venus' breezy skies, provide efficient energy solutions. That energy could also be put to use on large turbines that could offer some maneuverability for the station, keeping it in one place to study surface features below, or changing altitude to take advantage of directional winds. Surface science packages and telerobotics could be deployed for surface science research.

Figure 7.3 The blistering surface of Venus will be difficult to inhabit, but conditions 50 kilometers up are much more Earthlike: temperatures hover at about room temperature, and pressures are equivalent to those at Earth's sea level. The Venusian skies may be perfect for buoyant outposts or even small cities. (painting by the author)

While Venus offers pie-in-the-sky possibilities, many visionaries prefer locations where you can walk around on the ground. Mars and its moons, Europa, the asteroids and other sites have been bandied about as future sites for distant homes. As we establish settlements on Mars and other worlds, we will immediately commence mining operations, but it's not gold we'll be after; it's water. The most important resource on other worlds, by far, is good ol' H_2O. We humans drink it. It can be split into hydrogen, used for handy things like rocket fuel—and oxygen, used for even handier things like breathing. The poles of the Moon and Mercury have small inventories of the treasured compound, but as our exploration of Mars has progressed, we have come to realize that the desert world has vast amounts of it, often near the surface.

On frosty Mars, the water is frozen into ice even more solid than what we find in terrestrial arctic regions. As one Antarctic ice core-drilling engineer put it, "The ice here regularly breaks our diamond drill bits". But not all Martian water is solid; there are a few exceptions. Impacts melt or vaporize subsurface water ice. Evidence shows that past volcanic eruptions have triggered colossal floods, much as Icelandic eruptions beneath glaciers let loose massive flooding across the subarctic landscapes there. There may still be active geothermal hot spots on Mars today. Even sunlight on equator-facing slopes appears to occasionally free liquid water (see Chapter 6).

The Mars Direct approach

Mars offers other resources as well. Its carbon dioxide atmosphere can be gathered and processed into rocket fuel. This fact figures into most of the scenarios for human Mars exploration. Rather than bringing fuel for the return trip, many designers now propose manufacturing the fuel on the Martian surface, and then tanking up the Earth vehicle for its return. Engineering visionary Robert Zubrin proposed this scenario for his groundbreaking "Mars Direct" approach for human Mars exploration. The architecture of his mission—utilizing fuel manufactured at Mars rather than carrying it with the crew—saves so much weight and is so efficient that NASA has adopted a version of it in its Mars plans as well, and the concept is catching on in many corners of the aerospace industry.

Details of the Mars Direct approach first gelled at a series of conferences called Case for Mars.[8] Zubrin and his team envisioned unpiloted vanguard spacecraft landing on Mars, where each would begin to produce fuel to refill their tanks. Humans would only be dispatched when flight engineers could confirm two flight-ready transports waiting on the surface. Only then would human crews make the dangerous journey, with a primary and back-up craft waiting for them. The craft that transported the first crew would remain on the surface, slowly refilling itself from the processed Martian air. Using this scenario, infrastructure could be built up on the Martian surface, establishing a permanent settlement for human occupation. With every Martian launch opportunity, new crews and equipment would settle on the Martian surface, adding to the established village taking shape there already (Mars launch opportunities open up roughly every 780 days).

A second way to lower fuel requirements is by use of a cycling ship (Figure 7.4). An interplanetary stack of several linked spacecraft could be set on an orbital path than continually loops between the Earth and Mars. As it approached Mars, small landing craft would depart for the Martian settlement,

---

[8] Participants included esteemed NASA and industry personnel, as well as planetary scientists, engineers and researchers: Robert Zubrin, Chris McKay, Tom Meyer, Ben Clark, Carol Stoker, Penny Boston, artist Carter Emmart and others.

Figure 7.4 Left: A cycling ship is a creative transportation option offered as part of the Mars Direct scenario. (art by the author, based on sketch by Carter Emmart) Center: concept of an early Mars outpost, with buried habitats for radiation protection. (art by the author) Right: the Four Frontiers Corporation envisions advanced Mars settlements with living areas buried in the side of a hill and greenhouses, energy plants and gas processors outside. (art by the author, courtesy Four Frontiers)

while others would ascend from the surface, carrying a crew to the cycling ship and on back to Earth. The cycler would make its leisurely way back to Earth, and as it glided by, the Earth-bound crew would depart, while a new crew from Earth would launch to meet the cycler, hitching a ride to Mars. The cycler would not slow down, but would continue on its long orbit, ferrying groups of astronauts between the Earth and Mars, a sort of interplanetary shuttle service requiring only enough fuel for slight adjustments in trajectory.

At the time of the first Case for Mars workshops, fuel manufacture on Mars was a theoretical possibility, but the concept has now been directly confirmed on Mars.[9] (Figure 7.5) The Perseverance rover carries a toaster-sized experiment called the Mars Oxygen In-situ resource utilization Experiment (MOXIE). The tiny atmosphere processor successfully separated oxygen from carbon dioxide in a dress rehearsal for future human missions. The oxygen can be used for breathing as well as for rocket fuel. Future plants would need to be scaled up. Launching four humans from the Martian surface back to Earth will require on the order of 7 metric tons of fuel and 25 metric tons of oxygen. MOXIE can generate 10 grams of oxygen per hour, enough oxygen for an astronaut to breathe for twenty minutes. The gas processor heats carbon dioxide to 1470°F (800°C), splitting Mars' ubiquitous carbon dioxide into carbon monoxide and oxygen.

Another approach to creating Mars-made rocket fuel at a larger scale involves the use of biofuels. Carbon dioxide is one of the few abundant

---

[9] Fuel and oxygen manufacture from carbon dioxide was also carried out at several firms, including Zubrin's Pioneer Astronautics.

Figure 7.5 Left: Early experiments at Robert Zubrin's Pioneer Astronautics demonstrated the efficacy of generating oxygen from the air and, as in this case, the Lunar regolith. (Mars Society, courtesy Robert Zubrin). Right: technicians install the MOXIE oxygen generator into the Perseverance Rover. (NASA/JPL-Caltech)

resources on Mars, and microbes are exceedingly talented at converting it to other forms. The prime candidate for this biological magic is cyanobacteria (or blue-green algae). Engineers propose to robotically assemble 3-D printed "bioreactors", resembling greenhouses, which would cover the equivalent of four football fields. The cyanobacteria would thrive in the high carbon dioxide environment, and would be broken down into sugars. Those sugars, in turn, would be automatically fed to *E. coli* bacteria in another bioreactor, converting the sugars into methane for rocket propellant. Other creative approaches for fuel production are under study as well.

Mars appears to have mineralogical resources that could facilitate a permanent human presence. But it lacks some key elements. One of those is methane (although methane has been detected sporadically). Another is nitrogen. Earth plants and microbes are "nitrogen-fixing". They rely on nitrogen in the soil to take in nutrients. Future inhabitants will need to seed the soil with fertilizers before any plants will grow. Martian soil is also quite acidic, containing peroxides and other chemistry detrimental to crops and other plants. Future farmers will need to extensively process Mars dirt—which is regolith, or crushed rock—into soil, a process that has taken Earth billions of years to accomplish. But we have learned technologies that enable us to treat and transform soils quickly. All it takes is a whole lot of material and microbes transported from terra firma.

Home-grown Martian rice?

Aside from its acidity and lack of nitrogen, Martian dirt may be well-suited for growing rice. Rice requires very little preparation for making it edible (unlike corn or wheat, which must be peeled and processed). Researchers[10] at the University of Arkansas, Fayetteville, planted rice in several Mars soil simulants, essentially crushed basalt from the Mojave desert. The first crop, sown in the pure Martian soil simulant, grew weakened roots and thin stalks. Another rice crop grew in pure potting soil as a control. The cosmic farmers placed other plantings in soils with varying mixes. The combination of potting soil and Mars simulant yielded much healthier crops. The team announced that even when only 25% of the mix was potting soil, the rice thrived. Rice genetically modified to tolerate perchlorates may yield healthy crops for future Mars inhabitants.

An assortment of scientists and volunteers are getting practice for establishing Martian outposts. In the Canadian Arctic, the Mars Society has carried out a series of ambitious simulations under the heading Mars Arctic 365. The Mars exploration rehearsals take place at the Society's Flashline Research Station on Devon Island, a desolate outpost just 1450 kilometers from the North Pole. Completed in the summer of 2000, the two-story cylindrical Mars Habitat stands 7.7 meters tall. On the first floor two airlocks open on to a shower and toilet, spacesuit storage and a laboratory/work area. The level above accommodates six crew rooms with bunks, a kitchen and common area. Above this level, a loft storeroom can also host a seventh crewmember.

The Mars Arctic 365 program subjected researchers to the most extensive simulation to date of a full-scale mission in isolated, harsh conditions. Research included geology, microbiology, human physiology and psychology, nutrition and experiments involving remote rovers and field stations. While the last mission took place in 2017, the Mars Society plans to restart the facility. Crews will repair and upgrade the facility beginning in the summer of 2023.

Perhaps the most large-scale Mars simulation was not even designed for the purpose. The U.S. Antarctic Program's McMurdo Station will look familiar to anyone who has studied architectures for Martian settlements (Figure 7.6). McMurdo is a remote, isolated setting, with an infrastructure nearly independent of the rest of the world.[11] Like an outpost on the red planet, McMurdo must be able to function independently for long periods of time. It generates its own power and manages its own emergencies while enabling researchers to

---

[10] Results presented by Abhilash Ramachandran at the Lunar and Planetary Science Conference on March 13, 2023.

[11] Although McMurdo relies on shipments of foodstuffs and other supplies annually.

Figure 7.6 Antarctica's McMurdo Station, ca. 2017. The remote community is in the process of a multi-million dollar upgrade. (photo by the author)

perform work in its austere environs. McMurdo's inhabitants and visitors watch out for each other, remain vigilant for the safety and health of the community, and take precautions to preserve the delicate polar environment. The frontier beyond McMurdo's border is so treacherous that lines of flags have been emplaced to help people find their way home in the unplanned blizzards (and they are all unplanned, at least by the humans). Any travel beyond McMurdo's safe haven, whether on foot, snowmobile, helicopter or Snowcat, must be okayed with the center, where both air and ground radio transmissions are constantly monitored. McMurdo has its own general store, a gym, urgent care center, hair salon, coffee house, pub, lecture hall, and chapel. An expansive galley remains open 24/7 to serve people in all work shifts (complete with fresh-baked cookies, pizza bar and deli). One of the most striking Mars settlement analogs at "Mac Town"—as the inhabitants affectionately refer to it—is the arrangement of the doors: they are all configured like airlocks, with outer and inner accesses. Inner doors are not open until the outer ones are sealed. Door handles are typically the horizontal type used on freezer accesses, security against Antarctic winds that can reach hurricane-force. The difference with these freezer doors is that they keep the cold out. On Mars, airlock doors will keep the air in.

Form follows function, and McMurdo looks like a Mars outpost because it must function in a very Mars-like environment. Antarctica surrounds any visitors in bleak terrain and glacial plains. The Harsh Continent has a thousand ways to kill trespassers, all creative and often surprising in their lethality. Mac Town is a frontier village where inhabitants respect the hostile nature of their surroundings. And much like a Mars settlement, McMurdo is the hub for the many deep field camps set up by visiting scientists. The base oversees research on the vast ice plains where meteorite hunts take place, on the remote camps near the South Pole, and in the famous Dry Valleys (some of the most geologically and environmentally similar places to Mars that you will ever see on Earth).

Antarctica is the coldest, windiest, most arid place on this planet, and the Martian setting is even more sterile than Antarctica's. Like Mars, Antarctica is a desert. Although covered in frozen water, scant precipitation falls annually. Averaged over its entire surface, a measly 166 millimeters (6.5 inches) of moisture falls each year, which technically qualifies the continent as a desert. But the southern continent is a tropical paradise compared to Mars, where the only precipitation today falls as dry ice (frozen carbon dioxide) over the polar regions. The Martian air has enough water vapor to form clouds in some areas under transient conditions, and clouds of carbon dioxide ice condense at higher altitudes (Figure 7.7). Frost forms on the ground in the mornings, as first seen by the Viking lander on Utopia Planitia. Antarctica's frigid skies often display haloes, sundogs and other phenomena related to ice crystals suspended in the air. These elegant ethereal phenomena may be a common occurrence on Mars as well, since dust and ices are suspended high in the Martian atmosphere.

As we venture beyond the Moon and Mars, we're faced with new challenges, but also offered new resources. Jupiter's four Galilean satellites are intriguing worlds in their own right, but Io orbits deep inside of Jupiter's mighty radiation fields. Aside from the Sun, the king of worlds has the strongest magnetic field in the solar system. A human on Io would receive a lethal dose of radiation in minutes. Next out is the ocean moon Europa. Europa has a less formidable radiation environment on the surface, but it is still quite deadly. While Io orbits Jupiter at an unnerving range of 421,700 km from the planet's center, Europa circles at a distance of 676,800 km (about twice the distance from the Earth to the Moon. Nevertheless, the ocean moon is still well within the Jovian magnetosphere. At their highest, radiation levels on the ground are about 540 REM per day, 40 REM more than a fatal dose. Radiation levels vary across the Europan surface. They are lowest at high latitudes, and along the leading hemisphere. In these regions, less radiation shielding will be needed for a comfortable lifestyle.

Within and beneath the ice, radiation levels drop. The ocean itself may be inhabited. As we saw in our Titan chapter, two elements critical to life are the right chemistry and an energy source. Jupiter's magnetosphere provides a formidable energy source. Volcanism on Europa's rocky sea floor, some 100 km

Figure 7.7   Three views from the Curiosity Rover of Mars clouds. Top: clouds of carbon dioxide ice; center: feather-shaped cloud; bottom: noctilucent high-altitude clouds in the evening.

down at the bottom of its ocean, may also provide sites amenable to indigenous life.

The entire Jovian system is bathed in Jupiter's deadly magnetic fields, but this radiation may, in fact, turn out to be a resource. Studies by the European Space Agency show that a spacecraft in orbit around Jupiter, dragging twin 3-km-long cables through the Jovian magnetic fields, could generate a few kilowatts of power. Those cables could be scaled up and strung across the surface of Europa for dozens of kilometers, generating enough power to supply an entire outpost or village. Similar experiments in low Earth orbit generated 3500 volts of electricity on a cable several kilometers long.[12] Earth's magnetosphere is frail compared to the available energy at Jupiter or Saturn.

Engage shields!

Two kinds of radiation will assault explorers in deep space. The first comes from our nearest star. The Sun bombards everything around it with high-energy particles, especially during Solar Particle Events (solar flares). These episodes disrupt radio and television transmissions and even cripple satellites. They also have the potential of conveying lethal doses of radiation to a traveler outside of the Earth's protective magnetosphere. A second type of radiation, cosmic radiation, comes from beyond the solar system, out in deep space. Its specific sources are unknown, but may include such highly-energetic objects as supernovae and galactic cores. Cosmic rays pose a host of long-term health hazards, including various cancers and genetic and central nervous system damage.

Designers have come up with some creative solutions. While some types of metal can provide protection from solar flares, they are heavy. Instead, ships can be designed with "storm shelters", small chambers where crews can retreat during such events. Rather than using metals for insulation, these shelters can be surrounded by the ship's fuel, especially if it is hydrogen, or the walls can be filled with the crew's water supply. Several feet of water or pure hydrogen will form an effective barrier from even the most deadly of solar events.

Cosmic radiation is different in nature. Its dense particles are stripped of all electrons, so they pass easily through solid metal and even the cores of planets. Our biome is sheltered by the Earth's magnetosphere, the protective bubble encasing the globe. The magnetosphere shunts solar radiation away from the surface and cuts down on incoming cosmic rays.

Engineers have been exploring the possibilities of a similar solution for spacecraft. Could a crew be protected by an artificial magnetosphere generated by their own ship? Physicists from the UK, Portugal and Sweden have now demonstrated that the technology can be manufactured cheaply with light materials,

(continued)

---

[12] The largest-scale test was conducted in February of 1996, when the Space Shuttle Columbia unfurled nearly 20,000 feet of cable from an Italian-designed satellite.

to protect inhabitants in deep space. Like a planetary magnetosphere, these local fields would generate a charge in the space around the craft, deflecting the deadly particles away. Designers estimate that a spacecraft could be protected within a magnetic bubble roughly 100–200 meters across, using a system that could be readily carried into space.

Other recent studies are under way at CERN and several French, Italian and international European centers.

Researchers at Johnson Space Center have been experimenting with high temperature superconducting structures to generate protective fields. Their studies suggest that, "a combined system of active and passive radiation shielding constitutes the most promising solution..." JSC engineers envision large, ultra-light, expandable coils to reduce radiation within the spacecraft. Ion drives generate their own magnetic fields, and those fields might be modified to serve as crew protection.

Mission planners are also looking into a pharmaceutical approach. Medication may provide a bulwark against radiation's destructive power. A class of drug known as radioprotectors has been successful in patients who have been exposed to radioactive environments, and several drugs even show promise when administered ahead of time. Genistein, synthetic triterpenoids (compounds derived from plant mosses), Filgrastim, and Amifostine (used to reduce the side effects of radiation treatments in cancer patients) all make the short list for possible use in long duration, space flight in high radiation environments.

Europa's surface is laced with what are likely complex hydrocarbons in the form of tawny stripes that paint the ice surface for dozens or even hundreds of kilometers. The organic material seems to be issuing from the ocean beneath, although some is generated within the chemicals on the surface as they interact with solar energy and Jupiter's radiation fields. Organic material represents a resource that can be transformed into other substances for various settlement purposes.

Astrobiologists are focusing on these organics, as they may make their way into the sea below. Recent work[13] suggests that Europa's chaotic terrain—regions where surface ice has fractured and rafted before refreezing—marks locations where saltwater may have gathered in large quantities. These brines drain through the underlying ice, transporting oxygen and complex organics in pulses, ultimately bringing them into the oceans beneath the crust. There, biogenesis may take place at the sites of sea floor hydrothermal vents or within the ocean itself.

---

[13] "Downward Oxidant Transport through Europa's Ice Shell by Density-Driven Brine Percolation" by Hesse, et al; AGU *Geophysical Research Letters*, February 10, 2022

There may be other resources on the surfaces of the Galilean satellites. Dark regolith blankets Callisto, and it may have important components for future settlers. The dark brown sediment may contain elements essential for life support, such as nitrogen, phosphorous, and carbon. Some silicon and common metals may also be present, but they are likely mixed throughout a matrix of ice rather than in isolated outcroppings. The water ice on the surfaces of Ganymede and Europa appears to be more pure.

Searching for life on Europa will be a primary goal of explorers. Europa's crust may have thinner areas suitable for dropping a submersible through a borehole. An expedition such as this would require a pressure dome above the ice tunnel, as water would otherwise explode out into the vacuum above. Perhaps humans may one day even free-dive in pressure suits adapted to the alien, high-pressure environment within Europa's version of Davey Jones' Locker.

Ganymede and Callisto offer similar natural resources to those found on Europa, but they orbit in a much safer radiation environment. While they aren't anything like the Earth, they will play an important part in future human exploration of Jupiter's realm.

Beyond Jupiter, the ice freezes to the consistency of granite. This odd bit of trivia can be used to the advantage of off-world construction crews. Just as Inupiat hunters of the arctic use ice to build igloo shelters, so astronauts could use blocks cut from the icy surface to construct various structures. The frigid conditions in the outer solar system make ice an excellent choice for masonry. But as any person living in arctic regions can tell you, it must be insulated from the interior heat. Humans like their rooms warm, and at a temperature of 20°C, their walls and floors will begin to melt. Houses built in permafrost areas rest atop stilts, sheltering the delicate frozen ground below from the heat of their homes. Similar precautions will be made with buildings on the ice moons of the outer worlds.

Another home-building strategy on ice worlds involves the utilization of natural or artificial caverns. In an ice cave, inflatable habitats could be erected without being in direct contact with the frozen surfaces. The cave itself would provide radiation shielding and moderate temperatures. Insulation could also be sprayed onto ice surfaces that, once stabilized, could provide a permanent barrier for interior heat.

At Saturn, two moons appeal for resources, Enceladus and Titan. Enceladus is a case with parallels to Europa: a subsurface global ocean of salt water, an icy surface, and a really big planet nearby. But Saturn generates a more benign set of magnetic fields. The environment of Enceladus would be much easier to set up shop on. Its active geyser-like eruptions will provide astrobiologists with direct access to water from beneath the surface. Some astrobiologists argue that Enceladus is an even better candidate for finding indigenous life than Europa is.

As for exotic natural resources, Titan has them all beat. Hydrocarbons pool in the lakes and fall from the sky. Our future residents will want to use hydrocarbons for fuel, but they won't need to go to those lakeshores; they can dig up all they want in their hydrocarbon-impregnated front yards. Frozen water can be broken down into oxygen and hydrogen, elements that can combine with the abundant carbon and methane to make fuels. For manufacturing, the fundamental elements of plastics, and industrial society in general, are there for the taking.

Aside from modernized igloos, inflatable habitats provide another option at Titan and in the vacuum environment of other moons at Jupiter, Saturn, Uranus and Neptune. For over a decade, Johnson Space Center has been conducting formal studies on habitats—both inflatable and rigid—for the lunar surface. With some added insulation, these technologies are directly relevant to the harsh environs of the outer solar system.

For our human explorers, the planet-moon offers unique challenges for landing. Because of its extended atmosphere, the descent from entry to the surface will last nearly three hours. On Mars, that familiar landing sequence of rovers like Perseverance was all over in "seven minutes of terror." Titan is a far easier place to land because of its dense air and low gravity. When the Huygens probe touched down using its small parachute, it bounced lightly. Even if its parachute had fouled, the probe probably would have slowed itself down enough to survive landfall, simply because of the thick atmosphere and the probe's slow speed in Titan's 1/8 g. Landing on Titan has parallels to the gentle landings of our probes on Venus. In fact, Titan is the only other world besides the Earth with a substantial nitrogen atmosphere. Human survival on the surface is simpler than at other sites: all a Titanian will need is an oxygen mask and a good fur coat. Second to Mars, Titan is the easiest site to inhabit.

Every silver lining has a dark cloud in the middle. Where spaceflight and space stations are concerned, fire is a major worry. In the case of Titan, humans will need to be extremely vigilant because of the naturally occurring materials outside their door, materials like butane, methane and propane. These would prove a perilous mix to residents living inside a habitat filled with oxygen.[14] Titan's flammable hydrocarbon sands and liquid backyard grille gases pose a

---

[14] And what other kind of resident could there be?

real and present danger, although miners have dealt with similar hazards in coal mines for generations. Any airlock on Titan would likely include an ante-room where incoming inhabitants would hose themselves down and vacuum off their suits and equipment.

Getting there, and settling in

The rich natural resources of the worlds around us beckon us to venture out, but engineers, scientists, nutritionists and sociologists face the practical problems of how to get us safely to distant worlds. The assignment isn't an easy one.

For some time, engineers have been addressing the complexities of inter-planetary travel. Sending a several ton SUV-sized lander to Mars is one thing, but sending a 100-ton+ craft capable of making the crossing with delicate humans aboard is another prospect entirely. Scenarios like Zubrin's Mars Direct, which relies on *in situ* fuel manufacture at the Mars, show promise for travel to- and building of infrastructure at- the red planet. Other companies, including Lockheed Martin, Blue Origins and SpaceX, have their sites on the Mars as well. New propulsion technologies will make a difference, as we will see.

For its part, Jeff Bezos' Blue Origins company plans to spread its wings with a NASA initiative called ESCAPADE (Escape and Plasma Acceleration and Dynamics Explorers), a pair of robotic spacecraft destined for Mars. With plans under way for the fabrication of their Blue Moon lunar lander, Blue Origins is already tooling up to go farther. Blue Origins envisions the 2024 Mars launch as marking the beginning of interplanetary operations for the organization, which until this writing has been limited to suborbital flights. Its 100-meter-tall New Glenn booster will serve as launch vehicle for the dual Mars mission. But Blue Origins' view is wider, taking in eventual human flight to the Moon and Martian environs.

NASA's Artemis Moon program, which plans to return human explorers to the surface of the Moon by 2025, utilizes technologies that can be modified and adapted to deep space missions. The backbone of its operation is the Orion spacecraft, a vehicle designed to ferry humans into Earth orbit and then into cislunar[15] space. Orion has deep roots: it was the centerpiece of NASA's Constellation program, designed to return humans to the Moon. Constellation was shelved in 2009 and refocused in 2011. Work continued

---

[15] the region of space from the Earth out to and including the region around the surface of the Moon

on Orion, NASA's next-generation human exploration vehicle. Orion now has a crew module manufactured by NASA and a European Service Module—sporting communications, solar power and propulsion—built by ESA's prime contractor, Airbus. The ESM uses Space Shuttle Orbital Maneuvering System Engines for propulsion. And while the original spacecraft was to rely on circular solar arrays, the ESM now sports two x-wing solar panels, one on each side.

Orion's interior is a marked upgrade from any piloted spacecraft before it. The craft features an advanced "glass cockpit", says Orion designer Timothy Cichan. Cichan, who is a Lockheed Martin space exploration architect, explains, "Orion has switches and physical controls for key things like the power controls for computers and manual backup system controls, but most of the functionality is on 3 screens with buttons on the sides that are configured for each type of screen. This saves weight, allows for a simpler interface, and allows us to more easily update displays as we learn how best to operate the vehicle. And everything is designed to be used with space suit gloves on."

While Orion is designed for cislunar space, even astronauts taking part in the Artemis Moon project will be carried to the outer fringes of Earth's protective magnetic field. On longer voyages, radiation will become even more of a concern. We have seen that simple water can be used as an effective shield. Water stored within the walls of a spacecraft or habitat could double as shielding. The "storm shelter" option is also under consideration. Orion's designers have built one into the craft, the first of its kind. Apollo astronauts traveled to the Moon during "solar minimum", the Sun's quietest time in its eleven-year cycle. But future space farers will not have that luxury: their travel will not be restricted to the Sun's more quiet moments. Of additional concern is that radiation levels increase farther from the Earth. In the event of a solar flare or other increased activity, astronauts will climb into Orion's radiation storm shelter. Orion's main storage lockers are situated at the bottom of the vehicle, surrounded by heavy computers and life support equipment. In a radiation emergency, crew members can remove items in storage, climb into the lockers, and then cocoon themselves within the heavier storage items. Orion's structure, heat shield, computers and life support equipment provide dense material on hand for protection. It's a fast, low-tech solution, but these radiation events will be deadly for only minutes to hours. The spacecraft is designed to withstand the most severe solar storms on record.

The spacecraft has already been tried out (Figure 7.8). The first operational Orion capsule flew into deep space on NASA's Artemis I mission. Orion has the capacity to carry out thousand-day-duration missions. To get to farther

Figure 7.8 Photo taken from an uncrewed Orion spacecraft during the Artemis I mission, looking back at the Earth/Moon system from the furthest point in its travels. (NASA/Lockheed Martin)

destinations like Mars, it will need to. Missions of more than a few weeks will require recycling of key elements like oxygen and water. Completely recycling these elements with no loss or waste is called a closed loop system. "For long duration flight we would want to have as much of a closed loop system as possible," says Tim Cichan. "This would involve filtering the waste water, and capturing the carbon dioxide and converting it back to oxygen." Engineers realize that getting to a 100%-efficient closed-loop system is not possible, but the closer they can get, the less consumable mass they will need to pack. The International Space Station (ISS)—the closest thing we have to a long-duration deep space mission now—has been developing, testing, and using partial closed-loop systems, slowly increasing the amount that they can convert.

Earlier missions on the ISS pulled oxygen from water. Today, the European Space Agency's Advanced Closed Loop System (ACLS) recycles fully half of the station's carbon dioxide (exhaled by the astronauts), generating oxygen. The new system captures carbon dioxide from the air, and then shunts it through small beads created specifically for the process. Steam extracts the carbon dioxide from the beads, carrying the gas into a reactor that combines methane and water. Electricity splits the water into oxygen and hydrogen, and the waste methane is vented overboard. The system uses far less water than earlier equipment. The ACLS process conserves some 400 liters of water annually. ACLS is an important step toward sustainable life-support systems for deep space. One early application will be the lunar Gateway, part of the Artemis architecture for returning humans to the Moon.

While the atmospheric recycling is not quite a closed loop, the ISS has a fully closed water recycling system. Moisture is captured from wastewater such as urine, sweat and vapor from astronauts' exhalations. Filtration systems pull contaminants and impurities from the water, yielding fully potable water that is reused for drinking, rehydrating food, bathing, etc. The resulting water is actually more pure than most drinking water on Earth.

Recycling aboard the International Space Station, and aboard Orion and other flight systems to come, echoes similar processes invented billions of years ago on planet Earth. Less than two percent of all Earth's water is fresh water (including the water frozen at the poles and in glaciers). Processes similar to those being tested on orbit promise to enable us to recycle water more efficiently, preserving precious water resources. The ultimate goal is to more proficiently incorporate recycled water to irrigate crops, drive plumbing functions (like water for flush toilets or industrial pumping stations), and recharge underground aquifers.

Deeper into space

Orion designers have built in systems that are beefy enough to carry out lengthy missions of three years. Lockheed Martin's base scenario for getting humans to Mars (Figure 7.9) incorporates Orion into a larger interplanetary stack of modules, called the Mars Transit Vehicle, or the Deep Space Transport (DST). Orion will function as the command deck. From its cockpit, crews can control and monitor the entire Mars Base Camp vehicle from stem to stern. Orion also serves as an excursion vehicle, and an emergency return vehicle (roles similar to those served by Dragon and Soyuz spacecraft at the International Space Station). Two Orion spacecraft would be married to dual habitation modules, each with a propulsion system. Propulsion would be via either electric propulsion, conventional chemical thrust, or nuclear propulsion. In addition to living areas for the crew, habitats would house research laboratories equipped with physical science instruments, electron microscopes, small greenhouse or hydroponic chambers, and equipment for medical and chemical analyses. From their orbital perch, crews could operate

Figure 7.9 Top: Artist Kevin M. Gill envisions an advanced Mars application for SpaceX's Dragon spacecraft. (image courtesy Kevin M. Gill, ©2015) Bottom: Lockheed Martin's concept for an orbital "Mars Base Camp". The interplanetary ship uses two Orion vehicles at its core, and uses nuclear thermal propulsion. (courtesy Lockheed Martin Corp)

robotic sample return missions to the surface, with samples investigated in orbit aboard the DST habitat laboratories.

Lockheed Martin's "Mars Base Camp" Deep Space Transport would use massive solar arrays for its primary power, as the ISS does today. While fuel cells are an option, they would take substantial amounts of cryogenically cooled hydrogen and oxygen, something that may not be ideal on long flights. "We haven't explored nuclear options yet," says Cichan, "but for the surface of Mars we will need Fission Surface Power [stations] given the dust storms and the distance from the Sun, so it would be good to use those on the journey as well."

According to the Lockheed Martin website, "The concept is simple: transport astronauts from Earth, via the Moon, to a Mars-orbiting science laboratory where they can perform real-time science exploration, analyze Martian rock and soil samples, and confirm the ideal place to land humans on the surface in the 2030s." Solar electric propulsion craft would be dispatched to Martian orbit ahead of time, where they would store key supplies awaiting the astronauts. These would include fuel, food supplies and oxygen, but also robotic landers that could be tele-operated from orbit.

Although the Orion spacecraft has the capacity to carry six astronauts, baseline Mars Base Camp studies call for four (various Artemis lunar mission concepts involve six astronauts). Orion relies on redundancy for safety, both in having multiple copies of systems, but also in having dissimilar backups for the most critical systems, such as the flight computer or access to manual oxygen valves in a crisis. On missions to remote destinations—Mars and beyond—multiple sets of full modules will serve as backup. The Mars Base Camp envisions two entire sets of modules, either of which could be used to return home. In other words, crews could survive and return home in only half the modules contained within the interplanetary stack. (see Figure 7.9)

Spacecraft designers are also studying the possibility of pairing one of the two Orion craft with one of the Mars Base Camp's propulsion modules. This configuration would enable one of the Orions to leave the main vehicle once it was parked in Mars orbit. Orion could then venture to one or both of the Martian moons. Nominally, the Mars Base Camp assembly would return to the lunar Gateway in high lunar orbit, where it would be refurbished, its systems recharged, its furniture dusted, and windows washed in preparation for a return trip to Mars. But if the Mars Base Camp interplanetary assembly had a critical failure, Orion would then return the crew directly to the Earth. On return flights from more distant venues, Orion would likely carry out a direct return, as Earth-approach speeds and fuel constraints would make orbital entry of the entire stack more problematical.

Orion is too cramped for a deep space mission. Crews require physical space and interior variety for mental health. Two large habitats form part of the core of Mars Base Camp, affording crews exercise and relaxation areas. Additionally, crews would require good communication systems to keep them in touch with home, even given the light-time delays of transmitting over great distances. Crews would need to be selected and trained for the long duration and high-risk environment. Much has been learned in this area from our experience with space stations like ISS and the Soviet MIR.

Many elements of the Mars Base Camp are in the early conceptual stage of development. Because Orion already has built-in capability for deep space missions, including life support, radiation protection, communications, etc, this reduces the complexity of the habitats.

Building a craft that can provide a healthy environment for a crew over long periods is demanding. The specific architecture for a deep space mission to another world is a moving target. Technologies improve and focuses change. A "control deck" like Orion will need to be a small part of a much larger living space consisting of one or more habitat modules, as seen in the Mars Base Camp designs. Many designers favor rigid modules similar to those used on the ISS. In the industry, these are referred to as "tuna cans". But another approach is an inflatable habitat. In this scenario, a firm core houses supplies, electronics and other support, surrounded by an initially deflated habitat. Once in orbit or en route to another planet or moon, the core would pressurize the elongated donut-shaped structure around it, expanding to a large living and working area. A deflated habitat takes up far less room on the launch pad than does a rigid one, so can expand to create larger volumes for the crew. And in the space environment, where there is no up or down. Since there is no ceiling or floor, the physical elements of the habitat can be more efficiently arranged for the use of every surface.

Early development of inflatable habitats took place in the halls of ILC Dover Corporation, on a project called Transhab. Transhab was under study for use as habitat modules on the International Space Station. The inflatable habitat module would expand to twice the diameter possible with ISS's rigid modules. The project was eventually scrapped due to budget constraints. Bigelow Aerospace purchased the ILC Dover Transhab's main elements, continuing work on an inflatable habitat for a projected private space station. Inflatable habitats are seen as a solid option for extended interplanetary flights.

Johnson Space Center and Lockheed Martin have been testing inflatable habitats for some time. During the Constellation program, JSC engineers designed a pressure bladder—the kind that would make up the interior of an inflatable habitat—supported by a web of metallic/fabric straps similar to Kevlar. The pressure vessel was inflated, and strips were systematically cut to determine how well the structure held up under pressure. Lockheed Martin is carrying out similar failure mode tests on inflatable habitat scale models, over-pressurizing the structures until they burst. The results indicate that inflatable habitats can be operated safely under the harsh conditions of the lunar surface or interplanetary space.

Lockheed Martin's Orion has been under development since 2006. Other space agencies and aerospace companies have benefited from its design and engineering studies. Because the US taxpayers funded and "own" Orion, lessons from Orion's design, development and fabrication are available to the public. Aerodynamic data, parachute design and testing, and heatshield technology are just some of the examples of information available to the commercial space endeavors of private companies as they develop their own systems. The reverse is not true. Commercial crew vehicles are owned by private companies, whose research and inventions are proprietary.

As for Orion-class missions with destinations beyond Mars, says Lockheed Martin's Cichan, "We have not really discussed going further than Mars, but there's nothing in particular that would preclude those types of future missions." Elon Musk's SpaceX aerospace company is eyeing targets even more distant than Mars, and they are developing quite a track record to back up their visions. As primary contractor for the lunar landing portion of NASA's Artemis program, SpaceX is fabricating their fully reusable Starship Human Launch System (HLS). Using new engines three times as powerful as the Merlin engines used in its Falcon launchers, the craft will be able to loft 250 metric tons to low-Earth orbit, more than twice the payload orbited by the Apollo program's Saturn V. The Super Heavy booster uses 33 methane/oxygen "Raptor" engines. The Mars scenario version of the upper stage Starship sports nine of them, three optimized for sea-level and the others ideal for operation in a vacuum. Combined with the first-stage booster, the Mars mission would utilize a total of 51 engines (the Artemis lunar configuration has six engines in the upper stage, making a total of 39 engines on the entire vehicle). After launch into low Earth orbit, the upper stage would refuel before setting off for Mars. The interior of the craft is roomy enough to accommodate living and research areas for a crew of four to six over a roughly thousand-day mission.

Designers envision caching fuel at depots situated strategically among the planets, moons and asteroids of the solar system. With filling stations in place, missions could be carried out at a variety of targets. As CEO Elon Musk announced at a recent meeting,[16] "By establishing a propellant depot in the asteroid belt or on one of the moons of Jupiter, on Enceladus or Europa or even at Titan, this system really gives you freedom to go anywhere you want in the greater solar system." And with its hundred tons of cargo capacity, the

---

[16] International Astronautical Congress, Guadalajara, Mexico, 2016

Figure 7.10 The aerodynamic design of the SpaceX Starship space system would make it a natural vehicle for landing on Titan—if we could just get it there. (art by the author)

craft would have plenty of room for supplies and crew habitation space, even for missions more complex than those to Mars (Figure 7.10).

SpaceX's much smaller Dragon spacecraft has ferried crews and cargo to the International Space Station on a routine basis. It was the first privately-funded spacecraft to do so. Engineers are examining ways of modifying even this craft for Mars ventures. Dragon is 8 meters tall and 4 meters in diameter, with a launch-to-orbit capacity of 6000 kg and a return payload mass of half that. Like Starship, Dragon is built for propulsive landings, making it capable of landing on the Martian surface. In fact, it can land on any solid or liquid surface. With powered landing, the craft could, theoretically, be dispatched to asteroids or the Galilean Satellites. In a Mars mission architecture, Dragon would serve as the crew module in a similar role to Orion's with the Mars Base Camp scenario.

Designs like Starship HLS and Orion show promise for the short term, but a fundamental issue, analysts say, is that travel times to distant worlds must be decreased substantially. The length of interplanetary voyages may constitute the most daunting aspect of human exploration of the outer solar system. Our closest target beyond Mars, Jupiter, is 588 million kilometers at its closest approach, but we cannot make the trip directly. Traveling to another planet is a game of cat and mouse. A spacecraft must coast along an arc stretching across the solar system to a point in front of the target planet. In effect, the spacecraft places itself in a position for the planet to catch up to it.[17] Crossing millions of kilometers of the void involves great spans of time, and for human crews, those spans translate into long-term radiation exposure, subjection to extended microgravity environments, increased crew anxiety and boredom, and the consuming of vast quantities of supplies. The obvious solution is to trim down the travel time. Shorter trips mean reducing crew exposure to the hazards of the space environment, as well as reduction in every other consumable, from water to oxygen to food to fuel to power. As the character Ian Malcolm admonished in Jurassic Park, "Must go faster!"

Rockets with spark

Solar Electric Propulsion (SEP) is one option under consideration for the Mars Base Camp baseline study.

SEP has been successfully used on several long-duration robotic flights. NASA's Deep Space 1 spacecraft used an ion engine, powered by photovoltaic solar cells, on its mission to asteroid Braille and comet Borelley. Japan's two Hayabusa asteroid sample return missions also used the technology, as did the NASA/Jet Propulsion Laboratory's Dawn asteroid orbiter at both Ceres and Vesta. The European Space Agency's BepiColumbo Mercury mission is currently using the technology en route to an eventual Mercury orbit (see Appendix D). SEP's advantages over conventional chemical propulsion include safety, economy, and greater power over long periods. SEP uses a tenth as much propellant. Although the power of the engines is far less than chemical propulsion, they can

(continued)

---

[17] Assuming a direct flight. However, many large spacecraft make use of "gravity assists", using planetary flybys as slingshots to gain enough velocity to get to the outer system. The Saturn-bound Cassini mission made use of multiple flybys of Venus, Earth and Jupiter for gravitational assists, building up its velocity and traveling a staggering 3.4 billion kilometers before reaching its goal.

remain on for constant thrust, building up great velocities over long periods and cutting travel time, an important feature for piloted voyages.

Studies show that for a human-scale mission, an interplanetary craft would require 600 to 800 kilowatts of power. This compares to the Dawn spacecraft's 10 kW of power. The thrust for such a mission would require ion engines with a specific impulse of 2000 to 2500 seconds, comparable to Dawn's 3100 seconds of thrust.

More powerful versions of Solar Electric Propulsion are under development, including:

Hall effect thruster: uses a magnetic field to ionize propellants, accelerating the ions as expelled thrust.

Pulsed Plasma Thruster: First used in space by Soviet Zond 2 and 3 in the early sixties, these simple thrusters send an electrical arc through solid fuel (usually Teflon) to create plasma thrust.

Variable Specific Impulse Magnetoplasma Rocket (VASIMR): This variation on the SEP sends radio waves into an inert propellant to generate plasma. A powerful magnetic field then constrains and directs the plasma through the rocket engine. VASIMR engines can process far more power than other types of Solar Electric Propulsion systems, but the technology is still evolving.

Reducing the travel time means developing more powerful and efficient propulsion. Engineers, scientists, inventors and science fiction authors have been exploring many types of propulsion over the years. In addition to the familiar chemical propulsion used by most spacecraft today, advanced studies of nuclear propulsion, electric ion drives, and solar sailing are beginning to pay off with concrete solutions. To date, three kinds of propulsion have been used to send spacecraft on interplanetary missions, chemical rocket engines, solar sails and solar electric—or ion—drives.

Conventional, or chemical, rocket engines are the standard workhorses of today's space programs. Solid rocket fuel is often used for lower "strap-on" stages, while liquid fuel is far more powerful. From the Falcon 9 to the Ariane 5, from the Russian Proton to the Delta Heavy, conventional boosters carry fuel and oxidizer, each stored separately. When the fuel and oxidizer mix, they create a controlled explosion[18] that flows out the back end of the rocket through the engine bell. A rocket moves by ejecting mass in the opposite direction of its flight. Rocket fuel takes up a lot of physical space, and it is heavy. But it is reliable, and to get out of Earth's grip, it's the only game in town, for now.

---

[18] Sadly, exploding rocket fuel is not always controlled. Space exploration is a difficult and dangerous business, and some have given their lives for it.

Once out of the Earth's atmosphere, spacecraft are beginning to use an efficient form of propulsion that can provide steady thrust over the long distances involved in outer planets exploration: electric (or *ion*) propulsion. Instead of heating chemicals, ion engines heat gas into a plasma. An ion drive can go faster on a lot less fuel. The strength (specific impulse) of an ion engine is very high, but its acceleration is gradual. Unlike conventional engines, the ion drive can run over long periods of time. Geostationary satellites regularly use ion thrust to remain in proper orbit over long periods of time. The Japanese Hayabusa asteroid sample return missions, China's Tiangong space station, and ESA's BepiColumbo Mercury projects are a sampling of missions that have included ion propulsion. The Dawn dual-asteroid mission used xenongas ion drive to carry it on a looping trajectory to orbit the asteroid Vesta. It then departed for another orbital study of the largest asteroid, Ceres. At one point, Dawn fired its ion thrusters continuously for 270 days. For comparison, Dawn's ion drive is capable of accelerating from standstill to 60 mph in 4 days. This compares to a conventional rocket, which leaves the ground and reaches an orbital speed of 17000 mph in less than ten minutes.

Ion drives use electricity to accelerate propellant out of the back of the engine. Solar electric propulsion combines magnetism and electricity to propel a ship through space. Solar panels deliver a positive electrical charge to atoms inside a chamber. Magnetic fields pull the atoms toward the back of the ship, where they are pushed by an oppositely-charged magnetic field out of the ship. The steady flow of ions streaming out of the spacecraft provides thrust. The thrust is much weaker than chemical propellant, but the engine can be run for months rather than minutes, building up tremendous velocity.

A promising new concept in faster propulsion is the brainchild of former NASA astronaut Franklin Chang Diaz. Chang Diaz has been developing the Variable Specific Impulse Magnetoplasma Rocket, or VASIMR, engine. VASIMR uses microwaves to heat propellant, turning it into a plasma. Says Chang-Diaz, "It does not use chemistry, it does not use nuclear, it uses electricity. The electricity has to come from somewhere. So in our early applications of the VASIMR, we would use solar arrays that produce electricity. That electricity will be what we use to produce the plasma out of the working propellant."

VASIMR shares some common ground with other electric propulsion engines, but the VASIMR is able to process far more power. VASIMR is also able to work around a problem common to other electric propulsion systems: heat. When an engine runs for extended periods, hardware heats up and moving parts fail. But VASIMR generates a magnetic field that isolates the flow of the hot plasma. The field forms a pipe through which the plasma moves without touching the hardware generating it. VASIMR is able to bring its plasma to millions of degrees, thus increasing its power.

The output of a VASIMR engine is in the neighborhood of 200 kW, about a hundred times that of earlier ion engines. But the problem with a 200 kW engine is that it requires a 200kw power supply. This is one of the limiting factors of VASIMR's technology. Advanced solar panels are barely capable of producing 200kW. (The massive solar panels on the International Space Station are each capable of producing roughly 100kW). But solar panels, once extended, are ungainly and prone to damage. Some designers prefer another option: nuclear power. VASIMR is not a nuclear propulsion engine, but the electricity to make it work might come more efficiently from a high-yield nuclear power plant, with an output of thousands of times that of solar power. Power levels like that would trim a human mission to Mars from six or eight months to less than 2 months. But the issue is that no one has invented a 200mW space-worthy nuclear reactor. Not yet.

Sailing the solar seas

The year 2010 saw the launch of the first interplanetary solar sail mission, the Japanese Space Agency's Ikaros. Unlike other forms of propulsion, solar sails are passive. Deployed in the airless environment of space after a conventional launch, solar sails utilize the pressure of sunlight to move, just as a tall ship uses wind to traverse the seas. The kite-shaped Ikaros spread some 14 meters across. Solar cells embedded in the sail fabric generated electricity. Liquid crystal panels could change the surface reflectivity, enabling the craft to steer. JAXA also plans to send a 40-meter rotating solar sail with dual solar-powered ion propulsion engines to encounter Jupiter's Trojan asteroids.[19] Called the OKEANOS (Oversized Kite-craft for Exploration and AstroNautics in the Outer Solar system), the spacecraft would also carry a small lander to study one of the Trojans from its surface. Although passed over for selection in the early 2020s, engineers hope to resubmit the mission for a launch in the near future.

Other solar sail projects are in progress. The Planetary Society's LightSail-2 successfully launched in 2015. Over the course of the next three years, the 1/3 acre craft demonstrated controlled solar sailing.

While these solar sail vehicles have confirmed the practical use of the technology, another approach involves the use of a maser to boost a solar sail

---

[19] The Trojans follow at roughly 60° behind or in front of Jupiter within its orbit around the sun.

Figure 7.11 Left: An original NERVA nuclear propulsion engine at Marshall Space Flight Center. (courtesy Mike Jetzer, heroicrelics.org) Right: Test of a Hall ion thruster, similar to ion drives used by spacecraft such as JAXA's Hayabusa series and NASA's Dawn. (NASA)

composed of a mesh of wires with the same spacing as the wavelength of the maser's microwaves. Masers could also be used to power a solar sail coated with a layer of chemicals designed to evaporate when hit by the maser's radiation. The evaporating coating would provide thrust. But to date, payloads using solar sails have been very small. Human-scale vehicles are still a future prospect.

Another style of propulsion under study is called the thermal nuclear rocket (Figure 7.11). This type of engine uses the heat from a nuclear reactor to accelerate its fuel and create thrust. Unlike solar powered ion or plasma engines, thermal nuclear engines perform essentially the same process as chemical rockets, but they do so using only half the fuel.[20]

Since the design of NERVA engines, new, lighter materials have become commercially available that perform at higher temperatures and pressures. A new generation of nuclear propulsion under development is the Nuclear

---

[20] Early US experiments with nuclear propulsion culminated in the NERVA (Nuclear Engine for Rocket Vehicle Application), a nuclear thermal rocket engine program that progressed steadily for nearly two decades before its cancellation.

Cryogenic Propulsion Stage, or NCPS. Plans are on the books for an engine complex with a fueled weight of 40 tons. For safety, the NCPS would be brought on line only after successfully reaching space. In the case of deep-space probes, the NCPS would not be initiated until the craft is ready to leave orbit.

Creating micro-Earths for temporary homes: close to Earth

At dawn, a young girl walks her standard Poodle, Ralphie, on a long paisley lead. She crosses a bridge over a gurgling brook, letting the doggie off leash to splash through the cool water. In front of her, the ground slopes up to a patchwork of farmlands and low hills. Waterfalls spill from the tops of distant cliffs as hawks soar overhead. Bumblebees hover over spring blossoms. As Ralphie snaps up one of the rare butterflies flitting by, he ventures on across a landscape that curves up into the sky. The Moon is setting quickly in the west. The Earth is rising in the east. Yes, the idyllic paradise is a perfect facsimile of Earth, but it's not our home planet. And it's not Mars. The girl has set off on her perambulation inside a vast metal cylinder. She lives in a habitat stationed somewhere to the side of the Earth and Moon, at the L5 Lagrangian point, a location equidistant from the Earth and Moon.

Space colonies like this one have been dreamt of for some time. Jonathan Swift, in his 1726 masterpiece *Gulliver's Travels*, was certainly one of the first when he described the great floating island of Laputa powered by magnetic levitation. But Laputa was not really a space station: it floated in the air. The American minister Edward Everett Hale penned what is probably the first true description of a space station. In his story *Brick Moon*, he even described, in 1869 fashion, how the station could orbit the Earth.

> The plan was this: If from the surface of the earth, by a gigantic peashooter, you could shoot a pea upward from Greenwich, aimed northward as well as upward; if you drove it so fast and far that when …it began to fall, it should clear the earth, and pass outside the North Pole; if you had given it sufficient power …that pea would clear the earth forever. It would continue to rotate above the North Pole, above the Feejee Island place, above the South Pole and Greenwich, forever, with the impulse with which it had first cleared our atmosphere and attraction.

The inhabitants aboard Hale's Brick Moon carried out activities that anticipated global navigation, satellite communication and even remote ocean science. Just a few years later, Russian space science pioneer Konstantin

Figure 7.12 Left: The interior of an O'Neill cylinder, with landscaping patterned after the San Francisco Bay area. Note the lines of the solar windows in the background. The painting is by master artist Don Davis, and was done under the direction of Gerard O'Neill himself. (NASA/Ames/Don Davis) Right: diagram showing sunlight track reflected into the habitat. Windows reveal the landscaped walls inside the titanic cylinder. (diagram by the author)

Tsiolkovsky described space communities in his book *Beyond Planet Earth*. But with technology came paradigm shifts, and one of the most dramatic unfolded in 1977 with the work of Gerard K. O'Neill. O'Neill, a physicist at Princeton University, envisioned vast columnar space colonies stationed at the Lagrangian points L4 and L5, essentially parking lots in the Earth/Moon system (Figure 7.12). Material from Earth would kick-start construction, but the bulk of building material would be sent from the Moon via mass drivers. The column walls would be arranged as six strips. Landscapes would alternate with long windows. The titanic colony would point with its longest axis directly facing the Sun. Mirrored louvers would direct sunlight into the colony, and would close for the night portion of the day/night cycle. The cylinder would spin, enabling its inhabitants to live on its walls under normal gravity. Regolith from the Moon and minerals from Earth would serve as the landscaped soil for farming and forestation of wilderness areas. The cylinder would be so large that some planners predict the condensation of clouds in its centerline, creating an active water cycle of precipitation and evaporation.

Aerospace futurists have put forth many other arrangements for space colonies, including floating globes, T-shaped affairs, spidery constructions (think International Space Station) and turning wheels. The latter design was illustrated beautifully in the classic space station scene from Stanley Kubrick's

2001: A Space Odyssey (1968). But our assortment of space colony choices begs the question: what's the advantage of establishing cities out there? It's a good question, and advocates believe they have good answers. Proponents of space colonies contend that space construction on this scale will enable humankind to transfer most of its heavy industry to orbit, thus relieving the ecologically stressed Earth. Manufacturing in microgravity, use of lunar, asteroidal and extraterrestrial resources, and access to free solar power unencumbered by an atmosphere are all points cited by space city promoters. Another important aspect of the O'Neill colony model is the production of energy using solar power generators in orbit. These, they say, could beam energy down to Earth and decrease humanity's dependence on fossil fuel or nuclear power. Removing so much human activity from the Earth's delicate biome would, they suggest, preserve the planet while assuring a human presence elsewhere, in case a natural disaster or global conflict would lay waste to the genesis world.

Creating micro-Earths for temporary homes: Mars

For those who prefer to walk on firm ground, engineers and explorers continue to draft plans for future Earth-like homes. Perhaps even before we attain our quest for high-speed travel, we may well be on our way to deep space en masse. Dozens, and then hundreds of scientists, workers, entrepreneurs, adventurers and settlers will make their way Marsward, once our brave astronaut explorers have blazed the trail. The exodus will not be one of escape. Mars could never support the human population, and as we have seen, many view it as irresponsible to look to other worlds as "plan B"s rather than caring for our own world. But we will go and we will build. Hopefully, we will create our new homes responsibly. How then shall we live?

As we saw in Chapter 6, Mars can be a brutal world. It has just enough atmosphere to make landing difficult and dangerous, but not enough to keep a habitat from explosive decompression. The first action in the construction of a large Mars colony may well be something familiar to camping fans: setting up a tent.

It's hard to change the entire climate of a planet, but what if you could cover parts of the planet in tents instead? Engineers and biologists envision "tented" craters on Mars. A pressurized enclosure would inflate like a balloon, and environs inside could be brought into Earthlike conditions. The air would be pumped up and infused with oxygen and nitrogen, both available in various forms on Mars (the oxygen, for example, would come from water). Soil could be processed so that crops could be grown and forests or jungles could flourish. Planners foresee vast wildlife areas where rare species would multiply and thrive. Villages and, eventually, cities would rise from the Martian soil, with open avenues and parks where inhabitants could wander without the protection of a space suit.

Smaller domed enclosures can also include localized facilities like greenhouses and small settlements, and could also house energy and water-processing plants.

Richard Taylor of the British Interplanetary Society wants to take things even farther. *Why not put a tent over the entire planet?* He asks. Taylor calls his scenario Worldhouse.

According to Taylor and others, the inventory of volatiles on Mars has been depleted over time, leaving us with a planet nearly devoid of important elements and gases. Turning the Martian environment into something livable, he says, is more doable under a tented, pressurized enclosure, especially in light of key missing material in the Martian environment. Taylor points out that "the likely low availability of nitrogen may constitute a 'critical limiting factor' preventing the establishment of a gravity-bound terraformed atmosphere of terrestrial composition and surface pressure...Under such conditions a fully habitable environment could still be achieved by constructing a deliberately restricted ecospheric environment (DREE) on the planet. This biosystem, enclosed within a quasi-global 'Worldhouse', can mimic closely the free-running seasonal biological and geochemical cycles of the Earth and would function in an almost completely self-stable and self-regulatory manner."[21]

Taylor's globe-spanning pressure vessel would create an inhabitable ecosystem on Mars far sooner than a global engineering project designed to convert the natural Martian environment. Beginning as small clusters of domes and tented lowlands, the pressurized envelope would be immediately inhabitable. As it grew to enclose more of the landscape, towers would hold the pressure vessel's roof aloft at an altitude of several kilometers, with supporting cables forming a planet-spanning web. Sunlight would pour through the clear canopy for a natural day/night cycle. This would be important to agriculture and establishment of wilderness areas.

Not all of Mars would be draped in plastic. To preserve the natural climate and cycles on Mars, some of the more dramatic features would be left out in the open. The highest mountains (like the volcanoes of Tharsis and Elysium),

[21] Richard L.S. Taylor, British Interplanetary Society, *Advances in Space Research*, Volume 22, issue 3, 1998

the deepest canyons, and the atmosphere-linked polar caps would all be independent of the pressurized regions.

Advocates of the concept claim that the massive undertaking could be completed utilizing technology in use since the 1960s. Nevertheless, Taylor's report states that, " Constructing, creating and maintaining a habitable contained environment on Mars would be the greatest architectural and engineering challenge yet undertaken by mankind, although it would require little technology much in advance of that currently available."

Skeptics point out that a project on the scale of Worldhouse requires immense amounts of construction and maintenance. The vast acreage of pressurized envelope would also be susceptible to puncture by incoming meteors, which would cause violent depressurization. Orbiters have observed dozens of newly-formed craters on the Martian surface, all occurring in just the few decades that our spacecraft have been in operation there. Barriers within the tented areas could prevent total decompression, isolating various sections from each other. Finally, it is unlikely that a bona fide water cycle could be triggered in the air beneath such a low ceiling. Water may condense on the underside of the tent, but probably not enough to carry out independent agriculture. A costly irrigation system would need to be emplaced for crops as well as for forested wilderness areas (Figure 7.13).

Figure 7.13 Thousands of plant species thrive within the domes of Cornwall's Eden Project. (photo by the author)

The other biospheres

The Earth's complex, interrelated cycles keep its biosphere healthy. The planet seamlessly recycles oxygen, water and food across a variety of environments. Nature's balances are difficult to mimic, but we will need to learn to do so if we are to successfully establish outposts and settlements on distant planets and moons. Closed loop systems, also known as closed ecological systems,[22] are interconnected systems that sustain themselves without the introduction of outside energy or material. They are self-regulating, maintaining balances in chemistry, soil conditions, air, etc. Closed loop systems will be necessary on long interplanetary voyages as well; because of the extended travel times involved, mission planners will need to incorporate closed loop systems and advanced recycling methods for success. To that end, the Biosphere 2 experiment was conceived as a test of protracted closed systems involving humans, animals and plants. The controversial experimental facility covers over 3 acres of desert in Arizona (US). The habitat originally housed five distinct biomes, and is the largest closed ecological system every constructed. Biosphere's biomes included tropical rain forest, grasslands, desert, and a 600,000-gallon ocean with coral reef. The structure also has a "lung" to move atmosphere and stabilize day/night pressures. An agricultural area for growing crops and raising fish and livestock supported living quarters, common areas and kitchen, and research laboratories. Biosphere executed its intended experiments to evaluate closed-loop ecological systems on two separate occasions, from 1991-1993, and March – September of 1994. Both experiments provided valuable data, and both suffered from low oxygen output and mortality of both animals and plants. Still, crews did not suffer from malnutrition, as they stuck to a revolutionary high-nutrient, low-calorie diet. The second crew rotation actually achieved "total food sufficiency". Additionally, Biosphere 2's agricultural segment was among the most efficient and highest-yield in the world, surpassing famously productive farming in Indonesia and Southern China by five times. Although some outside material was introduced into what was supposed to be a closed system, valuable data was gathered in biomedical research, agriculture, and the growth and interplay of complex ecosystems.

Biosphere 2 was widely criticized for its perceived "alternate" scientific approaches, despite the fact that many of the players were bona fide researchers in their various fields. One reviewer referred to the crew and its overseers as "a clique of recycled theater performers." Ecology magazine characterized Biosphere 2's approach as "New Age drivel masquerading as science". Political intrigue on the outside and conflict among the "biospherians" tainted some results, but the lab continues to perform research. The University of Arizona now oversees the facility (after a several-year stint by Columbia University). The laboratory is now focused on its "Land Evolution Observatory", a $5 million project that studies the physics of watersheds in different conditions. LEO studies how ecosystems are driven by climate, and how they interact with the atmo-

(continued)

---

[22] Closed loop systems are also found in non-biological fields, such as mechanical or electronic

sphere. Biosphere 2 also includes a Moon/Mars greenhouse prototype that uses plants to recycle and purify water.

Biosphere 2 is in good company: organizations across the globe continue to do research into closed loop systems, isolated biological assemblies, biodiversity and climate change. The Eden Project in Cornwall (UK) consists of several domed biomes. The primary dome houses the largest contained rainforest in the world. Adjacent to it, a second dome emulates a Mediterranean climate. The Eden Project is largely an educational outreach center, but also works to preserve various rare plant species. Russia's BIOS-3 is a closed ecosystem designed to support three humans. Algae recycled a substantial fraction of the air in initial experimental runs, while food crops contributed about 25% of the oxygen. 10 human "flights" were carried out from 1968 to 1984, with the longest test lasting 180 days. Work continues there in cooperation with the European Space Agency. The European Space Agency has erected its own experimental closed system, called MELiSSA (Micro-ecological Life Support System Alternative). 30 European engineering and scientific organizations contribute, with the goal of developing regenerative life-support systems for long duration space flight.

Eventually, our technology will progress to the point where we can engineer planetary environments on a global scale. The process, known as terraforming, is a long way off technologically, but we are already getting glimpses into how to embark upon such a massive project. Terraforming is fraught with issues that transcend the engineering ones. It brings us face to face with philosophical and moral questions (we'll explore the possibilities in our last chapter). But what if we could find an Earth-like world just waiting for us out there? Would we have the capacity, or even the need, to go?

8

Distant Earths: exoplanets with potential

<u>That was then…</u>

At the opening of the twentieth century, Martian canals whispered with cool waters beneath blue skies (Figure 8.1). Carboniferous jungles thrived on the banks of carbonated Venusian swamps. Jupiter's Great Red Spot just *had to be* a vast cloud from a volcano. Armed with just enough knowledge to be dangerous, we had made the worlds in our planet's own image. But the Mariners and Veneras and Pioneer probes changed all that. Our solar system was a lonelier place than we thought, with nary another Earth to keep us company.

Beyond our solar family lay a myriad of other suns that begged the question: did any of those distant stars have Earths? Our best telescopes could not resolve other worlds orbiting even the nearest of stars. Was it possible that the Earth was alone, or at the very least a very, very rare case?

The "plurality of worlds", the idea that there are many Earths out there, is an ancient one. Some Greek philosophers saw the universe as arranged in an infinitely repeating pattern, with Earths at center, circled by a sun and planets. It was a view held by such greats as Democritus, Leucippus, Epicurus and Lucretius. These Greek thinkers viewed these Earth-centered universes as operating independently of each other, each arranged as concentric crystalline spheres holding various planets or stars embedded upon their inner surfaces, all on display for us Earth dwellers.

M. Carroll, *Planet Earth, Past and Present*, Springer Praxis Books, https://doi.org/10.1007/978-3-031-41360-5_8

Figure 8.1 Mars in Earth's image: our first estimates of conditions on other worlds were far too earthly. Mars was seen as an "older" world, a frigid desert planet similar to the Earth during what we now know as the snowball period. Before the twentieth century, the true nature of the Martian conditions—and even its moons—could only be guessed. What preconceptions do we have about Earthlike planets in other star systems? (painting by the author)

During the Industrial Revolution, philosophers resurrected the plurality of worlds concept. It was the "machine age", and it seemed that technology and manufacturing infused the air with exciting new ideas.[1] In 1817,

---

[1] The industrial age infused the air with coal dust, too, and its presence would have been a dead giveaway to extraterrestrial astronomers that this little blue planet has life.

Scottish scholar Thomas Chalmers suggested to his audiences that the planets are "the mansions of life and intelligence…there are other worlds, which roll afar; the light of other suns shines upon them; and the sky which mantles them is garnished with other stars."[2] Other lecturers followed his lead, adding to the theme of the Earth as one among many. In 1827, philosopher and theologian William Whewell preached, "the Earth is one among a multitude of worlds."

Despite the many champions of distant Earths, we had no proof of their existence. As the twentieth-century opened, science only added to our sense of isolation by showing us just how large the universe really is. Careful observation and advances in technology demonstrated that the Milky Way was not the whole show, but was one galaxy of many in a universe more vast than anyone had thought.

The modern understanding of vast cosmic numbers brought a sober reckoning of our place—and scale—in the universe, but with it came a new realization of possibilities. Wernher von Braun, whose engineering prowess led to the Apollo Moon expeditions, asserted that, "Our sun is one of 100 billion stars in our galaxy. Our galaxy is one of billions of galaxies populating the universe. It would be the height of presumption to think that we are the only living things in that enormous immensity."

It now appears that the majority of stars in our galaxy play host to planets of their own, and among these we may find hundreds, if not thousands, of planets similar to the Earth. How many are truly like our home world? It may be that our planet simply "hit the lottery", arising in the right location at the right time. Earth may simply have been lucky when it came to its star, its location in its star's habitable zone, its geological makeup, its huge Moon to stabilize its axial tilt; the list goes on (see Recipe for an Earth, below).

Our search for Earths of distant suns may be rooted in a sort of biological imperative, or pure curiosity, or simple survival. Whether our fascination for distant Earths is a Darwinian trait or something more profoundly emotional or transcendent, humans are driven to search the heavens for company in the great starry night. We now know something about the distant Earth-candidates out there, and we are developing techniques to find out what they are like and even whether they might host living biomes.

---

[2] Chalmers, T. (1817) *A Series of Discourses on the Christian Revelation Viewed in Connection with the Modern Astronomy*, Edinburgh, pp. 3-4

The Earth Similarity Index: What makes home "home"?

To better evaluate the habitability of exoplanets, scientists have devised a sliding scale called the Earth Similarity Index (ESI). The index measures the similarity of a planet or moon compared to Earth. Its constraints include the planet's diameter, density, gravity, and its span of surface temperatures. The scale traverses numerical totals from zero to one, with the Earth being one. On this scale, the ESI for Venus is 0.444, while the ESI for Mars comes in at 0.697.

Astronomers base the variables within the Earth Similarity Index on several factors, depending on the technique used for finding the planet in the first place. For example, a planet's density is influenced not only by its mass, but also by its size. A large, low-density planet is likely gaseous or covered in deep ocean. A small, dense globe will be made primarily of rock and metal, terrestrial in nature.

In addition to size and mass, the temperature of a planet's surface can be influenced by its surface albedo (brightness), the amount of heat falling upon it from the parent star, tidal heating and the type and structure of atmosphere (which can act as a greenhouse "blanket" to retain heat).

An ESI rating is not a direct estimate of a planet's habitability, although some of its variables influence that aspect of a planet. Worlds with high ESIs are most likely terrestrial, with solid rock surfaces and sizes fairly close to that of Earth. To date, the planet with the highest known Earth Similarity Index number is TRAPPIST-1e, checking in at 0.95 on the scale. Planet KOI-4874.01 is currently awaiting confirmation, but theoretically has a rating of 0.98. But as for stable planets with active biospheres, we are currently left with only one example: the Earth.

As we look to the heavens in search of other Earths, it helps to review what factors contribute to the benign nature of our own planet's environment.

The orbit: Perhaps the most important prerequisite filled by our planet is that it orbits in the Sun's habitable zone. Like the fairy tale, it is the "just right" Goldilocks zone. But in order to have liquid water on any surface, there must also be air pressure. Too much air, and the atmosphere will pump up temperatures to the point where water becomes steam. At the other extreme, if the air pressure is too low, water cannot remain a liquid, and turns into vapor or freezes to solid ice. The majority of the Earth's surface sustains liquid water.

The magnetosphere: Adding to the Earth's life-empowering list, the planet's core is large enough to generate and retain heat from radioactive materials.

This internal heat keeps the metallic outer core molten, and the Earth's rotation is fast enough to set up currents within that molten metal, creating a dynamo (essentially a giant magnet). The core's currents generate a "force field"—a magnetosphere—deflecting most of the solar radiation away from the surface.

The amount of radiogenic material captured by a forming planet may be a critical factor in its habitability. The radioactive heating from heavy elements like uranium and thorium helps to drive plate tectonics, important for carbon recycling and volcanic activity. The decay of radioactive elements also keeps the planet's core molten. Convection from the currents moving within the liquid metallic core generates the magnetic fields surrounding the planet. But not all planets are created equal: some planets gather less radiogenic material to themselves than others. If a planetary core lacks radioactive elements, the planet's crustal activity will die off, and the planet will become geologically dead. With too much of a good thing, too much radioactive material ends up in the outer layers of the mantle and crust. The heat in the outer layers prevents the core from losing heat fast enough to continue convection, and the planet's dynamo dies. This scenario also results in abundant volcanic activity that might lead to mass extinction events.

The right water/rock blend: The interplay of dry continental land forms with large bodies of water brings a rich diversity of life to our home world. The interaction between land and water may play a critical part in biodiversity and advanced life. (Still, among the water worlds, sea floors may host underwater communities of rich life.

Planetary companions: Judging by the exoplanet architectures we have mapped so far, the layout of our solar system seems rare. Our solar system lacks hot Jupiters close to the Sun —which would disrupt the orbits of terrestrials in the habitable zone, but we do have a massive gas giant farther out. Jupiter's influence tends to clear the inner solar system of large asteroids, creating a shield for the inner planets. The Earth has another significant planetary companion: its own Moon. The Moon is so large compared to Earth that the two essentially constitute a double planet. The Moon plays a key role in keeping the Earth's axis stable, and contributes other helpful influences to our planet (see Chapter 2).

Planet-scale recycling: The Earth has several innate recycling systems. The most evident one in day-to-day life is the water cycle. Rains fall into rivers, lakes and seas, the water evaporates, and clouds condense to begin the cycle again. Similar cycles are found on other planets involving other volatiles like ammonia (in the gas giants) and methane (at Titan). Early in its Noachian period, Mars, too, probably had an active water cycle. The Earth's water cycle purifies its water, keeping it balanced for biogenic processes.

But the Earth has other recycling operations, and these appear to be more rare among the worlds around us. Our plate tectonics find rough analogies on Mars, Europa and Enceladus, but nothing operating in the same way as terra firma does. The Earth's movable provinces crash into each other—raising mountain chains in the process of uplift—or slide under each other (subduct). Plate tectonics revitalize our environment in two important ways. First, they act as an atmospheric recycler. The stone on a terrestrial planet acts as a chemical sponge, soaking up gases and locking them into the surface. But as plates subduct, rock melts and the gases are freed, often venting through volcanoes and back into the atmosphere (Chapter 2). This rejuvenating was present on primordial Mars, where we find the largest volcanoes in the solar system. But when Martian crust ceased its movement and those volcanoes shut down, the atmosphere drifted away, eroded by the Sun's radiation and the planet's low gravity. The process of plate tectonics is responsible for recycling the Earth's minerals as well as air, cycling material from the sea floors to its highlands. Tectonics have been recharging the environment over eons.

If our own solar system is any indication, plate tectonics on terran worlds may be the exception rather than the rule. Venus is a twin in size to Earth, but lacks any global evidence of plate tectonics. The Venusian crust appears to have upwelled in great arenas and ovoid zones, with material emerging at the center and sinking back down into the crust on the outer rim. This form of tectonics may not result in the same revitalizing force that Earth's tectonics do (Chapter 4). Smaller Mars probably had a subdued form akin to the Earth's moving puzzle pieces at one time, but on the red planet, the process shut down early on.

Some models suggest that terrestrial planets larger than Earth are encased in crusts too massive to fracture into plates, so the recycling of minerals in the rock and the gases in the air may not happen at all, or may happen only at volcanic sources. One factor involved is the viscosity (thickness) of the molten rock, or magma, beneath the surface. Low viscosity results in a stagnant 'lid' of rock (similar to what Mars has today). A terran planet might have a few hot spots, but a frozen-in-place crust. Alternately, if the magma is less viscous, tectonics might abound, with thousands of micro-continents. The problem facing researchers is that they are finding many super-Earths—probable rocky worlds larger than Earth—and they have no idea what kind of viscosity is typical on super Earths. Those who model planetary structures have come up with two competing theoretical models. One envisages volcanic activity increased by a factor of a thousand, while the second foresees a quiescent crust, with volcanism reduced by a factor of a thousand.

The type of star that a planet orbits is also a critical factor. Early searchers looked to Sun-like stars as the only viable haven for other Earths. But with a new understanding of biology, atmospheric dynamics, and the nature of stars, we have expanded the search to other star types. The most common star type in our galaxy is the red dwarf, or M star. Nine out of ten suns in the night sky are M stars, many as small as Jupiter, burning feeble and dim. Planetary systems are common around these ancient, cool stars. But M stars are not as constant as our own. Although they are long-lived, they have a nasty tendency to jettison deadly flares (red dwarfs are also called "flare stars"). Planets orbiting such cool stars must remain close by to be in the habitable zone, but their tight pathways place them in the danger zone for deadly radiation. On the other end of the scale, giant and super-giant stars burn themselves out in a few million or billion years. These stars may not burn steadily enough to support stable conditions long enough for life to arise. On the other hand, our Sun's behavior is comparatively bland. Its output is stable and fairly predictable on an 11-year cycle, and it is hot enough that our habitable zone is fairly distant from its flares and prominences.

Type of galaxy (and our location within it): The stars of some galaxies are loosely compacted, drifting in amorphous, cloudlike flocks. Other galaxies resemble great whorls of suns in vast disks, many with spiral arms extending out hundreds of light years. In addition to the dangers associated with the densely packed stars at our galaxy's center, proximity to other galaxies poses a new set of problems. In provinces of the universe where galaxies are densely packed together, bursts of high-energy radiation from one galaxy can destroy life in adjacent ones. Some researchers suggest that galaxies inhabiting the quiet voids are more likely to host Earth-like worlds with life. Our Milky Way galaxy resides in just such a location.

The densest of galaxies discharge lethal gamma ray bursts (GRBs). These mysterious flares may come from matter falling into black holes, or they may earmark the fusion of a black hole and neutron star (Figure 8.2). Gamma Ray Bursts last from milliseconds to hours. In that short time, these high-energy explosions pump out hundreds of times the power of a supernova, enough to destroy all life in entire solar systems through vast regions.

Figure 8.2 Gamma ray bursts may come from matter falling into black holes, or they may mark the fusion of a black hole and neutron star, as seen here. (art by the author)

A gamma ray burst on July 9, 2005 took place a billion light years away. Despite its remoteness, its weakened wave of gamma radiation passed through our solar system. The High Energy Transient Explorer satellite, mounted by an international team from NASA, Japan, France, Italy, Brazil, and India, recorded the "short duration GRB" event. The fading wave of energy lasted just 70 milliseconds, but astronomers across the world teamed up to tease out details. The gamma rays marked a drastic event: a 25 km-wide corpse of an ancient sun (a neutron star) self-destructed as a black hole ingested it. Though the event's radiation wave was feeble at our distance, any nearby planetary system would have been sterilized by the cataclysmic blast.

So it seems that the Earth has dodged a few bullets over the eons, and a host of cosmic confluences came together to make our planet what it is. Many have commented on it, from theologians to astrobiologists. In their landmark book *Rare Earth*, Don Brownlee and Peter Ward made the case: "Our planet coalesced out of the debris from previous cosmic events at a position within a galaxy highly appropriate for the eventual evolution of animal life…[not] in the center of the galaxy, not in a metal-poor galaxy, not in a globular cluster…we became a planet where global temperatures have allowed liquid water to exist for more than 4 billion years…a volume of water sufficient to cover most—but not all—of the planetary surface…Earth received the right range of building materials—and had the correct amount of internal heat…alone among terrestrial planets…"[3] Astrophysicist Hugh Ross agreed, stating, "To make the existence of a planet with the characteristics of the primordial Earth possible required 9.2 billion years of…innumerable just-right events in the universe as a whole and in our Milky Way Galaxy in particular."[4]

Recipe for an Earth

The below list represents only a few of the aspects that make our planet well suited for life. It provides an overview of some factors that contribute to the benign, biologically-friendly environment of Earth. Although some of these factors may be unique to Earth or rare among other terrestrial worlds, and although not all may be prerequisites, as we search for Earthlike exoplanets this short list will help to inform that pursuit.

   Galaxy type
   Location in galaxy
   A stable star
   An orbit within the habitable zone (allowing for liquid surface water)
   A stable planetary orbit to keep the planet in the habitable zone for extended period
   A Jovian-class asteroid deflector
   Surface temperature between water's melting and boiling point
   Large companion moon to stabilize axis
   Plate tectonics (or similar tectonic processes) and volcanoes to recycle minerals and air
   Robust magnetosphere to serve as a radiation barrier
   Moderate axial tilt (obliquity) tames seasonal changes
   Few giant impacts after formation period, preserving life conditions
   The right amount of carbon dioxide and other greenhouse gases
   The right balance of continental and ocean surfaces

---

[3] *Rare Earth* by Peter Ward and Donald Brownlee (Copernicus Books, 2000), pp282-283
[4] *Weathering Climate Change* by Hugh Ross (RTB Press, 2020)

The first hunting expeditions for distant Earths

The Earth's list of factors contributing to its habitability seems long, but it may not be inclusive or essential. It was a place to start, but in the search for Earthlike worlds, early searches could not afford to be too picky. As astronomers became more familiar with the types of stars out there, and they pieced together a sketch of what alien solar systems might look like, the hunt for exoplanets matured. At first, telescopes seemed the best tool, and the first place to look was the nearest star system. The closest star to our own Sun is Proxima Centauri, a red dwarf star in the Alpha Centauri solar system. Proxima is about half again the size of Jupiter (209,000 km across). The cool red sun is part of an intriguing triple star system. Proxima orbits two larger stars that are similar in nature and size to our Sun, Alpha Centauri A and Alpha Centauri B. Proxima circles a distant 1/5 of a light year from the other two. It takes Proxima half a billion years to make the long, looping trip once around the system. Proxima is currently passing along the segment of its orbit nearest Earth, making it the closest star to Earth by a trillion kilometers. To the naked eye, Alpha Centauri A and B look like a single star, the third brightest in the night sky of our southern hemisphere[5]. Little Proxima is invisible without a telescope.

Alpha Centauri A is slightly bigger than our Sun, with a mass about ten percent larger. It is the same stellar type as our own star (G2). Surface temperatures reach 5500°C (comparable to the Sun's), but its greater diameter, 25% more than Sol's, makes it 1.6 times as brilliant. Alpha Centauri B is smaller and more orange. Its type is known as spectral type K2 (or an "orange dwarf"). Its lower temperature of 5000°C results in the star giving off only half the luminosity of the Sun. These two primary stars circle each other once every 80 years at an average distance of 11 AU. The M star Proxima simmers at just 2825°C, and glows like a dying ember, only 1/500th as bright as our Sun. As the closest nearby star system, Alpha Centauri seemed the logical first choice for an Earth hunt.

Direct telescopic observation can tell us very little about even nearby stars. It is a challenging technique: the light from a planet's star overpowers the planet's comparatively dim reflected light. In effect, any nearby planet is lost in the glare of its own sun, hidden in the brilliant light of day. An instrument called a coronagraph can block light from the primary star, but this is an exacting technique that is still evolving for exoplanet searches. A Jovian-sized planet has tentatively been found transiting the

---

[5] Only Sirius and Canopus are brighter.

star Alpha Centauri b in Hubble Space Telescope data. It has yet to be confirmed, and is controversial.

In the summer of 2016, researchers at the European Southern Observatory (ESO) brought another technique to bear as they studied the nearby Alpha Centauri system. Called the radial velocity method, the technique makes use of changes in light due to movement of the star. As a large planet orbits its parent star, its gravity causes the star to wobble from side to side. The star also moves directly at—and away from—the viewer. This movement can be sensed by a shift in the star's light. As the star is yanked away from us by its nearby planet, its light reddens. When the planet makes its way around to the other side of the star, it pulls its sun toward us, and the starlight becomes bluer. Finding a planet using radial velocity is a lot like listening to a symphony from a distance, straining to hear the cadence and tempo of the sonata. ESO researchers used the radial velocity method to confirm a planet orbiting Proxima Centauri. Known as Proxima b, its discovery was the finale of a sixteen-year study. Conditions on the new-found planet may well be "Earthlike". The beat of Proxima's music—the Doppler shift of its starlight—reveals that planet Proxima b weighs in at a minimum of 1.3 Earth masses. Its 11 day, 5 hour orbit places it on the inside edge of Proxima's habitable zone. With the data in hand so far, Proxima b may be anything from a rocky world—like our terrestrial planets—to a sub-Neptune behemoth. But its existence has been independently confirmed: the Earth, it seems, has a sibling world just 4.2 light years away.

The range of possibilities for Proxima b's nature faces researchers with each new planetary discovery (Figure 8.3). In the case of Proxima b, a survey of those possibilities will give us a deeper appreciation of the challenges faced by astronomers. Researchers do not yet have a measure of Proxima b's radius, and only a range for its possible mass, two factors that can tell us a lot about a planet's character. But if the planet has half again the Earth's mass, it may qualify as a sub-Neptune, a bizarre hybrid planet falling between a super-sized Earth and an ice giant gas planet like Neptune.

If the planet's radius is less than 1.6 Earth radii, Proxima b's temperament may be quite different from a sub-Neptune. Computer simulations indicate that on the lower end of the scale, giving Proxima b a radius of 5590 km, the planet would hide a dense metallic core comprising 2/3 of its overall mass, surrounded by a rocky mantle. Intermediate sizes in the computer studies suggested that some water would remain as liquid or break into hydrogen and oxygen atoms. Although oxygen and water are good news for life on Earth, oxygen is such a reactive molecule that it actually prevents the growth of many prebiotic molecules (Chapter 5). Astrobiologists suggest that the presence of

Figure 8.3   A deadly flare spews from the red dwarf Proxima, closest star to our solar system. If planet Proxima b is rocky, it may well have an Earth-like environment. If its core is molten, it may have a protective magnetosphere, enabling life to take hold. In this view, we see Alpha Centauri A and B above and to the left. (painting by the author)

oxygen early in a planet's environment could short-circuit life's operations. It could be that Proxima b has abundant water and oxygen but is not habitable by Earth's biological standards.

Alternatively, Proxima b may be a larger water-world with a 1.3 Earth mass, awash in global oceans hundreds of kilometers deep. Some studies imply that with a radius 1.4 times that of Earth (~8,920 kilometers) Proxima b would possess a rocky core making up 50 percent of its mass, blanketed by an ocean that makes up the other 50 percent.

Proxima orbits close to its sun, a red dwarf known to belch out deadly flares. But floods of radiation may not spell disaster for Proxima b's potential for life. An over-sized Earth may have an over-sized-core. If Proxima b accommodates a large enough molten metallic core, a strong magnetosphere could shield its surface from much of Proxima Centauri's radiation. The star itself would appear three times the diameter of the Sun in our sky.

With an orbit so close to its parent star, Proxima b may be tidally locked, keeping the same face toward its sun. In this case, only one hemisphere of the globe would ever see the sun, while the other hemisphere would endure eternal night. But that long dark night might shield life forms from Proxima's deadly radiation (see Eyeball Planets, below). Those who study celestial mechanics—the movement of planets and stars—suggest that Proxima b may take a less leisurely daily spin, perhaps akin to Mercury, which turns three times for every two revolutions around the Sun. If Proxima b moves in this way, with a day slightly shorter than its year, winds will trigger atmospheric circulation that could allow for quite moderate surface conditions on the planet. But that atmosphere might also simmer as a pressurized hothouse, similar to what we see on Venus today.

There are more stars—and planetary systems—out there. To find them, we must use some very clever techniques.

How to Find an Invisible World

The search for exoplanets is the search for the hidden. Directly imaging them is the most difficult way of revealing these distant worlds. Only recently have advances enabled us to find worlds directly. But there are other techniques, and they fall into five categories: radial velocity (which unveiled Proxima b), astrometry, timing technique, gravitational microlensing, and transits.

Direct Imaging of an exoplanetary system means getting an image of actual planets in orbit around another star. To directly image distant Earths, the glow from the planet's star must be blocked out. The star's scattered light that would normally blind a telescope's view can be cancelled out by a promising technique called interferometry. Astronomers can link two space-based telescopes "in phase," so that their view of the sky becomes a series of fringes, an interference pattern. With correct pointing, the star falls in a minimum of the fringe pattern, and essentially disappears. A nearby planet falls outside of this cancelled-out region, and its light gets through.

To directly image a distant planet requires a very big telescope. It is inefficient and expensive to build large, single mirrors, but when engineers craft
several smaller telescopes and combine their light, higher resolution can be
obtained. Telescopes chained together in this way can resolve objects equivalent to those seen by a single telescope covering the entire area *between and
including* the two tandem scopes. Some ground-based telescopes today are as
powerful as orbiting ones like the Hubble Space Telescope. However, telescopes that see into the infrared part of the spectrum do better in space, above
the Earth's warm atmosphere. The James West Space Telescope is an example;
the heat in the Earth's atmosphere would blind its powerful instruments.

Despite the challenges, directly imaging a planet has distinct advantages
over other techniques. A direct image provides researchers with a spectrum, a
chart of reflected light, from the planet itself. Spectra can tell scientists what
the planet is made of, whether it has clouds, its temperature and how it
changes from one observation to another. These important clues help us to
paint a picture of a complex world.

Early in its mission, NASA's James Webb Space Telescope directly imaged
a young exoplanet orbiting the star HIP 65426 (Figure 8.4). The planet has as
much mass as seven Jupiters, so is classed as a super-Jupiter gas giant. It orbits
92AU from its star. JWST blocked the light from the star in order to image
the planet in four different parts of the spectrum. While the Hubble Space
Telescope has directly imaged several exoplanets, JWST brings greater capability and resolution, and promises to dramatically enhance exoplanet studies
using the direct imaging approach.

Radial velocity measures a star's movement as its nearby planets tug upon
it. Astronomers used this technique to discover Proxima b. Radial velocity
measurements are well-suited for revealing the mass of planets and their
period of orbit (length of year). It has been especially successful in determining the masses of planets at least as bulky as several Earths, and in orbit close
to their parent star.

The *timing technique* works with stars that vary in regular ways. For example, pulsars are the corpses of collapsed suns. They shed radio waves as they
spin, like a radiation-beaming lighthouse. These waves are regular and predictable, pulsing out of the star as it spins. Since the rotation of a pulsar is so
uniform, researchers can use slight changes in the pattern of a pulsar's radio
beams to detect changes in its motion. Just as our Sun wobbles around a point
between itself and Jupiter, a pulsar will wobble back and forth as its planets
pull on it. The shift in timing of the pulses reveals the size of the orbit, the
planet's distance from the star, and the approximate mass of the planet. The
timing technique is so sensitive that it can measure the effects of a planet

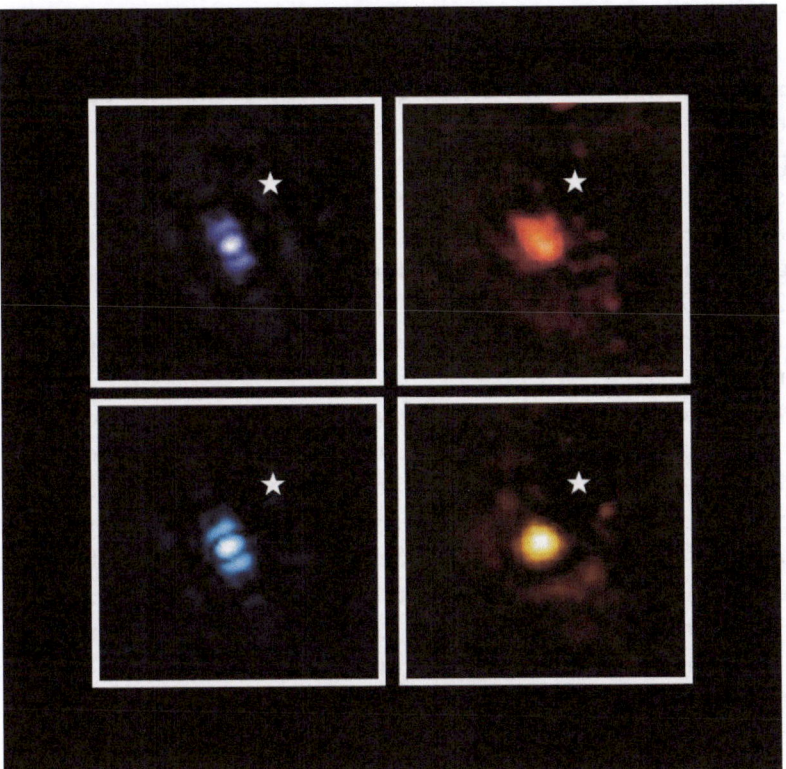

Figure 8.4 This James Webb Space Telescope image shows the exoplanet HIP 65426 b in different wavelengths of infrared light. The images each reveal varied data at different parts of the spectrum. Sunlight from the host sun—marked with a white star—has been blocked out so that the super-Jupiter planet can be seen. (NASA/ESA/CSA, A Carter (UCSC), the ERS 1386 team, and A. Pagan (STScI)

down to less than one tenth the mass of Earth. Researchers also use the method to detect multiple planets tugging on each other and the star itself. Unlike some other methods, the timing technique can detect planets that orbit far from their parent star. Some other types of stars oscillate in predictable ways. In all of these cases, the timing of a star's varying brightness or spin will appear to be slowed or sped up by a planet's passing.

Gravitational microlensing also looks for dimming starlight, but it takes advantage of a prediction in Einstein's General Theory of Relativity, the idea that mass distorts the space around it. Because of this phenomenon, light bends when it passes near a large mass. This warping of space bends images behind a heavy object like a star. Sir Arthur Eddington and his three teams of astronomers first confirmed the effect[6] in 1919. By tracking stars as they vanished and reappeared from behind the Sun during a total eclipse, the observers were able to demonstrate the bending of starlight around the Sun.

The same phenomenon can be applied to planets circling stars. When one star appears to pass near another (from the viewpoint of Earth), the closer star acts as a lens, increasing the brightness of the farther star. If a planet transits across the face of the star, the starlight will dim. Because a close star magnifies the light from the star being transited, that light can be charted in detail. By early 2020, 49 exoplanets had been found using this technique, one of which was, up to that point, the least-massive exoplanet discovered around a main sequence star.

Timescale or autocorrelation technique: Another technique that will aid in the detection of Earthlike worlds is still being perfected. Called the "timescale technique," it is a measure of gravity's pull at the surface of a star. Autocorrelation function enables researchers to more accurately characterize the planets affecting the star surfaces. Scientists can measure surface gravity of distant suns with an accuracy of about four percent. A star's gravity depends on its mass and diameter, but many stars are too distant to calculate these. With these remote stars, the dip in light of a transit may be the only hint of a planet, even though the size of the star is unknown. The new technique enables planet hunters to lock down the mass and size of distant suns, leading to a more accurate view of the planets transiting them. The timescale technique uses data already collected from orbiting telescopes like Canada's MOST (Microvariability and Oscillations of Stars telescope) and NASA's Kepler and TESS Observatories.

The transit technique: The most striking development in the hunt for exoplanets has come at the hands of the transit technique. By carefully observing the light level coming from a star, any drop in that level might betray the passing of a planet in front of it. It's a challenging process; light levels drop by only 0.01% to 0.1%, depending on the size of the planet in comparison to its star (Figure 8.5). A transit takes place once in each orbit of the planet. The duration of the transit can reveal the planet's distance from the star and details about its orbit. The size of the planet can be determined by how much light it blocks out. When combined with the radial-velocity method (which determines the planet's mass by how much it tugs on its star), observers can establish the density of the planet and gain insights into its physical nature.

---

[6] Eddington's observations were taken at the same time as those from teams in Brazil and an island off the west coast of Africa.

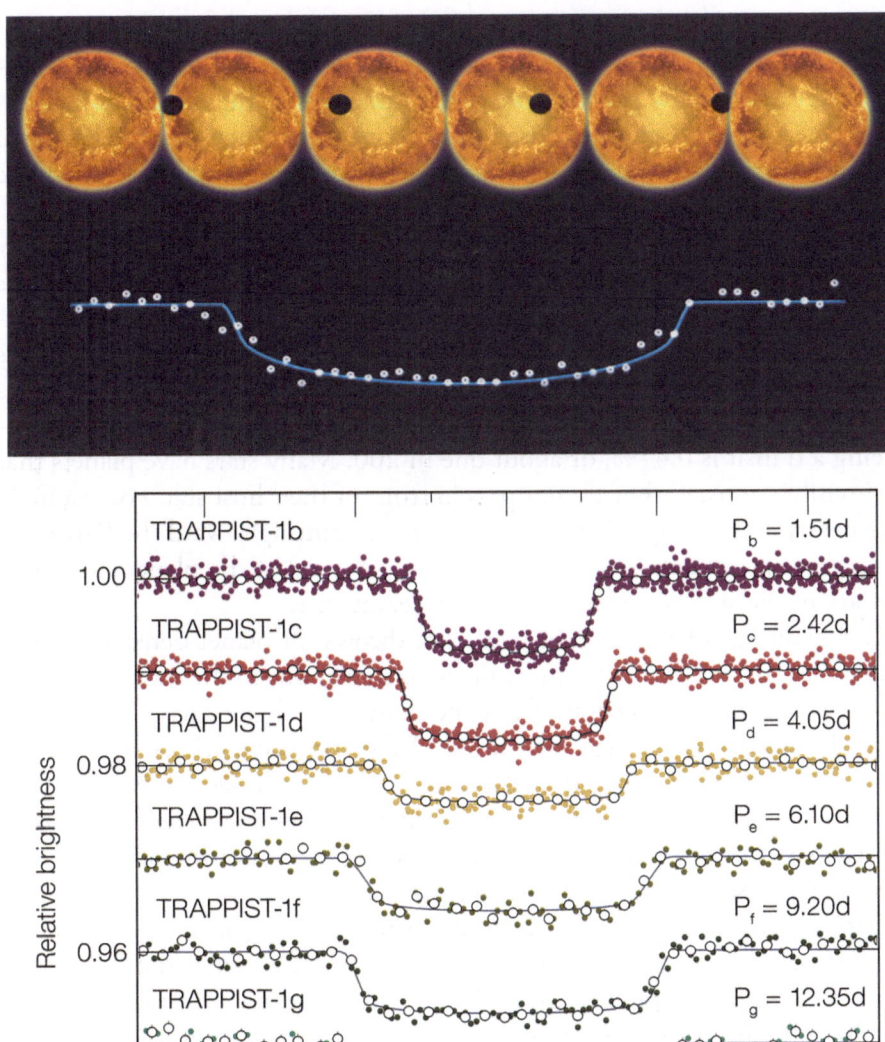

Figure 8.5 Top: as a planet transits (crosses) the face of its star, light levels decrease and increase in specific patterns, as seen by this series of simulated images above a typical transit light curve (art by the author). Bottom: Starlight from the red dwarf TRAPPIST-1 dims with each pass of a planet in front of the star. Here, each planet's light curve is shown against time. (Spitzer Space Telescope image courtesy European Southern Observatory/M. Gillon et al.)

A related planet-searching method is called the transit timing variation (TTV) technique. If one planet is discovered by the transit technique, multiple observations of its passing may reveal variations in the timing of its transits. These variations may be caused by other nearby planets that are not lined up to transit the star. This method is so sensitive that planets with Earthlike masses can be revealed.

The transit method has two major drawbacks. First, planetary transits only happen when a planet's orbit is aligned with their parent star from the Earth's viewpoint. Only about 10% of planets in close orbits pass directly in the line-of-sight to their sun as seen from Earth. The fraction decreases for planets with larger orbits. For a planet orbiting a Sun-sized star at 1 AU (Earth's orbital distance from the Sun), the probability of a random alignment producing a transit is 0.47%, or about one in 200. Many stars have planets that are invisible to us, as they do not pass in front of their host star. For example, a planetary system aligned with the star's pole pointing toward the Earth will present a solar system looking like a target, so none of the planets will cross the face of the star as viewed from our observatories.

Red giant stars have another issue that throws off planet detection: while planets around these stars are more likely to transit due to the stars' larger size, their effect is hard to separate from the main star's light curve, because red giants have frequent changes brightness. These can occur within periods of a few hours to a few days. M dwarf stars also "flare" at times, throwing off the light curve. But with enough repeat sightings, astronomers can determine if an observed transit is really from a planet (once in each of the planet's 'year') or simply a false reading from variations in starlight.

Astrometry measures the location and drift of a star against the background sky. Astrometry can tell observers the inclination of a planet found through other techniques, helping to ascertain its mass. The mass, in turn, provides insights into its composition and surface gravity, which reveals something about the nature of the planet itself: terrestrial vs. gas giant, extent of atmosphere, possibility of liquid water, etc.

Searching from space

With its launch and daring on-orbit repair, the Hubble Space Telescope racked up a list of spectacular images: Cepheid variables, supernovae, stellar disks and nebulae and even potential black holes. More important in the search for alien Earths, Hubble was able to image protoplanetary disks around stars, providing our first glimpse of the actual formation of planets. Later in its mission, Hubble was conscripted to search for exoplanets and confirm ones discovered by other means.

The Kepler Space Telescope ushered in a renaissance of exoplanet research. Launched in 2009, the SUV-sized Kepler Observatory carried 42 cameras. Those cameras combined forces to study a small patch of the sky in great detail. Kepler's view only covered an area of the sky equal to a fist held at arm's length. Surveying the same stars over and over again, Kepler watched for slight variations in starlight. The amount of decline in starlight revealed how much of the star's disk a planet blocked. With careful timing of the planet's orbit by further observation, scientists can determine the size and mass of a planet, along with its orbital traits.

The Kepler Observatory eventually confirmed over 2700 exoplanets, with another 3601 awaiting independent confirmation. The confirmed planets span a wide range of sizes and natures, from hot Jupiters to super Earths to ultra-heated terrestrials. Kepler and TESS team scientist Thomas Barclay points out that, "there were predictions before Kepler launched that the space-craft would find no super-Earths. We thought planets about twice the size of Earth just didn't exist. But now we find that the most common planets out there fall into this class."

In 2018, NASA launched the Transiting Exoplanet Survey Satellite (TESS). The advanced probe continued Kepler's work of charting the transits of plan-ets. TESS' increased capabilities enable it see transits of smaller planets than Kepler could, planets more the size of Earth. These planets will orbit nearby stars, close enough to also obtain radial velocities. TESS is like Kepler on ste-roids, quipped one science team member. While Kepler's main mission searched a specific area of space, TESS surveys the entire sky with an eye out specifically for super-Earths and Earthlike worlds. The spacecraft uses four wide-field CCD cameras. By October of 2022, the spacecraft had already clocked some 250 exoplanet discoveries.

The James Webb Space Telescope ushers in the next generation in space telescopes. Webb studies the atmospheres of those targets. Webb carries a coronagraph to block the blinding light of a star, revealing nearby planets for direct imaging. The spacecraft is also carrying out spectroscopy of those plan-ets, unveiling the composition of the atmospheres. Webb studies stars and planetary systems in the infrared.

As of 2022, world observatories on the ground and in space had confirmed the existence of 5220 exoplanets. Of these, only 192 were terrestrial. 1593 are considered super Earths, while 1617 are found to be Jupiter-like worlds. 1813 planets the size of Neptune make up the bulk of the list. Within these ranks, we have found bizarre and wondrous planets. Most bear no resemblance to the Earth, but they all give us perspective on our home world and the preciousness of Earth-like worlds across the universe.

The rarity of Earth-like planets is due, in part, to the arrangements of other solar systems. Exoplanetary systems around other stars differ markedly in their architectures, especially when compared to the planets around our Sun.

Strange Arrangements

We have seen how the organization of our planetary system affects the very nature of planet Earth. Based on the size and arrangement of planets in our own system, astronomers expected to find many rocky, terrestrial-sized worlds clustering around other stars, with massive gas or ice giants orbiting farther out. But our ground-based and orbiting telescopes reveal a surprising variety of exoplanetary layouts.

The Kepler Telescope's discoveries provided insights into how atypical our planetary system's architecture really is. Our own planets suggested that since temperatures are high near the star, the only planets that can form there are small, rocky ones. Farther out, where it is cold and large planets have an easier time forming, we you get Jupiters and Neptunes. But that is not what we have found. Instead, we have seen Jupiter-class worlds in tight, four-day orbits just next door to their suns, and multiple Earth-sized planets packed together cheek to jowl. In fact, the majority of planet-hosting stars that we've studied have an Earth or twice-mass-Earth planet close to their stars. "What we've never really found," says Caltech dwarf planet expert Mike Brown, "is something like the Solar System."

In 2015, researchers finally discovered a solar system that has a design resembling ours. HIP 11915 is a G type star similar in size and temperature to the Sun. The system is 186 light years distant, and it has a gas giant in orbit some 4.8 astronomical units from the star (Jupiter orbits the Sun at a similar distance of 5.2 AU.) The planet HIP 11915 b has a mass 93% that of Jupiter. Since its location within its star's outer planetary system is so similar to that of Jupiter in ours, the planet may have carried out planetary migration similar to that experienced by Jupiter. And since models of Jupiter's migration (i.e. the *Nice* or *Grand Tack* models) indicate that it laid the groundwork for the terrestrial planets here, a similar situation may exist within this distant solar system. HIP 11915 seems a rare example of planetary architectures like ours, and it underscores the uniqueness of our own planetary system's design among the thousands of planetary families mapped so far.

Not-so-Earthlike-worlds

There is a planet out there, circling a Sun-similar star 608 light years away, that's about the size of the Earth (1.4 times as far across). The star is Kepler 10, close in temperature to our Sun but nearly three times as ancient. Its Earth-sized planet, Kepler 10b, is a rocky terrestrial just like home (Figure 8.6). The bad news is that its orbit carries it 1/20 as close to its star as Mercury is from the Sun; surface temperatures reach 2839°F (1560°C), as hot as an industrial blast furnace. Its tight orbit carries the baking world once around its star every 20 hours. Kepler 10b is a heavy world, weighing in at over three times the mass of the Earth. Despite its similar size to Earth, the planet is an alien, scary world that provides us with a cautionary tale: other Earths may be quite difficult to find.

Kepler 10b has a sibling that reinforces this view. Orbiting at a distance of just 0.3 AU, Kepler 10c circles its parent star once each 45 days. At this range,

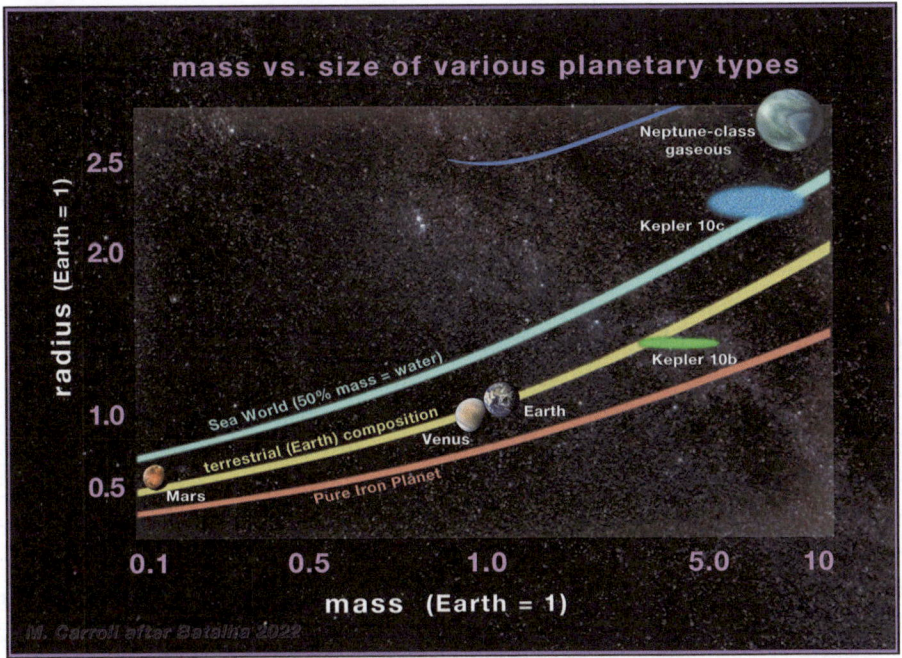

Figure 8.6   Kepler 10b is slightly larger and more massive than Earth. Here, the green oval shows the possible range in which the Earth-sized planet falls compared to Mars, Venus, Earth, and a typical Neptune-class gaseous giant. Above Kepler 10b is a blue ellipse showing the possible range of the much larger Kepler 10c, likely a hot sea world. (author art based on public domain NASA diagram by Natalie Batalha via https://commons.wikimedia.org/wiki/File:CompositionOfKepler-10b.jpg)

its surface temperatures are four times that of Jupiter, hovering at about 591°F (311°C). Its density suggests an ocean planet rich in water and other volatiles. The planet falls into the category of super Earth or sub-Neptune. The two worlds of Kepler 10 are just a taste of the exotic planets we have found in our search for distant Earths.

Sub-Neptunes

Weighing in at up to ten times the mass of Earth, the ubiquitous mini- or sub-Neptunes join the exoplanet parade with similar masses to our own Uranus and Neptune. Uranus holds 14.5 Earth masses, while Neptune encompasses 17. Super-Earths larger than 1.6 Earth radii appear to contain substantial amounts of hydrogen and helium, transitioning from terrestrial to gaseous.

Sub-Neptunes may represent a wide range of planetary natures. In fact, planets of the sub-Neptune class may turn out to be the most varied of any size worlds. Gas giants larger than Saturn all tend to be about the same size, because they are dominated by their hydrogen/helium atmospheres. The composition puts limits on their sizes, although their densities vary (Jupiter is only slightly larger than Saturn, for example, but has twice the density. The majority of planets close to the Earth's mass are rocky worlds with thin atmospheres. But across the range of sub-Neptune-mass worlds, there likely exists a wide spectrum of conditions between rocky terrestrials and gaseous globes.

The transition from rocky super Earth to gaseous sub-Neptune happens at about 1.5 or 1.6 Earth radius. Once a planet reaches 2 Earth radii, it complexion becomes more like a sub-Neptune.

Astronomers have been studying the elusive transition using Kepler data, radial velocity measurements and other techniques from ground and orbital facilities. Using the transit technique to determine a planet's diameter, astronomers merge the data with radial velocity or transit timing techniques to understand its density. The combined data offers astronomers a critical measure of a planet's nature. If a planet weighs twice as much as the Earth, but is the same size, it must be very dense, and so it is rocky. But if a planet of the same mass is ten times the size of Earth, it must be a low density, fluffy world more like a gas or ice giant. Sub-Neptunes present hybrid worlds somewhere between rocky exoplanets comprised mostly of iron and silicates, and worlds with extensive atmospheres made up mostly of hydrogen and helium with underlying ices or water.

The examples are stacking up, and their natures underline the variety of this planetary class. A prime example is the super-Earth BD+20594b, which has

16 times the heft of Earth, and roughly the diameter of Neptune. Its numbers come in right on the edge of all the models. Some researchers contend that BD+20594b's mass and radius require a terrestrial character; this would make BD+20594b the largest solid-surfaced exoplanet yet found. Its proximity to its star results in ground temperatures of approximately 128°C. Another sub-Neptune, super-Earth K2-3d, has a density greater than iron and a diameter half again that of Earth. Although its average density is that of the rusty metal, its surface may contain lower density silicates. Its orbit lies at the inner edge of its star's habitable zone, so the planet may have liquid water on the surface. But with hotter temperatures, its surface conditions might range more toward the Venusian, with only water vapor locked in a dense atmosphere. Its true nature is still unknown.

Like Neptune, many sub-Neptunes are gaseous, with dense, hydrogen/helium rich atmospheres. Unlike Neptune, these super-Earths tend to be warm versions of the gaseous worlds we know, as they orbit close to their primary star. These warm, gaseous worlds with no solid surface have no analogue in our solar system. Because solar radiation at such close range can strip an atmosphere fairly quickly, sub-Neptunes must form in the outer regions of their planetary systems, beyond the snow line. As happened in our own system, they must have migrated in from the cold later. Now, close to their suns, their atmosphere is being stripped away by the relentless solar wind of their stars.

Giant planets and Earth-moons

Exoplanet hunters have found a multitude of Neptune- and Jupiter-class giant planets. Many orbit remarkably close to their host star, with searing temperatures to show for it. Others circle at a distance, in the outer darkness, probably in the same region that their formation took place. But other gaseous giant planets orbit within the habitable zone of their parent sun. These behemoths will be fundamentally different from the Earth, but they undoubtedly carry moons with them. The natural satellites circling giant worlds may be able to have Earth-like climates. Jovian planets possess immense gravity fields, enabling them to hold on to large moons in stable orbits. We find an example in our own planetary family: the seventh-largest moon in our solar system, Triton, orbits Neptune, which is on the smaller end of the gaseous worlds in the Sun's planetary family.

Recent work reveals that moons weighing a fraction of the Earth's mass can retain a respectable, warm atmosphere within a habitable zone, even when they orbit within the fierce radiation of a gas giant planet. Moons this size or larger could easily bear similar terrain to the Earth or Mars, and even the smaller ones, if subjected to the kind of tidal heating we've seen in the moons of our own system, could have active geology with volcanoes and terrestrial-style tectonics. Such small worlds may enjoy the same kind of feedback loops so important to life on Earth, like atmospheric recycling in the rocks and hydrological cycles.

The Earth-planets appear: Earths, Almost-Earths, Mega-Earths, and Super-Earths

Smaller on the scale of Earthlike worlds are the ExoEarths, rocky planets similar in size and mass to the Earth. Their orbits also vary widely, creating surface conditions ranging from hellish to possibly life-sustaining. Thanks to the overall data from the Kepler and other observatories, researchers estimate that there may be as many as 40 billion Earth-sized planets orbiting stars within habitable zones. While many orbit the common red dwarfs or other stars different from our own, our galaxy may host up to 11 billion Earths in orbit around stars very much like our Sun.

Early in our search for Earth-sized and Earth-like planets, observers stumbled upon the unusual world Corot 7b. It is half again larger than Earth, and probably a terran planet rather than a gaseous one. The French CoROT observatory—the first spacecraft earmarked specifically for exoplanet research—unveiled the planet after several years of studies culminating in 2009. Corot 7b sprints around its sun—a star slightly smaller and cooler than our Sun—once each 20 hours, 29 minutes. It is not the only planet circling its sun: another super-Earth, Corot 7c, orbits farther out, taking just 3.7 days. Neither are promising targets in our search for life, as temperatures on both would vaporize any biological material, and even some metals! The surface of Corot 7b is likely molten rock.

While Corot 7b seems a bit toasty, it has nothing on a blistering planet orbiting the star 55 Cancri A, a G type star similar to the Sun. With a mass nearly nine times that of Earth, super Earth 55 Cancri e is nearly twice as far across as our own world. It orbits so close to its star that its sun-facing hemisphere cooks at 1700°C (3100°F). The surface of this magma ocean world simmers at the vaporizing point of rock. The Hubble Space Telescope was able to detect hydrogen, helium, and hydrogen cyanide, but no water. Across its skies drift clouds of sapphires and mineral-rich gases. Some clouds may well be made of corundum, which forms semi-precious gems like rubies and sapphires.

Generally, the term super-Earth applies to planets that are larger than Earth but still have a rocky surface and an atmosphere thinner than the giant, hydrogen-rich planets. The term sub-Neptune refers to a small gaseous giant. However, there isn't a clear boundary between these two classes of planets, and the terms are often used interchangeably. The line between the two is blurred even further by the fact that often, size and mass are the only established constraints, so conditions are not known well enough to rank a planet as a small sub-Neptune or a large terrestrial super-Earth.

Super-Earths comprise the most common exoplanet type. Although roughly three out of ten exoplanet discoveries to date fall into this class, these worlds have no analog in our solar system. Astronomers class them by their mass, not by any conditions that the planets themselves might exhibit. Super-Earths may orbit near or far from their host sun, with a wide variety of configurations and dispositions. The vast majority discovered so far orbit close to their star. Some may be terrestrial and rocky, while others may lack solid surfaces, resembling gaseous sub-Neptunes. The largest super-Earths fall in a mass range of ten Earth masses. The lower limit is more debatable, ranging from 2 to 5 Earth masses.

Super-Earths fall into four possible categories. Low-density planets contain large amounts of hydrogen and helium, and are referred to as dwarf Neptunes, mini-Neptunes or sub-Neptunes. Medium-density super-Earths probably contain water as their major component, making them ocean worlds. Some of these may be completely covered by deep ocean. A third type has a denser core than a sub-Neptune, but like a sub-Neptune, it has an extended atmosphere. The extent of that atmosphere will depend upon how distant the planet is from its star: the farther away and the cooler it is, the more atmosphere it can retain (Ganymede-sized Titan, for example, is cocooned in a dense atmosphere while warmer Ganymede is airless). Finally, on high density super-Earths, known as Mega-Earths, rock and/or metal likely comprise the bulk of the planets.

It is unlikely that super-Earths can arise in the same place they end up, making super-Earths a prime example of planetary migration models. For example, Uranus and Neptune have solid cores. But if a Neptune migrates into the orbit of Mercury, its hydrogen and the helium will be blown away by solar heating. Beneath their delicate atmospheres are frozen mixtures of carbon, hydrogen, oxygen and methane, much heavier than hydrogen and helium. These materials will also be vaporized into space over time. What's left is a large core of iron, nickel, etc. It is this pathway that may result in the super-Earths 2-3 times the size of the Earth.

Just how "Earth-like" is a super-Earth? Those features that contribute to the uniqueness of our own world provide us with a good yardstick. The Earth's plates play an important part in our ecosystem, but they may be a rare occurrence. But some astronomers contend that super-Earths may not benefit from plate tectonics, for several reasons. The first has to do with chemistry: it takes just the right mineral mix to drive the shifting plates seen on our world. As one plate descends beneath another, it is initially too light to sink all the way down into the mantle. But as the plate plunges down, a transformation takes place. 40 kilometers down, increasing pressures reorganize atoms within the plate, making the rock denser. Without this atomic alteration, plates would stall out, paralyzing the entire tectonic process. The change in density is dependent on the plate's composition. While scientists cannot yet directly measure the mineral composition of exoplanets, they can chronicle elements within their stars. This data reveals the building blocks of those nearby planets. Planets rich in silicon cannot maintain the conveyor-belt of plates seen on Earth. Aside from composition, the crust of super-Earths may simply be too thick to carry on tectonics. Thick crusts may be common among super-Earths, raising a physical barrier to plate tectonics. Some models demonstrate that the circulation of heat and minerals on a super-Earth class planet is too sluggish to spawn such life-critical phenomena as volcanoes. A super-Earth with ten times the mass of the Earth grows a thick rock crust incapable of continental shifting. But we have never seen a super-Earth up close, and the universe is full of surprises. Some researchers assert that the increased heat within super-Earths might be enough to drive plate tectonics after all.

One of the most Earthlike worlds yet discovered is an exoplanet just 12% larger than our own. Kepler 438b checks in with an Earth Similarity Index number of 0.88. It orbits within the habitable zone of the red dwarf Kepler 438, circumnavigating its sun every 35 days. Kepler 438b is probably terrestrial in nature, with a mass of 1.4 times that of Earth. Surface temperatures on the distant Earth range from 0°C to 60°C (from the freezing point of water up to a sunny 140°F).

The planet has the disadvantage of orbiting close enough to its parent star to feel the fallout of solar flares so common to red dwarfs. In fact, Kepler 438 is very active, unleashing flares of radiation and plasma (called coronal ejections) approximately every hundred days. The storm of charged particles from the star is strong enough to shred the atmosphere of close planets. M stars also have a more energetic magnetic field, and they emit more X-ray and UV radiation, which create greater erosion on a planet's atmosphere. But in the case of worlds like Kepler 438b, atmospheric ozone and a strong magnetic field from a molten rocky core could shelter the planet's surface, allowing life to thrive even there. Many super-Earths probably have molten cores generating a protective magnetic field. Even super-Earths in close orbits to their suns will find magnetospheric protection from sterilizing solar radiation.

Another addition to our list of Earth-like planets is Kepler 452b, the first roughly Earth-size planet discovered in the habitable zone of a G-type star (similar to the Sun). With an ESI rating as high as 0.83, Kepler 452b is roughly half again as large as Earth. It takes 385 days for Kepler 452b to orbit its sun, a year comparable to Earth's 365-day year. Although the planet is slightly farther from its star than Earth is from the Sun, its star is brighter, so the planet gets slightly more energy from its sun than Earth does from ours.

Kepler 452b's atmosphere is probably denser than Earth's. With its increased heat, the planet may have developed a cloudy, hot atmosphere, perhaps even as hot as that on Venus. A rocky world that size may well have active volcanoes. Kepler 452b was once in the center of its star's habitable zone, but as the parent star has aged and temperatures have risen, its habitable zone has migrated out, stranding this planet on the inner edge. Any oceans it once had are evaporating into its thick atmosphere. (Our Sun's habitable zone has migrated outward, so that both Venus and Earth are warmer than they once were.)

Rounding out our list of Earth-like exoplanets of note is Kepler-186f. The planet is just a tenth more massive than Earth. The planet gets only about a third of the energy that the Earth does from its M dwarf star, which is half the mass of the Sun. At high noon, daylight would be as bright as the Earth's an hour before sunset. Its distance from its star gives the planet an advantage: it is probably not tidally locked, so has a day/night cycle to even out temperatures and drive circulation of the air. Kepler 186f is not alone; its four known comrades all fall in the range of earthlike or Super-Earths.

Assorted Super-Earths

Three super-Earths may circle the star Gliese 581, a red dwarf twenty light-years from Earth. The closest in, Gliese 581c, orbits at the inner edge of Gliese 581's habitable zone. Gliese 581c orbits so close to its star that it suffers a runaway greenhouse effect much like that found on Venus. At five times the mass of Earth, the planet likely has an extensive atmosphere.

Another world in the system, 581d, is significantly more massive than Earth, at roughly 16 Earth masses. The huge size of the planet – five times that of a super-Earth class – caused astronomers to add a new class to exoplanets, the mega-Earth. When first discovered, estimates put the planet at the outer edge of the habitable zone, but recalculations in April of 2009 show that it orbits closer in, with a period of 66.87 days. This places the planet in the heart of Gliese 581's habitable zone.

The nearby star system Gliese 667 (or GJ 667) contains three stars some 22 light-years from Earth. Gliese 667 A and B are both K type (orange dwarf) stars orbiting around each other, completing their tango once each 42 years, at a distance ranging from 5 to 20 AU. Gliese 667 C is a red dwarf that orbits the pair at a distance of 230 AU, nearly eight times the distance between Neptune and the Sun. Around the K star Gliese 667C, we find at least three planets. This is surprising: orange dwarfs like Gliese 667's larger couplet are somewhat depleted in the kinds of heavy elements that terrestrial planets use to assemble themselves. Gliese 667 may harbor three super-Earths close to – or within – its habitable zone.

GJ 667Cc has a mass approximately 1.5 times that of Earth. The world may be a rocky terrestrial, but most studies place the planet in the class of a dwarf Neptune. GJ 667Cc circles its red dwarf sun at a furious speed, completing a circuit in just 28 days. But its orbit is far enough from its sun for liquid water to exist on the surface. Gliese 667Cc collects about 90% the light that Earth receives from the Sun. While it is in the habitable zone, it is still within reach of the red dwarf's energetic solar flares, which may periodically fry the surface in deadly radiation. The over-sized Earth is so close as to be tidally locked to its parent star. GJ 667Cc may have siblings. The sub-Neptune GJ 667Cb resides at less than half the distance of GJ 667Cc's orbit, circling its sun in just over 7 days. The planet is at least 5.6 times as massive as Earth.

At a distance of 600 light-years away[7], the Sun-like star Kepler 22 safeguards a family of planets that includes the super-Earth Kepler 22b. Kepler 22b has the distinction of being the first habitable-zone planet discovered by the Kepler Observatory. Its density implies that the planet is a rocky terrestrial. Kepler 22b is larger than Earth, weighing in at 2.5 times the mass. It may have a denser atmosphere, and since its orbit carries it through the inner region of its star's habitable zone, its climate may resemble Venus more than Earth. But a host of factors (rotation, cloud cover) might moderate temperatures there, including the planet's rotation, axial tilt and cloud cover. Some models indicate that the surface may sit at a "room temperature" of 22°C (72°F). The planet is large enough that it may be a gaseous sub-Neptune rather than a terrestrial, with a deep, cloud-banded atmosphere.

The super-Earth HD 85512b has 3.6 times the mass of Earth. This rocky Super Earth orbits the orange dwarf star Gliese 370, some 36 light years from

---

[7] At its distance of 600 light years from here, any astronomer on Kepler 22, viewing our planet today, would see the Earth as it was in the Renaissance.

here. The planet is barely within the inner edge of the habitable zone, with temperatures probably above the boiling point of water.

Two super-Earths inhabit space in the Tau Ceti star system. Their parent star, Tau Ceti, is similar to the Sun, but smaller and slightly dimmer. Tau Ceti e orbits within the hottest region of the habitable zone. With its large size and likely dense atmosphere, its surface conditions may resemble those of Venus. Tau Ceti f—6.5 times the mass of Earth—orbits farther out, close to the outer edge of Tau Ceti's habitable zone. Its climate may allow for liquid water, especially if its atmosphere presents a substantial greenhouse effect to warm its environment. If so, Tau Ceti f may well be an ocean world.

Sea World: a new kind of planet

The "ocean planets" represent an entire class of world not seen before the Kepler mission. These watery globes probably form in the outer regions of their solar systems as ice giants similar to Uranus and Neptune, or as big-brother-versions of Ganymede or Titan. Some time later, they may then migrate into the inner system, where their dense atmosphere and ices either condense into liquid or are stripped away by the solar wind. What's left is a planet completely blanketed in a vast ocean hundreds of kilometers deep. The planet OGLE-2005-BLG-390Lb may be a sea world. Investigators estimate its bulk to be five times that of Earth, but some analysts point out that not all ocean worlds are warm enough to remain liquid on the surface. They describe OGLE-2005-BLG-390Lb as an ice globe orbiting its red dwarf sun.

No blustery volcanoes or crashing surf pound through the waves of sea worlds. Any geologic activity lies far below the surface, leaving a susurrating liquid planet. The quiet splendor of such sea worlds may be only skin-deep. The enormous pressures near the sea floor could generate unusual ices of a type not seen on Earth. This might be bad news for life, say some researchers. Because of super-Earths' massive bulk and increased gravity, great pressures would build at the sea floor where mineral-rich rock meets water. Ices under such pressures might form a barrier between the sea floor and ocean, effectively blocking off the carbon cycle. But astrobiologists still have little understanding about how chemical cycling would play out on such worlds.

The geological record indicates that oceans on the Earth have been present for most of its 4.54 billion year lifetime. Our first oceans probably gathered at the close of the Hadean Eon some 4.0 billion years ago, when the surface cooled enough to solidify and water vapor began to condense into the first rains. But would hundred-kilometers deep oceans last as long?

Recent studies indicate that the oceans of water worlds may be long-lived affairs, lasting for billions of years.

Two other exoEarths that fall in the class of ocean worlds orbit within a habitable zone of the orange dwarf Kepler 62 (Figure 8.7). Both are roughly half again as large as Earth, putting them at the border between Earthlike and super-Earth. Each is a possible candidate for having a rocky, terrestrial surface. With the right combination of atmospheric greenhouse gases and cloud cover, both worlds could sustain liquid water on their surfaces. Kepler 62f has an ESI rating of 0.67, and may have an atmosphere denser than Earth's, perhaps nearly as dense as that of Venus, but cooler. Nearby Kepler 62e, with a higher ESI rating of 0.83, is likely a water world, swathed by a global, 100-km-deep ocean. Sibling Kepler 62f may also have a large component of water, but may be far enough out in the habitable zone to have congealed a frozen surface, at least at its poles. The twin worlds are probably too big to be a typical rocky planet like ours.

Sea worlds are larger than Earth, and as such may have some intrinsic problems in the area of habitability. The Earth's mantle—the hot rock layers beneath the crust—holds enough water to fill our oceans multiple times. Plate

Figure 8.7 Two sea worlds: Kepler 62e (left) and 62f are both larger than Earth, with a large contingent of water. Because of their differing locations in the habitable zone of the red dwarf Kepler 62, their surfaces may be remarkably dissimilar. The two planets are large enough to possess multiple moons. (art by the author)

tectonics continually pull the water into the planet's deeper layers. Descending water recycles back into the environment through volcanic activity, primarily at the mid-Atlantic and Pacific Ocean ridges. Since super-Earths may have much thicker crusts than Earth does, researchers at the Harvard Center for Astrophysics wondered if that thick shell would short-circuit the kind of ocean recycling that keeps Earth's environment stable. The Harvard study modeled planets up to five times Earth's mass and up to 1.5 times its volume. Results showed that oceans on planets with 2 to 4 times Earth mass are even more long-lived than those on Earth have been, lasting at least ten billion years. The biggest planet studied in the model had five times the Earth's mass. In this case, the thick crust staved off early volcanic activity. But once eruptions began to recharge the atmosphere, stable oceans were established. The study suggests that older super Earths are better bets for living biospheres than younger ones.

Super-Earths in habitable zones need not all be water worlds. Studies show that even if a super Earth possesses 80 times the amount of water in Earth's inventory, the planet's surface may resemble Earth's continent-and-ocean mix. The stronger gravity of super-Earths forces the added water into the mantle, leaving room for dry land peeking through the global seas.

Some sodden super-Earths may orbit outside habitable zones, where global seas would freeze into a deep ice crust covering the ocean beneath. Planets that orbit inside the habitable zones of their suns might flaunt bubbling seas hovering at the boiling point. These worlds may exhibit a bizarre, twilight-zone transition from liquid to vapor. Undulating waves would stretch from horizon to horizon. At their crests, boiling water would tumble into the air, turning to vapor and drifting into incandescent fog overhead. Dense air would keep the humid mix aloft, blending it with low-lying clouds. These hot water worlds would be truly alien places, akin to an ocean-clad Venus.

The remarkable worlds of TRAPPIST-1: the case for new, distant Earths

Nights are eternal on TRAPPIST-1 e. Oddly enough, so are the days. Like so many of the planets discovered in orbit around red dwarf suns, TRAPPIST-1e must be tidally locked, with one hemisphere always facing its parent star. The reason has to do with gravity: TRAPPIST-1 e is so close to its sun that the gravity of sun and planet have locked each other in a grip where the planet turns exactly once for each revolution around its star.

The planet's awkward-sounding title comes from the name of its sun, TRAPPIST-1. The star seethes like a cool ember, just 1/12th the Sun's mass. As stars go, it is small, just a forth larger than the gas giant planet Jupiter. As such, TRAPPIST-1 numbers in the majority of stars in the Milky Way, one of the galaxy's multitude of M stars (re dwarfs). The cool star and its planets are ancient, perhaps as old as 9 billion years. The TRAPPIST-1 system lies 39 light years from our own. The planetary system surrounding TRAPPIST-1 is what makes the star truly remarkable: seven Earth-sized planets make up the TRAPPIST-1 system, and three may be in the habitable zone.

Scientists estimate that out of every 100 exoplanets, 20 to 30 are Earth-like in size and temperature. That's tens of billions of Earth-like worlds in our Milky Way galaxy alone. So far, the majority we've found orbit red dwarf suns. As of September 2021, researchers at Puerto Rico's Arecibo Observatory who maintain the Habitable Exoplanets Catalogue, list 59 "habitable worlds" (from a total of roughly 4000 confirmed planets). 38 of them are "superterran", falling between the Earth and Neptune in size (super-Earths). 20 are roughly Earth-sized, and another one is "subterran", measuring a comparable diameter to Mars. Though not Earth twins, these red-dwarf-companions could be distant cousins, and the list is growing.

Red dwarfs offer unique possibilities in our hunt for Earth-like exoplanets. It is easy to find planets around them, as their close orbits in habitable zones make their appearances frequent. Red dwarfs are abundant and nearby. It appears that red dwarfs host a wide variety of planets, from tidally locked planets to ice and gas giants. Exotic worlds like hot Jupiters and super-Earth water worlds circle in tight orbits. Because planetary systems orbiting red dwarfs are closely spaced around their dim stars, nearby planets often cross their skies, companions periodically passing by at close range. Some of these brief meetings would offer spectacular views, with the passing planet many times the visual size of a full Moon in Earth's sky. But nearness to the star itself comes at a price: such an orbit locks one face of a planet toward the star, placing the night side in eternal darkness.

In our search for other Earths across the galaxy, candidates in a red dwarf system offer good targets, as many occupy the habitable zones of their cool stars. They must orbit their suns at short distances due to the coolness of their star. This makes their year—the length of time the planet completes a circle around its star—very short compared to our own home world. The Earth's year of 365 days is determined by its distance from the Sun, but our star is hotter than the typical red dwarf, so its habitable zone is set back farther away. Planets of the cool M stars like TRAPPIST-1 orbit so closely that their years are on the order of days or weeks rather than months or years.

TRAPPIST-1 smolders at a temperature of 4160°F (2295 °C). Like other stars of its kind, its low temperature makes the M star reddish, lending it the nickname red dwarf. And while it is dim in the visible spectrum, red dwarf planetary systems don't need as much energy output to heat up the habitable region, because this type of star is brighter in the infrared part of the spectrum. As a planet moves closer to its star, any water in its atmosphere slowly evaporates. Since water is a good infrared absorber, and the star itself is a good infrared emitter, it's a perfect feedback loop. The planet gets increasingly warmer, which means it can remain at a distance and still retain water in a liquid state. In other words, the habitable zone of a red dwarf can remain farther away from the star and its deadly flares, all because of the type of light given off by the red dwarf.

The planets surrounding TRAPPIST-1 cover a range of diameters, all similar to Earth's. The largest is planet b, which spans 14000 kilometers (compared to the Earth's 12740 kilometer diameter). The smallest, planet h, measures 9860 kilometers across.

Earth-*sized* does not necessarily mean Earth-*like*. The closest planet of the TRAPPIST-1 family, TRAPPIST-1 b, simmers at a surface temperature of about 126°C. All the worlds of TRAPPIST-1 generate tidal heating as they pull on each other in their tight orbits. Their star contributes even more tidal friction. All planets in the TRAPPIST-1 system may be volcanically or cryo-volcanically active. TRAPPIST-1 b is very probably covered in a magma sea. The next world out, TRAPPIST-1 c, is not much better in the Earth similarity department. Conditions there are likely to mimic Venus. If Earth orbited TRAPPIST-1 at the same distance, our oceans would vaporize. Next out, TRAPPIST-1 d, has surface conditions somewhere between those of Venus and Earth. Its surface may be too hot for liquid water, but conditions may be more temperate, as its orbit hugs the inner edge of the habitable zone.

Then we come to TRAPPIST-1 e, a very promising Earth-like world. Although the world has less mass than Earth's, it is nearly identical in size. It is dense enough to consist of a rock/metal mix. Its orbit carries it around TRAPPIST-1 once each six Earth days. Its surface is capable of supporting liquid water; estimates say that average surface temperatures remain below 30°C. As of June 2022, TRAPPIST-1 e's ESI scale number was a remarkable 0.85. If planet d (closer to the star) has the right atmospheric conditions, its ESI rating may be even higher, at 0.9. But researchers feel it likely that planet d has a dense atmosphere that increases temperatures beyond water's boiling point.

Beyond TRAPPIST-1 e, things get chilly fast. Temperatures on planet 1 f drop to between -50 to -23°C. Planet g is even colder. If the planet is airless,[8] it is probably encased in an ice crust similar to our solar system's ice moons

---

[8] If TRAPPIST-1 g has an atmosphere, it may be a sea world with a dense steam atmosphere.

Figure 8.8 The outermost planet in the TRAPPIST-1 system, TRAPPIST-1h, has a diameter two-thirds that of Earth, and a density similar to that of Mars. With these factors in hand, astronomers estimate that the planet is rocky with a thick ice crust. Here, we see a version of the planet with a mix of ice and rock, its sky filled with all the other planets of the TRAPPIST-1 system. TRAPPIST-1g is eclipsing the star, a common sight in a planetary system like this one. (art by the author)

such as Europa or Enceladus. Planet h is colder still. Orbiting 9 million kilometers from its star, TRAPPIST-1 h's temperatures plunge to -104°C (Figure 8.8).

Computer simulations show that within 500 million years of their formation, all planets in the TRAPPIST-1 system had become tidally locked, keeping the same face pointed toward their sun. The system formed in the same way that our own did, emerging from a disk of dust and gas. Planets coalesced from TRAPPIST-1's solar nebula much as our solar system did, spinning as they solidified. The drag of their sun's gravity eventually slowed their rotations, finally relaxing them into the dance they perform today, turning once for each revolution they take around their star. Our solar system has no planets tidally locked to our Sun.[9]

When the M star was very young and extremely active, these planets were already locked. Young stars go through an adolescent phase of brightening and extreme activity, spewing flares and radiation before settling down into stable adulthood. For M dwarfs, this "T tauri phase" can last for billions of years. The nearby planets would be susceptible to atmospheric erosion, so it may be that TRAPPIST-1 b and c have no atmospheres at all.

Eyeball Planets

As exoplanet discoveries came in, observers began to see patterns. Within the overall population of exoplanets, a common form arose: planets near enough to their stars to become tidally locked. These worlds keep one face toward their sun. The result is one illuminated hemisphere with a hot spot in the center. For many water worlds in this situation, oceans freeze in arctic environments along the terminator, but thaw in daylight, with a bright, hot region facing the sun. Theoretically, this would result in a planet with the appearance of an eyeball (Figure 8.9).

At the sub-solar point directly under the sun, clouds will likely burn away and water temperatures may rise close to the boiling point. Even with a thick atmosphere to dampen the effects, these super-Earths may suffer temperature extremes. Atmospheres transfer heat from the dayside to the night side – and the thicker the atmosphere, the more effective this is. The day/night temperature disparity depends on many factors, including the amount of radiation falling on the planet, the density of the atmosphere, and the reflective quality of the atmosphere itself. To provide an idea of just how extreme the temperature swing might be, researchers applied a global circulation model to a model super-Earth. The example they used was Proxima Centauri b. Their results revealed a possible day/night temperature range of -130C to 0°C (-202F to 32F).

---

[9] Mercury comes closest: it turns three times for every two revolutions around the Sun. Mercury and Venus will both eventually be tidally locked.

Figure 8.9 Seen from a nearby moon, an ocean-covered world exhibits features that led astronomers to coin the term "eyeball planet." Here, we see frozen oceans along the terminator, giving way to open water toward the sub-solar point. (art by the author)

Figure 8.10 Evening on a volcanic moon: Tidally locked terrestrial planets may develop eyeball facades. Seen from a nearby rocky moon, a terran world displays desert conditions toward the sub-solar point and cooler land with standing water near the terminator. If the atmosphere is quiescent, constant arctic conditions will assault the night side. (art by the author)

Smaller, rocky worlds will have similar eyeball traits (Figure 8.10). On the hemisphere facing the star, even if the planet orbits within the habitable zone, desert conditions likely occur where the sun continually shines overhead. The environment will become more clement closer to the evening side, perhaps allowing for seas and lakes within that twilight zone between dark and light. The night side is another matter: if the planet has an atmosphere robust enough for wind currents, the night side may remain at temperatures benign to life. But if the atmosphere is thin or stagnant, night time temperatures will

Figure 8.11 The hot Jupiter exoplanet HAT-P-12b compared to Jupiter. Note the chevron of cloud banding, quite different from the patterns on our own Jupiter. (art by the author)

drop to well below freezing, even if the planet is on the inner edge of its star's habitable zone.

Eyeball planets come in an assortment of sizes. The planet HAT-P-12b orbits a K-type star slightly cooler than the Sun (Figure 8.11). The exoplanet falls into the "hot Jupiter" gas giant class. The colossus is tidally locked, always pointing the same face toward its star. This spin, slower than the giants of our own solar system, likely results in belts that are far more subtle. Jupiter's well-defined belts form because of the planet's rapid spin, which takes just under ten hours. Golden Saturn—nearly as large—rotates every 10 hours 42 minutes. But since HAT-P-12b is tidally locked, its daily spin matches its course around its sun, making a turn once every 3.2 days. Its leisurely turn will not drag the atmosphere into the kind of tight bands seen on our Sun's gas and ice giants. Instead, equatorial jets of flowing air develop chevron-shaped forms like an arrow pointing in the direction of the rotation. Computer atmosphere models show lobes coming out from the equatorial jets. Heat at the planet's equator generates cells that circulate material poleward. HAT-P-12b is the least dense of any Jovian world found so far.

Giants inside the habitable zone: the promise of the moons

A churning host of moons surround all four of the giant worlds in our own Solar System. Some are as large and complex as planets. But within the life-friendly habitable zones of other stars, it is the gas giants that make up the majority of the known planets. In fact, moons of giant planets could number among the greatest population of habitable worlds. Three large moons in our own system are considered candidates for life or life's precursors: Jupiter's Europa, Saturn's Enceladus, and Saturn's Titan (see Chapter 5). If we took these moons and brought them into the habitable zone of our Sun's system, they would become sea worlds filled with the prospect of life. Since many exoplanet gas giants circle their stars within the habitable zone, they provide places to search for exoEarths tagging along. In the case of exoplanets circling red dwarf stars (by far the most common Earth-like worlds found so far), a moon offers something that a planet lacks: a day/night cycle. Habitable zones are so close to M dwarf stars that the planets within them tend to be tidally locked. But moons orbit those planets, turning on their axes far more quickly than their parent planet makes a circuit of the star. This provides moons with periods of day and night, circumventing some of the problems of a tidally locked planet. Some biologists suggest that day/night cycles provide more opportunity for biogenesis, as the planet or moon can host more varied conditions of light and temperature when the sun is crossing the sky rather then sitting in one spot.

Several candidate moons of exoplanets are awaiting confirmation, and researchers continue to pursue new techniques for finding small worlds orbiting giant ones. One of the new candidate moons orbits a giant planet circling the Sun-like star Kepler-1625. Kepler 1625 has a mass roughly equal to our Sun, but the star is older, so has expanded to 70% larger than our star. Its planet, super-Jupiter Kepler 1625b, is about the size of Jupiter, but the dense world weighs in at nearly twelve times the mass. The bulky globe resides along the inner edge of the star's habitable zone, with a year lasting 287 Earth days. Around this planet, in turn, orbits a large moon. Kepler 1625b-i doesn't look much like a moon compared to the natural satellites of our own system. Its size rivals that of Neptune, so it is likely gaseous. Data revealing the moon's existence was hidden in Kepler Observatory data, and confirmed later by the Hubble Space Telescope.

But what about moons with solid surfaces and more benign environments? Astronomers continue their search among the giant exoplanets. Typical of the giant worlds orbiting within habitable zones is Kepler 47c. Kepler 47c holds the distinction of orbiting two suns. The Kepler 47 system is binary, with the primary star similar to—but slightly dimmer than—our Sun. Its companion star is a small red dwarf a third the size of the Sun. Kepler 47c orbits within the habitable zone encircling both stars. But the shared habitable zone of this two-star system is complicated. The primary star circles a point between itself and the smaller star once every 7 ½ days. Its location dictates that its habitable zone precesses, wobbling like a child's toy top around the primary star. Kepler 47c remains within this shifting zone. It takes the planet 303 days to trace its path around the stellar couple. The planet itself is a super-Neptune, gaseous and probably not hospitable for life. However, any moons around the planet, if large enough to hold atmosphere and water, may well be Earthlike.

A second planet in the system, Kepler 47b, revolves closer in to the stars. Its 50-day route keeps conditions on its surface parched, with an atmosphere that may be more Venusian than Earthly.

A Neptune-sized moon orbits a planet belonging to the Sun-like star Kepler-1708. Kepler 1708b is a gaseous giant. Its moon is smaller than Neptune, and may be in the right size range to have a solid surface. The moon may also be oceanic. Its host planet dwells within the habitable zone in an orbit similar to that of Mars around the Sun. Conditions may be cool, but the moon is large enough to possess an atmosphere that imbues the surface with greenhouse warming.

Researchers have found another gaseous goliath within the habitable zone of a binary star system called Kepler 16. The primary star, Kepler 16A, is an orange dwarf slightly smaller than the Sun. The star orbits around a point between itself and Kepler 16B, a red dwarf. Like Kepler 47b, planet Kepler 16b orbits both of the stars in their shared habitable zone, making the yearly trek once each 229 days. Because the planet was spotted transiting two stars, more data was available than it would have been in a single star system. The planet's mass is known to a precision of 0.3%. Kepler 16b is similar to Saturn in mass, plenty large enough to muster a swarm of moons. The planet is likely composed of half gas and half rock and ice. Kepler 16's habitable zone extends from 55 to 106 million kilometers from the dual stars. Kepler 16b orbits in the zone's outer region, at about 104 million kilometers distance. Temperatures on the planet should range from -100 to -70°C. Computer models suggest that sometime in the Kepler 16's past, other planets may have perturbed an Earth-sized inner planet from another location in the habitable zone, causing it to migrate out of its orbit, where Kepler-16b could have captured it as its moon. If the scenario played out as proposed, Kepler 16b may be one of many worlds with Earthlike moons in tow.

Beyond Europa: deep sea astrobiology

Since water worlds make up a large proportion of exoplanets, astrobiologists naturally turn to the question of life on them. While some water worlds orbit within the habitable zone, many do not. But considering an expanded microbial habitable zone, our narrow view of life's niches has expanded. Our discoveries of cryophiles—extremophiles adapted to extreme cold—adds fuel to the possibilities of biomes in even the frozen super-Earths of distant planetary systems.

Planets like Kepler 62f underline the fact that water worlds come in assorted varieties. Just because a planet has water does not ensure a life-bearing environment. Large super-Earths may have dense atmospheres that preclude habitable surfaces, whether those surfaces are made of land or water. Some oceanic planets orbit in the inner edge of the habitable zone, so temperatures would be high. However, research shows that global seas could moderate temperatures as long as the ocean is deep enough. Studies show that even on a planet with an axis tipped over (like Uranus), habitability is possible, barely. Oceans of 50 meters or deeper can moderate an atmosphere, keeping temperatures warm enough for an active biome, even at the poles. Polar temperatures would remain above 10°C. But if the ocean shrinks to a 20-meter depth, it reaches a tipping point. Models imply that if global seas develop even a thin shell of ice, the entire ocean system could freeze solid in a marine version of a reverse runaway greenhouse effect (something akin to a snowball Earth; see Chapter 6).

A world awash with crashing waves may be a fine candidate for life, if those waves interact with a shoreline. The supposition comes not from the field of astrobiology, but paleontology. Gas bubbles trapped in Earth's sediments represent shorelines some 3.2 billion years ago. At that time in Earth's development, the atmosphere had no free oxygen, so it lacked a radiation-protecting ozone layer. But in this deadly environment, life was able to hang on. Microbes mixed their single cells with grains of sand, forming "biofilms", carpets of gooey biological material.

Paleontologists found some amazing fossils in 3.22 billion-year-old sandstone from South Africa. Microfossils from that epoch insinuate that colonies of microbes may have inhabited trapped gas bubbles, where they could endure Earth's austere primordial conditions. Long microscopic channels in crystalline quartz betray the colophon of microbial life that cloistered within the interior of gas bubbles. The ancient life forms also left a chemical calling card: a ratio of light and heavy types of carbon that's usually associated with living organisms. The creatures left faint impressions of cell chains similar to modern bacteria. If microbes could survive the primordial Earth's harsh conditions hunkered down in bubbles, it's possible that life survives on the bleak surfaces of high-radiation planets near M stars, or on planets of even more exotic suns like white dwarfs or neutron stars.

Living Worlds

What kinds of biological prerequisites will various exoplanets and their moons possess? Can conditions quite dissimilar to Earth's lead to life? After all, life on Earth began in an environment quite different from the one we enjoy today. Opaque clouds blanketed the sky for hundreds of millennia, and that sky had no oxygen, but instead seethed with a mix of elements that seem toxic today. Within this alien setting, the first life took hold, life that would have expired in the oxygen-rich environment of modern Earth. Anaerobic bacteria inhabited the oceans and lakes. More complex life such as colonies of microbes formed mats on the sea floor, incorporating minerals into a matrix that exists today in the form of fossilized stromatolites. Photosynthetic life emerged, using sunlight to make energy. Their waste product was oxygen, and as the oxygen piled up, creatures learned to make use of it. Life on Earth relies on carbon and water to operate. We humans are carbon-based life forms, as is all terrestrial life around us. But do other environments offer alternate construction materials for life, perhaps like those we see on Titan?

Biosignatures: life's footprints on the landscape of astrobiology

In our search for Earths beyond our own solar family, we naturally ask ourselves about life. But many of the planets we've discovered appear as points of light or dips in starlight. How can we detect an active biome many light-years away? One promising approach is through the light cast out by various chemicals on a planetary surface or atmosphere: any terrestrial world with a vibrant ecosystem will give itself away chemically. In fact, living organisms leave an assortment of clues to their presence. Life, it seems, has engraved its monogram across the face of our terrestrial globe. Its message, written in minerals and water, gas and chemistry, is called a biosignature. If life exists elsewhere, it, too, will leave such a calling card.

As we saw beginning in Chapter 2, nature constantly attempts to bring environments into equilibrium. But the spectrum of light coming from the Earth—the only example we have of a living planet—shows increased levels of oxygen in our atmosphere, one fairly reliable 'tell'. Oxygen is a sociable gas: it doesn't stay around for long periods, because it tends to combine with nearby elements. Left on its own, the oxygen blanketing our world should be soaked up by carbon-silicate rocks, part of nature's ongoing attempt at balance. The elevated levels of the gas, at least on Earth, are due to the metabolism of life forms. In the same way, if Earth had no methane-producing microbes, atmospheric methane levels would drop as the methane oxidized into the oceans and lakes. Methane also combines with carbon dioxide to become more stable. While methane can also be generated by volcanism, volcanic methane will ebb and flow globally as the environment soaks up and processes erupted gases, while the action of living systems will raise methane levels to a steady level over time.

With the advent of space telescopes like the James Webb Space Telescope, we can tease out the spectra from exoplanetary atmospheres. Astrobiologists[10] advise that we should be actively hunting for compounds floating around in atmospheres that organisms manufacture, specifically to protect themselves from toxic elements. We might be fooled by gases like oxygen and methane that can occur naturally. Methane often accompanies non-biological volcanism, hot springs, etc. Oxygen may be produced in the steam atmospheres of hot sea worlds, where ultraviolet radiation from a star breaks water into hydrogen and oxygen. The light hydrogen escapes into space, leaving behind heavier oxygen. But one class of gases stands apart: methylated gases. These gases are generated when organisms link carbon and hydrogen to toxic elements like bromine and chlorine. The new combinations (compounds) evaporate, and as they do, they carry away the toxins. Various microbes, algae, fungi and some plants carry out this protective linkage. Researchers have found that methylated gas production comes—almost exclusively—at the hands of living creatures, making it a good life signature. On Earth, methylated gases are most abundant in swampy river deltas, wetlands and estuaries. Even life forms not based on DNA might generate methylated gases in response to elements like chlorine and bromine, which appear to be quite toxic over a spectrum of conditions. This makes it a good agnostic tool for seeking life.

The Earth's atmosphere is not its only biological signature. The albedo, or brightness, of our land masses and oceans is radically affected by the hue and pigmentation of living things (Figure 8.12). Plants and forests paint the land in the hues of living color. The algae, diatoms and other microbial life add their own shades to river and sea. The hydrological cycle itself can even be modified by living things. For example, plankton influences local climate by pumping out dimethyl sulfide gas. This gas pumps sulfur particles into the atmosphere, increasing the condensation of water vapor. Large plankton populations generate their own cloud systems over tropical seaways.

It follows that our search for extraterrestrial life must involve a search for conditions that are out of balance, for levels and light curves that are not in equilibrium. In fact, living communities not only keep things out of balance,

---

[10] Michaela Leung, oral presentation at the American Astronomical Society, January 9, 2023

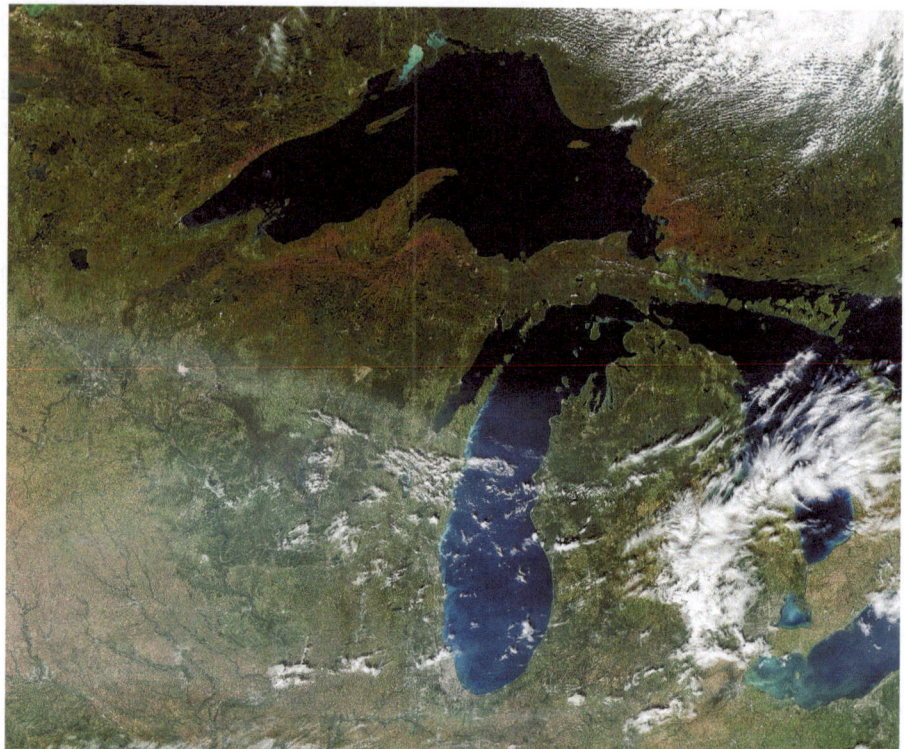

Figure 8.12 Autumn comes to the Great Lakes of North America. Seasonal changes in color such as these may be one indicator of active life on exoplanets. (NASA Earth Observatory image by Joshua Stevens from NOAA 20 data; NASA EOSDIS/LANCE and GIBS/Worldview and the Joint Polar Satellite System)

but they also respond to variations in their star's energy in ways that are more complex than a sterile, stable world would. An example from our own world is the Earth's carbon dioxide level. The Earth's carbon dioxide varies consistently with the change of seasons. As winter visits on hemisphere, vegetation there falls dormant, and atmospheric carbon dioxide rises. But as summer brings the flourishing of vegetation, photosynthesis blankets the forests, plains and jungles of our biosphere, and carbon dioxide levels drop. Long-term changes in carbon dioxide could come from volcanic or other geological activity, but seasonal changes would point exobiologists to a biological explanation.

Seasonal changes in exoplanet atmospheres might also provide clues about the surface of a distant Earth-like world. In the case of our own planet, it would make sense that seasonal carbon dioxide levels would cancel each other out as summer moves from the southern hemisphere to the north and back again. But because of the configuration of our continents, most of the real estate appears in the northern hemisphere. Seasonal change on planet Earth is felt most strongly north of the equator.

We have seen how oxygen, an unstable gas, could not build up in primordial Earth's atmosphere until the advent of life put it there. But one study cautions that false positives from high oxygen levels may occur where strong radiation strips oxygen atoms from carbon dioxide. If radiation is strong enough (for example, on an Earth-sized world in a tight orbit around a red dwarf), oxygen atoms may be freed from the carbon dioxide. The process can outpace the natural balance (normally, oxygen tends to recombine with carbon over time, turning back into carbon dioxide).

Some ocean worlds in close orbits around red dwarf stars may offer another false positive. Red dwarfs may bathe nearby planets in radiation that vaporizes surface water. In this case, the atmosphere would become saturated with water vapor. Hydrogen atoms in the upper atmosphere can be stripped away by radiation, leaving behind a fog of dense oxygen, resulting in another biological mirage. But on the other hand, a lack of oxygen might lead to a false negative. Just because a planet has low oxygen levels does not rule out the possibility of an active, global biome (the Earth had life without oxygen for some time, after all). The possibilities of both the false positives and false negatives serve as cautionary tales in our search for living systems.

In the quest for life beyond, biologists have two primary modes of *in situ* investigation. They can either collect samples and culture them in an attempt to grow microbes from the sample, or they can hunt for biomolecules (carbon-based organic molecules, amino acids, etc). But in their remote search for life on exo-Earths, astrobiologists have neither option. Instead, investigators must look for the waste products of life.

Life's pollution and consumption may make themselves known, even from light-years distant. Life often has associations with color. Photosynthesizing plants are green. Algal-seeded salt ponds are candy-apple-red. Life tints the color around it. This "pollution" of the spectrum may be a positive sign to search for.

A third test, one that is almost within our current capabilities, is to look for chirality. Most life forms known to science have a twist to them. This turning format is asymmetrical in a specific way: if the form is flipped, it does not become a mirror image of itself. The phenomenon can be seen in a person's hands, from which the Greek word comes.

All known living organisms exhibit homochirality, biological "twisting" in one direction. The vast majority of organisms use only left-handed L-amino acids in proteins and right-handed D-sugars in nucleic acids. A chiral life form contains an asymmetric center, forming mirror-image structures. This property of twisting in mirror image may be a necessity for self-replication. If it is, it may well be a universal characteristic of all biochemical life.

The polarized light reflected by chirality could supply observers with a convincing biosignature within the spectrum of distant planets. Astrobiologists advocate three foundational steps in the search for exobiology. First, look for oxygen (or other imbalances in the gases of an atmosphere). Second, look for polarized light from chiral-selective microbes or large-scale vegetation. A third test has to do with the colors associated with photosynthesis. Chlorophyll exhibits a red edge in its spectrum (a peak just at the near-infrared). Vegetation absorbs incoming visible light, but reflects back infrared wavelengths of the spectrum. Finding this unique tint within the light of a distant exoplanet may point to active photosynthesis taking place. If vegetation on an exoplanet carries on processes akin to photosynthesis, taking in the sunlight of the red dwarf star and converting it to food energy, their pigmentation will differ from the green chlorophyll we see on Earth. M dwarfs emit less visible light and more near-infrared light (redder) than our Sun. While terrestrial plants absorb the Sun's spectrum, leaving behind green, plants adapted to the light of a red dwarf might appear dark yellow or even black.

Our advanced observatories, along with techniques learned so far in our search for Earths beyond our solar system, afford us new capabilities for not only locating Earth-like exoplanets, but also understanding them as new worlds in the cosmic neighborhood.

Generation Ships

Just as our future civilization might build cylinders or torus-shaped structures to house colonies in orbit around our home planet, interstellar architects may construct vast rotating ships to carry humans and wildlife, gardens and forests and farmlands, to other worlds. Like the O'Neill colonies, the massive structures would spin so that their contents would be forced onto the walls, creating artificial gravity. The ship's central axis would always appear to be "up", while the inner walls would serve as the ground.

Sending robots on a 50-year mission to Alpha Centauri is one thing, but human passengers create complications on an interstellar mission. They must eat and breathe and live. Near-relativistic speeds slow time, aiding in the problems of the long journey, but to voyage to another star will take generations. If we are to use conventional engineering, even of the kind we can reasonably dream of with projected technologies, interstellar travel is a long-term effort. We may need to be content with sending a ship on a journey of hundreds of years. Such a ship would arrive with the descendants of its creators.

These innovative vessels have been called generation ships, because they would require many generations of travel to reach their ultimate destinations.

As we have seen earlier, going faster solves problems along the way. A ship traveling at a substantial fraction of the speed of light will encounter a big problem at the far end of the journey. After centuries or millennia of acceleration, how do we stop the thing? One obvious solution is to turn the ship around at the halfway point in the voyage and decelerate for the duration. It works, but this approach lengthens the trip, and it requires a lot of fuel. Spacecraft-designing physicists have stumbled upon a solution using a natural resource found around every star: the magnetic field. A generation starship can be thrusted under the power of lasers beamed from Earth. The lasers would be shut down as the ship approached the target star system. That's when the star's magnetic field comes into play: a huge loop of superconducting tether can become a magnetic brake, or parachute. When current races through the tether, it creates drag against the background magnetic fields intrinsic to the star up ahead. Loops of conducting wire tend to force themselves into a circle because of their self-generated fields, so deploying this type of assembly would be easy, as it would naturally expand into the required shape.

Theoretically, a generation ship could extend dozens or hundreds of the metal cables, each loop stretching hundreds or thousands of km out. The crew would infuse the cables with a high-voltage charge (~800,000 volts). The current's interaction with the magnetic fields would cause the ship to make a huge, gradual rotation around the target star so that it is eventually approaching from the far side (now moving toward the Earth). Years or decades before the starship is in position, lasers or microwave beams can be turned on again, timed to reach the distant planetary system at just the right time, slowing the ship as it approaches from behind. The magnetic "sea-anchor" concept enables a ship to drain off roughly half of its velocity – even if its speed is nearly relativistic – without the use of weighty propellants.

The subject of generation ships has been explored—in various forms—for over half a century. In 1956, E.C. Tubb wrote *The Space-Born*, a dystopian adventure told from the perspective of a 14th-generation chief of police who must carry out death-sentences prescribed by the ship's computer. 1973 saw the debut of Arthur C. Clarke's multi-award-winning classic novel *Rendezvous with Rama*, a tale of astronauts who explore a vast hollow cylinder of alien origins. More recently, visionary author Kim Stanley Robinson describes the successes and ominous failures of an interstellar journey in his book *Aurora*. Robinson's star voyagers endure a journey of nearly two centuries – seven human generations – during which the ship's 2000 inhabitants, quantum computer and robots must tend to all the ship's systems. The Aurora carries several distinct habitats, including ecosystems named Sonora, Amazonia, Olympia, and others, each supporting a population of wild creatures. Robinson delves into how the descendants feel about their fate, and about the challenges of creating a home on an Earth of another sun.

Despite the lure of finding life on exoplanets, when humans seek a second home among the stars, indigenous biomes may be bad news. The first reason is simple: human settlers can't afford to risk infection from an alien source. The consequences could be devastating. When the Apollo missions returned from the Moon, astronauts were quarantined, and geologists handled lunar samples in an environmentally secured laboratory. While the precautions may seem provincial now, they were prudent for our first visit to another world. We're faced with similar challenges when we return samples from Mars in the near future.

Secondly, no matter how much care we take, travelers from Earth will carry a wide assortment of microbes with them, and these could ravage any *in situ* biome we visit. Some astrobiologists caution that mission planners need to exercise extreme care concerning human travel to the planet Mars, suggesting that if life forms are found there, all piloted exploration endeavors should cease immediately, leaving scientific study to our sterilized robots. Still, the ultimate home away from home might be found among the planets of other star systems.

In the 1981 book *Illustrated Encyclopedia of Space Technology*, author Kenneth Gatland makes the point that in discussions about interstellar exploration, "it is implicitly assumed that there is both somewhere to go, and a motive for going." At the time of his writing, the search for exoplanets was in its infancy. Today, we have found thousands of planets circling other stars, and dozens of them qualify as Earth-similar (a subject for our next chapter). But the issue he brings up still stands today. In concluding remarks, Gatland says, "a decision to travel to the stars will involve mankind in a profound analysis of human motives, of the purpose of life and of our relationship with the Universe." And isn't that what science, at its best, does? The scientific exploration of the cosmos, in all its glory, causes us to look outward, but also to look inward.

9

Earth=Venus, part II: future Earth?

We have seen how the primordial Earth—during its Hadean epoch—resembled a toned-down version of Venus, with its high levels of carbon dioxide and its volcanoes belching toxins into a sky devoid of oxygen but full of falling metal and stone. We have witnessed our snowball Earth, enduring massive glacial epochs with parallels to the frigid desert world of Mars. And we charted the prebiotic Earth, its biochemistry percolating across our primordial ponds and seas, its chemistries similar to those occurring today on Saturn's moon Titan.

And we can't leave the Earth there; it is a dynamic and changing world still. As long as a radiogenic fire burns within the heart of our planet, its continents will continue their tectonic dance. Land masses will parade across its face—moving about as quickly as a fingernail grows[1]—pulling apart to form new oceans, bumping into each other to lift up new Everests and Denalis. With all that random moving and shaking, predicting how the continents will move in the future is quite difficult. But orbital imaging and advanced computer models based on surface stations gives us some idea. Africa appears to be rotating clockwise and drifting northeast. This movement will eventually fuse it to the sprawling Eurasian Plate to the north. The Mediterranean and Red seas currently separate the two plates, but may vanish in the collision. A new mountain range will rise up to fill the Mediterranean

---

[1] The speed differs with location. One of the most rapidly-moving sites lies within the rift valley of Iceland's Þhingvellir, where the Eurasian plate meets the North American one. Here, plates are spreading apart at about 2 cm/year as new rock is created. An Icelandic friend once told me, "This is the place to buy real estate. You can get a small plot, and in a few million years you'll have a whole farm."

M. Carroll, *Planet Earth, Past and Present*, Springer Praxis Books,
https://doi.org/10.1007/978-3-031-41360-5_9

Figure 9.1 The red death: Even the most advanced civilizations could not survive in the face of an expanding, red giant Sun. By then, human civilization will hopefully have moved on to other frontiers across the cosmos. (art by the author)

basin. Northern Europe will travel toward the north polar region. The Great Rift Valley of Africa will split open, forming a new ocean as the African continent fractures. The island of Madagascar will accompany eastern Africa on a voyage east, burying (subducting) the Indian Ocean Plate on the way and obliterating the Himalayas. The Atlantic Ocean will continue to expand while the Pacific narrows. Southern California will splinter off from the

North American continent along the San Andreas Fault, sliding northward. Australia and the Indonesian archipelago are destined to ram into Southeast Asia. 100 million years into our future, Australia and Antarctica will unite as Africa and Eurasia merge into a new supercontinent. Some work indicates that the Americas will drift east, and in 250 million years will combine into a new supercontinent that they call Pangaea Ultima. But other models predict the fusion of the Americas. In this scenario, the land masses will end up as the eastern seaboard of the African-Eurasian-Australian group, forming a colossal new supercontinent Melonia. Whichever route our continents take, any space traveler visiting the Earth in a few hundred million years will see an unrecognizable arrangement of continents on a very alien world. But if our visitor had come one hundred million years in the past, we could say the same thing about our dynamic world.

Why, now, return to Venus? Many of the changes brought to our world over the ages can be seen through the lens of climate and climate change. Those changes may be spurred by volcanism, by changes in obliquity (our axial tilt), or by perturbations of our orbit (remember those Croll-Milankovich cycles?). Changes in climate are more than academic; they affect the life and quality of the entire terrestrial environment in drastic ways. Biophysicist Ben Clark points to a drastic cooling event in Earth's recent past as an example: "[A drop in world temperature] may have caused the medieval period. India and Europe went through a very bad several hundred years where crops failed and civilization sort of collapsed. You had to wait for the Renaissance for things to get better. They think that may have all been climate induced; cooler weather."

As the 16th century dawned, the warm days leading up to the medieval period gave way to an ominous chill across northern Europe. The period has come to be called the "little ice age". With colder winters, coastal seas and rivers froze over, crippling shipping, communication and commerce. Crops wilted, livestock starved along with their keepers, and massive rains devastated summer harvests. In England, the River Thames—main artery of travel though London—froze over periodically from 1400 to 1815. The unprecedented events were not all gloom and doom, however. Entrepreneurial promoters disseminated flyers celebrating "frost fairs" on the river ice, where jugglers and puppeteers entertained, and partiers danced, played games and skated across the remarkable frozen surface (Figure 9.2). Even King Charles II and the royals came to the Thames to visit "Freezeland" as promoters were calling it. But amidst the celebrating, fuel prices (for wood) soared, river workers lost their livelihoods, and the economy nearly collapsed. Across Europe, the change in climate imprinted its cost in political revolutions, citizen rebellions, plagues and blame shifting.

Figure 9.2 Artist Luke Clennell painted "The Fair on the Thames" in February of 1814. Said one account of the great "Frost Fair", "The Thames, from London Bridge to Blackfriars, was for nearly a fortnight completely blocked up at ebb time…there were counted more than 6000 people at one time on the ice, chiefly skaiters." Painting ©Trustees of the British Museum, used under Creative Commons License (CC BY-NC-SA 4.0)

Really old climate records

After the discovery of fossilized giant turtles in Dorset, 17th-century British polymath Robert Hooke suggested that England had once possessed a warmer climate, perhaps due to the shifting tilt of the Earth's axis. He was right about changing climate, but researchers have only been able to take direct measurements of Earth's axial tilt for a few centuries. Climate patterns before recorded history, called paleoclimates, can be studied in various clever ways. Techniques include the analyses of tree rings, ice cores, rocks and sediment layers, coral reefs, the fossil record and seashells. We have seen how arctic ices trap gas bubbles, providing insights into the ancient atmosphere. Layers of volcanic ash within deep ice cores mark eruption events, and the thickness of ice layers tells us something about temperatures and precipitation. Coral reefs and tree rings provide similar data hidden within their strata. Tree ring sequences have been dated back to climates some 13,000 years ago.

For the more ancient of nature's chronicles, researchers must turn to the stony subsurface of the Earth. Fossils embedded in the layers of sediment show changes from desert to alpine conditions, or even changes from dry land to marine environments. Levels of various chemicals can yield estimates for temperatures in the past, and isotopes provide insights into changes in ice thickness and ocean salinity, which in turn tells us about ocean temperatures. Finally, radiometric dating uses radioactive materials within the sample to estimate age, a valuable tool in deciphering the complex history of our planet's climate. Dozens of different radiometric tests can be used to confirm results.

The Earth's unique plate tectonics have played their part. As we saw in our discussion of Rodinia, the arrangement of continents has a profound effect on our planet's weather and climate. As plates bash into each other, causing uplift, mountain chains build. Mountains cause changes in temperature, precipitation and airflow. As continents move toward higher latitudes, glaciers form at greater elevations, causing global sea levels to drop. As longer coastlines appear and more land is exposed, surface erosion and weathering bury organic material more quickly. (Note that weathering differs from erosion: weathering is a change in rock, soil, chemistry and minerals due to its interaction with water, gases, and organisms. Very little movement is associated with weathering. Erosion, on the other hand, is the movement of surface materials due to interaction with rain and flooding, ice and snow, wind and transport of material via events like landslides.) Increased land area lowers overall temperatures, and oxygen levels increase. Over the long term, these changes affect the climate, with lower global temperatures and changes in weather patterns. Some biologists see the changes as critical to biological processes, as new continental arrangements bring new niches for life.

Continents may have played their own part in the expansion of life. As plate tectonics gave birth to the continental crust, weather began to erode the newly-formed landmasses. Rivers and floods carried minerals from the high country into the seas, enriching the water with new chemicals for use by living systems. The rise of the continents also created inland lakes and, perhaps more importantly, shallow ocean waters along the continental shelves. The flow of nutrients encouraged growth of plant life in the shallows, and oxygen surged into the air, ushering in the GOE.

The rise of the continents brought forth a delicate equilibrium, a critical balance between water surface and dry land. The planet's water buffers our temperatures. Oceans cover 70% of the Earth's face. Ocean moons like Saturn's Enceladus or Jupiter's Europa offer examples of oceans too deep to allow for dry land above. In fact, both moons have a "decoupled" crust, a rind of ice that floats on top of the subsurface ocean, independent of the rocky sea floor. If our oceans had been just a bit deeper, the continents would have remained beneath the surface, as they must within the ocean moons. Those life-aiding minerals would have no place to wash down from, starving the water of critical life elements. Without exposed rock, limestone would not be present to moderate carbon dioxide levels, and temperatures would skyrocket. Taken far enough, this trend might have resulted in the entire planetary inventory of water boiling away.

If, on the other hand, our oceans covered less surface area, leaving more land exposed, temperatures would swing between extremes. Vast areas of dry real estate also mean less carbonate development, and less absorption of carbon dioxide, which tends to take place mostly along the ocean borders and in deep rock exposed to the seawater. The great expanses of mountain and plain would reduce carbon dioxide even further as the chemicals in the rock draw down the greenhouse gas more. Less carbon dioxide—and less warming—could lead to global glaciation, leaving the Earth looking like those ice moons we mentioned.

The balance is further affected by glacial periods. Massive glaciations result in lowered ocean levels, while interglacial periods result in a rise in ocean levels, covering the continental shelf with deeper waters.

How much land is too much? How much is not enough? We don't yet know, but we're getting some insights from models of exoplanet environments. One study suggested that without the chemical buffering of continental weathering, oceans on Earthlike exoplanets would remain too acidic for life to thrive. Wherever the balance lies, the Earth hit the sweet spot throughout the development of its planethood. In those days when prebiotic chemistries resembled those on Titan, life took hold, and it has tenaciously held on, morphing into varied forms through glaciations, acid-raising events, hot and cold spells and the rest.

Slippery habitable zones

In the 1970s, astrophysicist Michael Hart realized that since the Sun was, at first, dimmer than it is today, its habitable zone was closer in to the star that its current spread. As stars advance through different stages of development, their habitable zones must migrate, he reasoned. As the Sun aged and warmed, the habitable zone moved outward. Within that moving habitable zone was a region that remained habitable throughout. Hart called this region the permanently habitable zone (PHZ). Four billion years ago, during the Earth's infancy, the Sun was 70% as bright as it is today. As our star continues on this warming path, it forces our habitable zone outward. Hart's analysis showed that the Earth is in a central-enough location that it should remain in the habitable zone throughout this shift. It's a sort of sweet spot within the habitable zone where conditions remain stable for liquid water, despite the outward migration of the habitable zone around it. But this zone is just barely stable, perched on a razor's edge. Astrophysicists have been able

habitable zone
continuously habitable zone
animal habitable zone
homo sapiens habitable zone

Figure 9.3   The different precincts of our solar system's shifting habitable zones. The continuously habitable zone is but a small subset of the whole. (art by the author)

to demonstrate that had the Earth formed only 5% closer to the Sun, our planet would have experienced a runaway greenhouse effect, making us more of a twin to Venus than we already are. Had our planet arisen a scant 1% farther away, Earth would have suffered a runaway glaciation, with the surface and oceans freezing over in a permanent snowball Earth condition (see Chapter 6). Hart's models indicate that both of these situations—frozen or baked—are irreversible.

The permanently habitable zone has important subdivisions (Figure 9.3). The "microbial habitable zone" may overlap the previously understood habitable zone, expanding close enough to the Sun to benefit from its energy, but far enough away to dodge high radiation levels and temperatures. The microbial zone is more vast than the classic habitable zone, because it extends to the ice moons in the outer solar system. These "ocean worlds" make up a substantial portion of the Sun's family. Astronomers estimate that subsurface seas exist at Jupiter's moons Europa and Ganymede, a number of Saturn's moons, including Titan[2], Rhea and Enceladus, several of the Uranian satellites and Neptune's Triton. Pluto may also have a global ocean of slush, a mix of ammonia, methane and seawater trapped beneath its exotic surface. Another subset of the habitable zone is that of the "animal habitable zone". This region is

---

[2] In addition to Titan's famous methane seas on its surface, the massive moon hides a vast water ocean inside (Chapter 5).

more narrowly constrained, spanning a distance at which a planet can host liquid water, but where temperatures do not exceed 50°C. Within the biomes of Earth, this seems to be the upper limit of temperatures survivable for animal life (in some locations on our planet, extremophiles can exist at much higher temperatures). An even more limited precinct might be called the *Homo Sapiens* zone, a region that engenders conditions enabling agriculture to feed billions of humans.

Another intriguing implication of a migrating habitable zone is that a frozen planet on the outer edge of the zone (think Mars) may become habitable as its star ages and warms. If we're in it for the long run, this will have profound implications for the habitability of a future Mars. Conversely, habitable planets closer in may become too hot to support life as their parent stars brighten.

The continuously habitable zones in other star systems are quite unlike our own. The Sun's continuously habitable zone is fairly narrow, but for smaller, long-lived stars like red dwarfs, stable regions within habitable zones last longer, as they migrate over a longer period of time.

Several factors that have recently come to light that may signal a greater spread of continuously habitable zone territory. As we've seen, factors that stabilize the climate appear to be built in to the design of Earth. The most influential of these is the chemical recycling of carbon dioxide and rock. Venus has proven to us the power of carbon dioxide as a greenhouse gas. Without it, the Earth's surface would average 40°C colder than it does. The complex feedback loop between carbon dioxide and the chemicals in rock stabilizes our global temperatures (see Chapter 6). As Earth warms, increased weathering removes carbon dioxide from the air. That reduction in the greenhouse gas leads to a global drop in temperature. As the temperature sinks, the weathering processes subside, allowing the carbon dioxide levels to increase again. The second stabilizing influence involves plate tectonics. Remember that the stony surface of the Earth acts as a sponge, chemically locking various gases into the rock. The Earth's plate tectonics bring the gas-permeated rock down into the mantle where it melts, freeing the chemicals and minerals. Then, these materials are recycled back into the environment through mountain uplift or volcanism (Chapter 2).

Atmospheric expert James Kasting calculates that the Sun's continuously habitable zone spans a distance of .95 AU to 1.15 AU from the Sun. While this is more optimistic than Hart's estimate, it is still a thin portion of the current habitable zone. Earthlike planets with surface conditions that remain habitable over long periods may be rare, which provides us with yet another item on our list of the Earth's unique features among worlds.

Terraforming: engineering on a planetary scale

The act of transforming an alien world into a second Earth presents us with as many moral issues as technical ones. It may be a good thing that our technology is a long way off from what we would need to terraform worlds. But when the time comes, we will grapple with the moral implications as well as the nuts and bolts.

Two camps of thought generally arise when considering the moral implications of terraforming: intervention and preservation. The interventionist perspective holds that we must embrace our human innate nature as pioneers. Some argue that the spirit of exploration is built in at the genetic level, a feature that has helped the human race to not only survive but spread and flourish. Terraforming, they say, is a natural expression of the human character of exploration and curiosity.

This concept dovetails with the second interventionist talking point, that we should increase our species' chance of long-term survival. Terraforming Mars or Venus, for example, would provide another world on which to live, but it would also offer increased resources to improve life on Earth.

A third rationale for terraforming Mars, in specific, is for rehabilitation. The idea put forward is that it is our responsibility to restore a Martian biome similar to what the planet originally possessed, enabling indigenous life on Mars to flourish. This perspective would only be valid should native life be found on Mars.

An alternate paradigm regarding terraforming is that of preservation. This perspective falls into three general assumptions. First, the supposition is offered that we should preserve Mars' value as a unique world, and a subject of scientific interest in its own right. Terraforming will transform the surface of Mars into a new and very different world.

Which brings us to our second point: preserving the integrity of the Martian wilderness. The act of turning Mars into an Earthlike world will undoubtedly devastate its more fragile territory. It would fundamentally change formations at the poles as ice gives way to water. Lakes and rivers would quickly erode the delicate Martian landscape as permafrost vaporizes and ground collapses. Increased water vapor will scour the skies, clearing dust and fundamentally changing the view. No more blue sunsets. No more rusty daylight skies. New floodplains wiping out majestic craters.

The philosopher Holmes Ralston III[3] sums up the philosophical situation in this way:

"Respect exotic extremes." (i.e. the delicate formations on the margins of the polar caps).

"Respect places of historical value." (i.e. the Viking landing sites, Mars rovers, and such famous landmarks from observational history as Syrtis Major, the first feature recorded on another planetary surface).

"Respect places of aesthetic value" (i.e. dramatic canyons like Kasei Valles and Coprates Chasma, and the volcanoes of Tharsis and Elysium, valley networks of the southern highlands).

The third point in the preservation perspective declares that Mars planners should avoid expressing "colonialist views". Many Europeans considered the new world as an extension of Europe, a resource to be utilized. Their explorations often conflicted with the indigenous peoples already living across the western continents. Past ideals of "empire" should serve as a cautionary tale when planning future management of new worlds, they contend.

The generations to come will define our philosophical approach to the worlds around us, but will our civilization mature at the same rate as our technology? Will we husband the worlds around us responsibly, or view them merely as resources to be consumed? Is there a happy medium?

Beyond the moral considerations lie the technical ones, and location will matter. The closest planet to Earth is also the hottest one around. The list of differences between Venus and Earth is daunting, and it forms a to-do list of what would be needed to terraform Venus (Figure 9.4). The place is a blast-furnace. How will we cool it? The air is more oppressive than Atlanta in the summer. How will we thin it? Toxins lace the clouds. Can we get rid of them? And the water on Venus is gone, long boiled away. How will we make our second Earth moist enough to live upon?

---

[3] Distinguished professor of philosophy at Colorado State University.

Figure 9.4 Left: View of Venus today without its clouds (Magellan map; NASA/JPL-Caltech); right: same view of Venus after terraforming. (art by the author)

Our first problem is all that atmosphere, because if we can get rid of some of it, that action will help with the heat as well. One early concept was to focus incredibly powerful lasers on the planet from orbit. The lasers would blast away the atmosphere. This prospect would take patience, as the process would take many millennia to have much effect. And the real problem is the "incredibly powerful" part. We do not yet have orbital power plants powerful enough to do the job. Work on space reactors and advanced lasers does show promise.

Another option would be to sequester the Venusian carbon dioxide. Magnetically enhanced mass drivers[4] installed on Mercury could shoot freshly mined magnesium or calcium to Venus, where these elements would combine with the carbon to create non-harmful carbonates. But the scale of this blissful operation is mind-boggling: it would take several hundred billion tons of Mercury minerals to do the job.

Thinning the air will help lower the temperature, although not enough. Terraforming experts have hatched other plans to cool things off. What about sunshades? they ask. A natural sun block could be created by fracturing asteroids into a ring around the planet (Figure 9.5). The shadows of the ring would drop atmospheric temperatures, perhaps fairly quickly, and

---

[4] Mass drivers are theoretical structures built like long tunnels. Payloads can be accelerated, using a series of magnets, to escape velocity.

Figure 9.5 Rather than the technical approach of orbiting sunshades, some terraform visionaries propose bringing asteroid material into orbit around Venus, creating a ring system similar to that of Saturn's. The ring shadows would cool the planet, just as Saturn's ring shadows cool the hemisphere where they fall. (art by the author)

would require very little advanced technology. A more complex solution would be found in a titanic umbrella, a sunshade deployed between Sun and Venus to cool the atmosphere. The shade would be stationed at the Lagrangian point between Venus and the Sun. At this location, gravity of the Sun essentially cancels out the gravity of Venus, so objects remain in one place relative to the planet. The problem with a giant parasol, according to some models, is that the object would need to be farther across than Venus itself! But extensive constellations of smaller shades, floating independently, could block out enough light and heat to make the difference. Either way, lowering temperature is critical to our terraforming project. In this scenario, for the first few decades the Venusian surface feels very little change. The atmosphere cools, but it remains dense. Soon the carbon dioxide/water clouds clear. Initially, this lets in more sunlight, perhaps raising surface temperatures even further, but within a few decades (some estimate 60 years), the temperature reaches the tipping point of 31 °C, the point at which carbon dioxide will turn to rain.[5]

The Venusian skies thunder with carbon dioxide torrents, and the atmosphere floods the surface in a Noachian deluge. Both pressure and temperature plummet. Alien lakes and oceans of liquid carbon dioxide appear within centuries as temperatures continue to drop. With an atmosphere seven times the density of Earth's, and temperatures reaching -63°C, carbon dioxide snows fall and the lakes freeze into dry ice. The remaining atmosphere reaches about three times the pressure of Earth's at sea level. It consists mostly of leftover nitrogen, similar to our own. It's time to warm things up again, but now we have a problem: if all that carbon dioxide ice melts again we'll have a return to the bad old Venus. The ices could be launched off planet and stored in orbit as a new moon. Frozen carbon dioxide could also be sealed underground, but this plan leaves us with a sort of ticking time bomb.

Our next step is to water the planet. Water ices can be mined at the poles of the Moon and Mercury. We can redirect comets to build our oceans, or send water ice from the moons of the outer planets. The latter ices would be transported from the surface by mass drivers. Once in orbit, they could be sent on their way using space tethers as slingshots, catapulting the frozen water across the solar system to their new home. Water has an added advantage: it combines with the carbon dioxide to lock the greenhouse gas into the surface as carbonate rocks (Earthly examples include limestone and dolomite).

A more advanced approach using technology in our distant future involves transporting one of the ice moons of the outer solar system all the way to Venus, in its entirety! In addition to adding water, the moon might be targeted for a blow that would increase the Venusian spin, shortening its day to something more akin to an Earthly one. But as we saw with the creation of our own Moon, this can be a violent affair, and might result in a molten surface for some time.

Warming up Venus again is a tricky prospect, because of the planet's long day (116 Earth days). If we simply remove our rings or mirrors, we'll fry one hemisphere. Instead, we can leave our sunshades in place and control surface temperatures by opening and closing the sunshade flotilla, or with the use of other mirrors. At the right temperature, bioengineers can seed the planet's oceans with cyanobacteria to pump oxygen into the air. The entire process of terraforming—on any planet—will tax the patience of many generations. But long-term projects with future benefits have been undertaken before. Ancient Egypt brought forth an entire industry keyed toward building tombs that

---

[5] At Venusian pressures

would last for "eternity". The stonemasons, sculptors, glaziers, carpenters and architects of the medieval times knew they would not see the full fruition of their vision as they built the great cathedrals of Chartres and Santa Maria del Fiore. Instead, they were working toward a more distant goal, toward a legacy for their children and grandchildren and many generations to come. The soaring gothic arches, towering belfries, and glistening stained glass masterpieces were bigger than themselves, something that would last beyond their lives. As we care for our own world, we husband its resources and beauty for generations to come. If we are ever to terraform, it must be with the same selfless attitude.

The terraforming process at Mars may be somewhat more benign than that for Venus, but it won't be a walk in the park. While Venus holds too-massive quantities of air to work with, Mars has the opposite problem. Its skies are filled with a diaphanous blanket of thin atmosphere, devoid of many resources. Our list for Mars differs starkly from that of Venus. The air is too thin—even in the deep canyons, pressures are equivalent to Earth's at 100,000 feet. How will we build up the pressure? The gases on Mars are toxic to terrestrial animal life. How will we transform it? The surface of the red planet is parched. How will we bring back the seas and rivers that once graced its landscape?

Two major factors are to blame for the thin Martian atmosphere. First, the little planet has low gravity, just 38% the strength of Earth's. This means that Mars has a more difficult time holding onto air. The second culprit for thin air is a lack of a magnetosphere. We've seen how a planet's magnetic field—generated by a molten core—acts as a shield against solar radiation, which can strip an atmosphere away. The combination of low gravity and lack of magnetic "force field" have contributed to Mars' lack of air. Unfortunately, there is nothing we can do about either of these two factors!

Can anything be done? The loss of an atmosphere is a gradual process, and it's likely that if a Martian atmosphere can be built up enough to sustain life, it will be stable for some time. Mars has plentiful raw materials to help us out. Evidence strongly insinuates that vast inventories of carbon dioxide ice in the south polar cap, and within the Martian subsurface globally, await terraformers just beneath the surface. Studies by Mars Odyssey indicate that this carbon dioxide may reach as much as 12,000 cubic km, equivalent to the water in North America's Lake Michigan. If planetary engineers could heat this carbon dioxide so that it sublimated (turned from ice to gas), the atmospheric pressure of Mars could reach more than a quarter what it is at sea level on Earth. This is comparable to the pressure at the crest of Mt. Everest, breathable by some very athletic humans.

Several scenarios show promise for warming of the Martian climate. One approach would be to pepper the polar ices with dark material from meteors. As the polar ices darken, water and carbon dioxide ices would melt, increasing the pressure. As the pressure increases, so would temperature. This cycle would continue as more and more gases sublimate and water vaporizes, increasing and warming the atmosphere.

Drones with compressed greenhouse gases such as sulfur hexafluoride or chlorofluorocarbons could seed the atmosphere during this warming process. As the atmosphere builds up, these gases would dissipate, hopefully leaving a more pristine environment. One study estimated that a steady train of deployments could dramatically warm the environment in just a few decades. Another type of greenhouse gas factory is far less mechanized: microbes. Specific strains of bacteria produce methane and ammonia, both of them powerful greenhouse gases. Bacteria of these types require liquid water, so the planet would need to be warmed by other techniques before the microbial factories were introduced.

As at Venus, orbiting mirrors have also been proposed to play roles in terraforming here. In the case of Mars, the mirrors would reflect sunlight toward the poles to melt the arctic ices. Constructed in synchronous equatorial orbits, long processions of these giant reflectors would beam sunlight onto the polar caps. On Mars, this will be a significant operation, because vast quantities of atmosphere are sequestered as frozen carbon dioxide within the polar caps. Gas and water vapor heated in this manner would pour into the atmosphere, increasing pressure and temperature.

Once a thicker atmosphere is established, planetary engineers envisage genetically engineered plants like cyanobacteria that could begin the process of transforming the primarily carbon dioxide atmosphere into a breathable one. Our rovers and landers have demonstrated, firsthand, that the Martian soil contains a wide variety of minerals, along with some nitrogen. Water is available as ice and vapor, and may exist as liquid in large underground aquifers. The planet's day/night cycle is just 37 minutes longer than an Earth day. Seasons on Mars are similar to terrestrial ones, as the axial tilt of the planet (which causes seasons) is very similar to our own. However, the seasons on Mars are nearly twice as long. These differences in natural cycles will have an unknown effect on plants and wildlife.

One major problem with the Martian environment is that it is highly acidic. The Phoenix lander confirmed suspicions from earlier Viking experiments that Mars dirt is rich in toxic acids. Phoenix and later rovers found perchlorates in the soil. Planting a forest on Mars today would be akin to seeding soil drenched in hydrogen peroxide. But as the atmospheric conditions warm and pressures rise, many of these volatiles will combine with other elements in rock, air and soil to become more stable.

This, at least, is the accepted baseline for the terraforming of the red planet. Bruce Jakosky, associate director of the Laboratory of Atmospheric and Space Physics, teamed up with planetary scientist Christopher Edwards to reexamine the inventory of resources on Mars. Their paper's abstract says, in part, "We want to answer the question of whether it is possible to mobilize gases present on Mars today in non-atmospheric reservoirs by emplacing them into the atmosphere… so that plants or humans could survive at the surface. We ask whether this can be achieved considering realistic estimates of available volatiles…" The researchers looked at data from NASA's Mars Atmosphere and Volatile Evolution Mission (MAVEN) and ESA's Mars Express. They combined this information with measurements of carbon-bearing minerals and carbon dioxide inventories in the polar ices (using data from the Mars Reconnaissance Orbiter and Mars Odyssey). Their conclusion is that Mars no longer holds enough carbon dioxide within its ices and rocks to provide "significant greenhouse warming" or a substantial increase in pressure using today's technology.

NASA/Ames Research Center terraforming authority Chris McKay responds that, "the question of the total CO_2—as well as nitrogen and water—on Mars has been a long standing question in the study of terraforming Mars. We still don't know the answer because we don't know what is under the surface. It is premature to rule out the possibility that there is enough CO_2 in subsurface layers to terraform Mars." McKay points to the recent discovery of possible subsurface carbon dioxide layers in buried polar deposits, and the subglacial lakes on Mars beneath the southern pole (revealed by Mars Express' MARSIS radar data). McKay concludes, "The subsurface could surprise us."

Mars is half again as far from the Sun as Earth is. Although bolstering its atmosphere could, theoretically, bring water to its dried sea basins and warmth to its climes, Mars will never be a tropical paradise. Martian settlers would find themselves at home in places like Tierra del Fuego or Oslo, but not Palermo or Nassau. And another aspect of terraforming is seldom discussed: collateral damage. If you like Martian mountains and canyons, think twice. Much of the Martian landscape is held up by subsurface ice, just as the tundra in Earth's arctic regions is. As ice melts and vaporizes on a global scale, some regions of the Martian surface will be transformed. Still, one day it just may be possible to gaze across a river-filled Martian canyon under the evening light of two moons, sipping a cup of tea and watching raptors spiraling along the updrafts over a newly forested world.

The Earth's Moon is close by, but its gravity could not hold on to a livable climate out in the open. A more distant moon, Titan, is replete in resources but too cold to carry out any serious terraforming today. That will have to wait, but a more Earth-like Titan may be in store in the far future. The future will bring changes throughout the solar system as our star ages and alters in character.

The Venusification of Earth (terraforming in reverse)

Whether by intent or not, we've been carrying out planetary engineering on Earth for centuries. The "terraforming" we're doing to our own planet may not bring positive changes—at least as concerns humans—in the long term. Since the advent of the Industrial Revolution (beginning in the second half of the 1700s), human activity has dumped a remarkable and nearly unprecedented[6] amount of hydrocarbons into the atmosphere. David Grinspoon sums up the situation by imagining beings from a distant star system coming upon the Earth and its sibling terrestrial planets:

> ...they would see the clouds of Earth becoming acidified, the unique stratosphere
> being chemically eroded, and the anomalously low CO_2 level being slowly
> pumped back up toward "normal" for this solar system. The conclusion
> is obvious. The inhabitants of the third planet ...are doing their best to
> transform their world into one that more closely resembles their sunward twin.

Just how far have we traveled down the road toward transforming our Earth into a new Venus? How far might that process go, and how likely is it that our living planet might end up as a new Perdition? What future does our effect on the climate spell for planet Earth?

The Earth's environment will never be a doppelganger of Venusian climes. We can never produce enough carbon dioxide—even if we wanted to—to trigger a run-away greenhouse effect (remember that Venus has 90 times the pressure of Earth's atmosphere at sea level). But we can still carry out actions that profoundly effect the climate on our home world.

The concept of climate change is one laced with political and moral undertones. Terms like "global crisis", "climate collapse" "urgent" and "looming ecological disaster" appear in the modern press with increasing frequency. No

---

[6] rivaled only by massive volcanic eruptions of the past, as we saw associated with the Deccan Flats or the Permian extinction.

matter which end of the spectrum the narrative issues from, it is clear that something is afoot. Global temperatures are on the rise, as are global sea levels. Weather patterns are shifting as extreme storms brew and ever more severe temperatures hit both hot and cold regions. Summers are hotter and winters are cooler, and the pattern fits with computer predictions of what meteorologists expect with global temperature rise. Various groups debate the source of the dramatic changes our planet seems to be undergoing (some even debate its existence). Is this a natural occurrence? After all, they point out, the Earth has moved through many changes in climate, long before there were humans around to blame. At the time of the dinosaurs, the Earth essentially had no ice at the poles! Others point out that the changes we are seeing have appeared more abruptly than those recorded in the fossil record, ice cores, tree ring sequences, coral reefs, etc.

The Earth is in constant change, as is evidenced by the preceding chapters. Change is the norm here, which makes tracking the long-term effects of phenomena like climate change difficult. In fact, only a century ago, many thought the Earth was headed toward another ice age. In the early 1800s, paleontologists and meteorologists began to understand the dynamics of past ice ages. Many projected that increased air pollution from industrial manufacturing had the potential of cooling the atmosphere as it blocked sunlight. In 1923, Sir Arthur Conan Doyle remarked that American newspapers were "full of alleged change of climate, encroachments of ice, and general signs of a glacial epoch."[7] But with a plethora of more precise data in recent years, scientific consensus is that the planet is headed in the opposite direction, as we will see.

While change may be the norm, the issue is one of speed. Proof for fairly abrupt global climate change comes to us from several fronts. Specifically, two substantiated facts lead many environmentalists to the conclusion that the planet is undergoing a change that may become as drastic as the Earth's transformation at the end of the Permian or Cretaceous eras. The first fact, something we've learned for the worlds around us, is that carbon dioxide is a powerful greenhouse gas. The second fact is that human activities, especially those involving the use of fossil fuels (oil, coal, natural gas) is contributing more carbon dioxide to Earth's atmosphere. The unavoidable conclusion, as astrophysicist Jeffrey Bennett puts it, is that "we should expect the rising carbon dioxide concentration to warm our planet, with the warming becoming more severe as we add more carbon dioxide." And while there is no doubt

---

[7] *Memories and Adventures* by Sir Arthur Conan Doyle (Little, Brown and Company, 1924)

within the scientific community that global warming is currently occurring, there is legitimate debate as to how extensive and serious an issue it is.

One of the most high-profile scientists skeptical of the threat of global warming is Richard Lindzen, a professor at MIT. In a speech to the British House of Commons in 2012, Lindzen said, "…the debate over climate change is…most certainly not about whether climate is changing: it always is. It is not about whether CO_2 is increasing: it clearly is. It is not about whether the increase in CO_2, by itself, will lead to some warming: it should. The debate is simply…how much the increase in CO_2 can lead to, and the connection of such warming to the innumerable claimed catastrophes."

It's an important debate, and one that deserves a look on both sides. Let's start with several major reasons for skepticism, and why those arguments may or may not hold up. First, some skeptics (a clear minority) submit that the world is not warming up. This argument assumes that the data only shows a "blip", an incongruity in the global temperatures. Temperatures do fluctuate, so we must look at the bigger picture, at overall temperature averages. Even so, judging global averages is a difficult prospect. The total number of weather stations changes over time, as does their location. The heat coming from cities (affected by vegetation, parking lots, building materials, etc.) evolves, and weather satellites have differing systems for taking data. Researchers must average out data coming from disparate systems, and this variety serves as a check. The overall data are in close agreement. The combined information does, in fact, show a clear trend of rising temperature from the 1970s to the present. Considering the trends, each five-year period since 1980 has set new records for high temperatures.

A popular argument against a heating climate is that global warming has paused or stopped since the latter half of the 1990s. Global temperature records prove this to be false. While variations in temperature are expected by nature (and the year-to-year numbers confirm this variability), averages over multiple years do show a slowing of the rise in temperatures since the late 1990s, but the trend continues upward. Since 1978, temperatures have been consistently above the average, and rising every year. In fact, 2014 and 2015 set back-to-back records for annual highs.

Another argument put forth by some is that while the Earth may be heating up, its warming is a natural one. Under this scenario, the obvious suspect might be the Sun. As we have seen, stars vary in intensity as a natural course of their development. The Sun's energy output varies, but the variation is subtle, at less than one percent from one year to the next. We even know that fluctuations in the Sun's output may well have contributed to the onset or extent of some ice ages. However, we have accurate data for the amount of

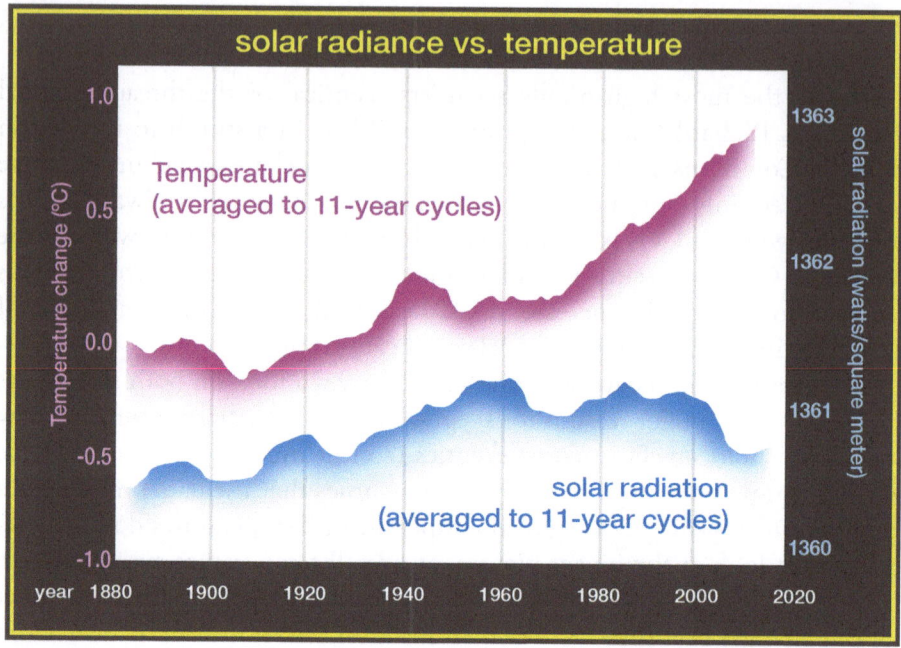

Figure 9.6 A comparison of global temperature to solar irradiance (amount of sunlight falling upon Earth's surface). Global temperatures are averaged over 11-year cycles, the natural cycle of the Sun. (art by the author, adapted from BigKidScience.com)

sunlight reaching the Earth's surface, and those records go back to the 1800s. If we compare change in Earth's temperature since 1880 to the change in sunlight levels over the same period, we see an interesting pattern (Figure 9.6). At first, the curves track each other fairly well. But by 1950, the two trends split, moving in opposite directions. Solar energy has been dropping while, at the same time, temperatures have been rising. And scientists have another set of measurements to reinforce the idea that greenhouse gases, not solar energy, are behind the temperature increase: the upper atmosphere has been cooling off, while the lower atmosphere is warming. This is consistent with greenhouse gas warming, not solar heating (see Appendix L). According to these lines of evidence, the Sun cannot be the cause of the planet's current warming.

Skeptics remind us that the Earth's climate is affected by many influences, both human and environmental. Some assert that the world is warming, and that humans are causing that change, but the change is not a cause for concern. The arguments generally fall into three categories: (1) the climate has gone through natural variations in the past, and our planet's ecosystems are still intact. What we're experiencing is part of a natural cycle; (2) future temperatures will rise less than the models predict; and (3) warming may even be beneficial to the planet. Let's examine each proposition in turn.

(1) Some suggest that the climate shift we see today is part of a natural course of events. The poles of the Earth were largely devoid of ice at the time of the dinosaurs (the Mesozoic era). The Earth has cycled through ice ages as well, and isotopes within the layers of ice cores have enabled us to chart the Earth's average temperature over the past 800,000 in some detail. And as an independent check, the increased temperatures correspond to higher levels of carbon dioxide. Is there, then, cause for concern, or are we simply in a naturally occurring planetary cycle? The Earth's current temperatures are already reaching the highest that they have been in over the course of the past 800,000 years. Today's carbon dioxide levels are roughly 40% higher than at any other moment in that 800,000-year record, and they continue to rise. The temperature changes chronicled by the ice cores all appear to have taken place gradually, over centuries. But the climate alterations we mark today are transpiring in timescales of decades.

(2) It may be that our climate models overestimate how the Earth responds to shifts in the carbon dioxide levels, and also overestimate future warming trends, say other skeptics. In his book *A Brief History of Earth*,[8] Harvard geologist Andrew Knoll addresses this issue head-on: "In the past, scientific predictions about…climate change have sometimes been off the mark, but mostly, it turns out, because they have underestimated the pace of change. Scientists are inherently conservative…" Additionally, as science accrues more and more knowledge and data, our models become more accurate.

(3) Some assert that there may be benefits to climate change.[9] One such benefit, they suggest, would be that increased carbon dioxide will encourage plant and crop growth. They point, as evidence, to the Mesozoic Period, when both temperatures and carbon dioxide levels were higher than today, and jungles blanketed much of the world. But creatures and the plants they dine upon adapt to their ecosystems over millions of years. The vegetation of today is adapted to the modern environment. If that environment changes drastically, extinctions will occur. Some in this camp also call attention to limited laboratory experiments that have been carried out in greenhouses and other confined areas. These studies show that some crops (soybeans, rice) benefit from higher doses of carbon dioxide. But they have missed the bigger point of detrimental effects on the

---

[8] *A Brief History of Earth: four billion years in eight chapters* by Andrew H. Knoll (HarperCollins, 2021)
[9] The editors have asked that I not use exclamation points here.

ecology at large. If temperature and carbon dioxide levels change gradually, species can migrate or adapt in other ways. But if climate changes take place at a rapid rate (as they are threatening to do), so will changes in the ecosystem, to the detriment of its inhabitants (human, vegetable and animal).

The Problem With Clouds

We've seen the complex effect that clouds can have on a planet. On Titan, they billow with organics as they shelter the surface from sunlight. On Venus, they reflect sunlight away while, at the same time, holding heat in. On Mars they mark the condensation of carbon dioxide or water, but are too weak to hold in any appreciable warmth. Clouds would seem to introduce baffling variables to our understanding of climate change on Earth. If the globe is warming overall, this would increase evaporation from the seas, forming even more cloud cover. But those increased clouds should reflect more sunlight away, cooling things back down again. Don't clouds set up a feedback loop that actually *prevents* further global warming?

This convincing-sounding argument against global warming misses two important factors. First, the evaporation that produces those extra clouds means that there is more water vapor in the air, and water vapor is an efficient greenhouse gas itself. Which wins out, the cooling of the air by clouds or its warming by the water vapor? Most studies and models cast their vote for the effects of heating winning out over those of cooling.

Secondly, the clouds aren't the only things reflecting sunlight away. As we saw with our snowball Earth, increased ice on the surface reflects more sunlight, continuing a cooling trend. Our planet narrowly escaped a runaway permanent deep freeze because of its brightening surface. But the opposite is also true: although fluffy white clouds may be on the rise, the surface beneath grows darker with warming, because surface ice melts and shrinks, and ice is far more reflective than clouds are. Climate researchers are in general agreement that the warming effects of escalating water vapor, combined with the reduced ice coverage on the surface, will outweigh any cooling effects that additional clouds might have.

Alternatively…

There's an ancient[10] Japanese proverb that says, "The reverse side of the coin also has a reverse side." So what about that rise in carbon dioxide? Rather than coming from human activities, could it be purely natural? After all, carbon dioxide seems to be just about everywhere we've visited.

---

[10] Japanese coins *are* ancient, dating back to the late seventh century.

One way to answer the question is to look at the chemical signature in the air. For more than half a century, researchers have been charting not only the amount of carbon dioxide in our atmosphere, but also the isotopes associated with it. Remember that isotopes affect the weight of an atom but not its nature (the nuclei of an isotope holds the same number of protons but extra neutrons). Using two stable isotopes—Carbon-12 and Carbon-13, we can compare the differences to pin down just where the carbon dioxide comes from. Scientists find that gases issuing from volcanoes or dissolved in the world's oceans have the wrong type of isotopes to test the increase in levels. But organic matter created during photosynthesis (including natural gas, oil and coal) has the right telltale signs. Using only the ^{12}C and ^{13}C comparison, the carbon dioxide added to the air might come from fossil fuels, but it also might come from forest clearing or other sources. But the researchers have another yardstick, a third isotope called Carbon-14 (^{14}C). ^{14}C is radioactive, predictably transforming into nitrogen at a steady rate over thousands of years. It shows up in living creatures (who ingest it from the environment) but not in fossils (or fossil fuels) formed millions of years ago. The ^{14}C levels clearly demonstrate that the primary source of the *increase* in atmospheric carbon dioxide must come from the burning of coal, natural gas and oil products.

Adding greenhouse gases to the mix of Earth's atmosphere should also warm the surface, and this is another variable that we can track. With the advent of Earth-observing satellites, it's much easier today than it was. For past measurements, we must base our estimates on old oceanographic and meteorological logs. This introduces some uncertainty, but the consensus of the scientific community is that over the past critical century, the mean temperature of the Earth's surface has increased by just under 1°C. Throughout this time, the poles have warmed more quickly than the equatorial regions. Current trends imply that another degree increase is reasonable to expect over the next three decades unless we can curb emissions substantially. An increase of one degree might seem slight, but it would bring Washington D.C.'s currently climate to Toronto, with less snow and longer summers. The northernmost regions of the continental U.S. may benefit economically from this warming trend. But on the southern border, states will lose an estimated 15% of their annual income on various fronts, including loss of land, damage to agriculture and reduction of tourism.

Holding disaster at bay

Rising temperatures will continue to bring tidal extremes as sea levels rise. Already, we have seen some ocean communities swamped with changing tides. In the UK's Norfolk, the towns of Happisburgh and Hemsby are fighting a losing battle with the sea, which is pulling buildings to the beach as cliffs decay. The medieval Corsican city of Bonifacio has held court over the Mediterranean Sea for centuries, but is now threatened as its sandstone cliffs erode away. Coastal towns in Alaska are collapsing as supporting permafrost melts and heightened tides eat away at their foundations. Jakarta, Indonesia has been called the "fastest-sinking city in the world". Half of the city is built below sea level, and some experts caution that major portions of Indonesia's capital could be submerged by 2050. Italy's Venice, Ukraine's Odessa, and New South Wale's Womberal add their names to the growing list of threatened cities. At a 2022 conference in Fiji's capital city of Suva, representatives of 15 Pacific Island nations declared climate change as their "greatest existential threat." Since 1900, global sea levels have risen by over six inches, and may rise another two and a half feet by the end of this century.

In a rebellion against rising tides, the Dutch have held back the sea for centuries. One quarter of the Netherlands lies beneath sea level, but since the Middle Ages, Dutch engineers have held the North Sea at bay with a series of dykes, levees and pumps powered by wind mills. The web of canals, waterways and dams has reclaimed 1650 square kilometers from the sea in the twentieth century alone. The Dutch have transformed these lowlands into farmland and settlements. Designers have modernized the system with the addition of 63 storm gates, each standing 7 meters tall.

Italian engineers are also using modern technology to hold back the mounting seas (Figure 9.7). 78 floodgates wait on the sea floor just outside of Venice, where the frequency of high-water events has increased from two per decade (at the beginning of the twentieth century) to a current forty per decade. When floods are in the forecast, the gates rise to form a temporary sea wall. The metal buttresses prevent rising waters of the Adriatic Sea from pouring into the shallow lagoon surrounding the famed city.

As temperature changes, so will the global pattern of precipitation. The planet's various biomes are all tied together. One affects another. In fact, the deforestation of the rainforest of the Amazon may influence the rainfall and temperatures some 15,000 km away, on the Tibetan plateau. The Amazon is known as one of the world's most efficient carbon sinks, and is seen by climatologists as a climate tipping point, in that the Amazon could transform from tropical rain forest to savannah beyond a certain threshold of global warming. A team at Beijing Normal University in China analyzed global climate data from 1979 to 2019. They found that high temperatures in the Amazon paralleled raised temperatures over the Tibetan plateau. They also found a similar correlation between the Amazon precipitation levels and a decrease in snowfall in Tibet and Antarctica. Some researchers warn that more study is needed

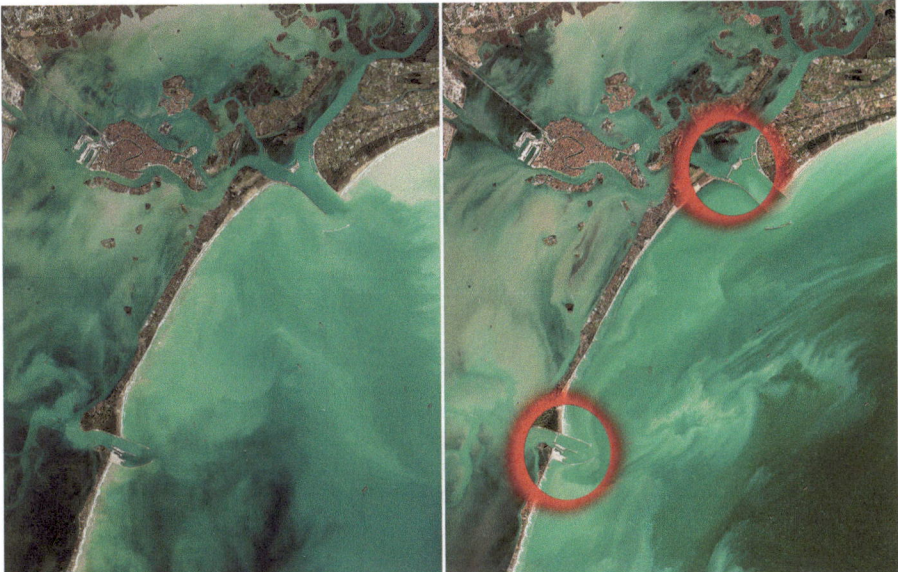

Figure 9.7 Gates rise from the sea floor to hold back the rising waters of the Adriatic. Venice is surrounded by the shallow lagoon to the left. Sea gates are circled in red. (NASA public domain image; NASA Earth Observatory/Landsat and ESA Sentinel image processing by Joshua Stevens)

to confirm the results of the new work. But if the Beijing team's conclusions are corroborated, the Amazon is yet another domino that might fall in the globally-linked climate picture (Figure 9.8).

Global warming: a sporting link

A study by a team at New Hampshire's Dartmouth College[11] contends that global warming is responsible for more than 500 home runs during America's baseball seasons from 1962-2019. This unexpected climate link pans out in data collected from more than 100,000 Major League Baseball games. The team of climate scientists applied a complex climate model to temperatures from decades of game days, controlling for greenhouse gas emissions, wind speed and humidity. Other factors, including stitching on the baseballs and steroid usage of the players, were accounted for in the calculations as well. The model's results show that temperature increases added an extra 58 home runs *per season* during the decade of the 2010s alone. As air temperatures rise, air density drops. An overall increase of just 1°C increases home runs by roughly two percent.

---

[11] Global Warming, Home Runs, and the Future of America's Pastime by Christopher W. Callahan, et. al., *Bulletin of the American Meteorological Society*, April 7, 2023

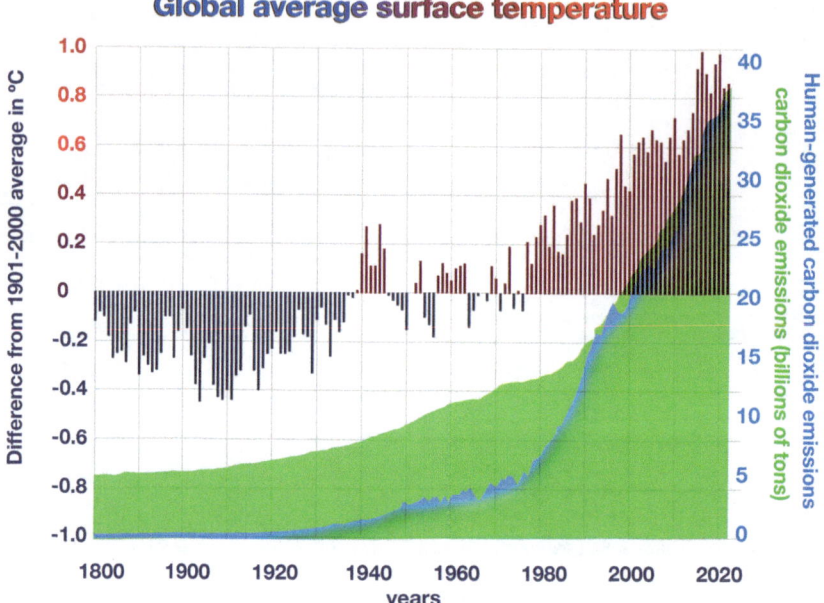

Global average surface temperature

based on data from the National Centers for Environmental Information

Figure 9.8 This National Oceanic and Atmospheric Administration chart shows yearly surface temperatures from 1880 to 2022, all measured in the month of May. Blue lines denote colder than average years, while red ones are above the average norm. Overlaid on this graph is an indicator of carbon dioxide levels, in green. Estimated emissions from human activity are in blue. Note how all lines track. (NOAA background based on data from the National Centers for Environmental Information, art by the author)

In the western United States and—to a lesser extent—Canada, water has always been a point of contention, a precious resource to be husbanded, shared, sometimes squandered and often fought over. The forecast for the desert southwest does not bode well. Climatologists project a decrease in precipitation for the southwestern US, Africa (especially in the southwest), the Iberian Peninsula, areas of the Middle East, and other sites around the globe. The United Nations' East African Subregional Coordinator, Chimimba David Phiri, announced that, "The current food security situation across the Horn of Africa is dire after four consecutive rainy seasons have failed, a climatic event not seen in at least forty years…"[12] Two billion people on the planet depend on the springtime melt of glaciers and snow in mountains, but the

---

[12] *UN News*, July 22, 2022

glaciers are shrinking and the annual snowfall is melting earlier in the year, causing water shortages and droughts in late summer.

Another prediction of higher temperatures (and the overall climate change they bring) is the occurrence of extreme weather events. Recent wildfires, formerly limited to wilderness areas, have recently devastated inhabited areas of California, Colorado, and New Mexico in the US, in wilderness areas of Australia, in the Canadian North Woods, and in other regions, many of which do not have a history of drought. Hurricanes appear to be increasing in severity and intensity. During the period of 1978-2017, Category 4 and 5 hurricanes (the strongest) have increased, while the number of smaller storms has decreased. Warmer ocean water increases wind speed and the amount of water within the storm. Meteorologists predict an increase in hurricane-related rainfall by 10-15% over the next decade. Storm patterns associated with tornadoes are also shifting. The famous "Tornado Alley" is shifting from the plains states to the south, and regions hosting frequent twisters are now referred to as "Dixie Alley". Tornadoes are also becoming more frequent there, where past tornado-spawning storms rarely formed in winter.

Just a few decades ago, skeptics claimed that global warming was not taking place at all (we'll examine some of the claims below). As data—and our understanding of climate on Earth and other worlds—grows the human role in global climate change can no longer be contested. Even so, some still hold to the perspective that current trends in our climate may not be cause for concern. The majority of meteorologists, climatologists and planetary scientists categorically reject this position. More than 97% of researchers[13] join these ranks as we continue to carry out studies of our Earth's environment (Figure 9.9).

Running on Empty: The Starving Of The Colorado
  The Colorado River has been called the "Nile of North America". It wanders, meanders cascades and gushes through 2300 kilometers of the western United States and northern Mexico. Indigenous peoples have inhabited its shores for 8000 years. 40 million people today rely on its life-giving waters, including such downstream centers as Las Vegas, Tucson, Saint George, Flagstaff, Phoenix, and Mexicali in Baja, California.
  In 1922, the Colorado River Compact was ratified. It governed the use of the river's waters. Colorado, New Mexico, Wyoming and Utah made up the Upper Basin, and agreed to use amounts of water that would leave some flowing

(continued)

---

[13] For details, see www.skepticalscience.com/global-warming-scientific-consensus-advanced.htm

down to the Lower Basin states of Arizona, Nevada and California. But there was a problem with the calculations. The water usage guidelines were based on accounts from the early 20th century, a time period that saw far more rain than usual. As early as 1928, one of the River Board commissioners warned that more water was being split up than was actually available.

The trend has continued to this day. Populations in the southwest of the US have exploded. Agriculture has, too, consuming roughly 75% of the river's precious resources. Compounding the problem, a nearly decade-long drought is visiting the western states whose water drains and feeds into the Colorado River. In the upper basin region, half the reduction in flow is being ascribed to climate change. The winter of 2022/2023 yielded a bumper crop of precipitation. The snowpack began to build back, and reservoirs to fill. But experts warn that to make up for the drought conditions, seven years of equal or more precipitation will be needed to recharge the Colorado basin and river course. The issue is partly one of rainfall, but the predicament comes down to a more complex series of variables. Warmer spring temperatures melt the snowpack in nearby mountains more rapidly, keeping it from making its way to the river or to aquifers along the way. The drought conditions have dried the soil, so more water is soaked up before it gets to the Colorado. Climate models point to the river's moisture reduced by 30% by 2050 if the warming trend continues. And while the legal papers say that inhabitants of the desert southwest can use 20.3 billion cubic meters, the river currently has only 15.1 billion to give.

With all the dams and recreational reservoirs dotting the Colorado, the landscape along the Colorado has changed, and wildlife has had to change with it. But the added effects of drought have brought the delicate river biome into crisis. To keep the river's ecosystem alive, hydrologists have instigated a series of "pulses", sending water through the system in an effort—somewhat successful—to preserve the natural balance. A 2014 release from the Morales dam at the Arizona-Mexico border sent 130 million cubic meters of water toward the Gulf of California. Other pulses have been freed, and the fish, cottonwood trees, and other life is coming back to some of the watercourse.

One other challenge that the Colorado faces is simple evaporation. Lake Powell loses 470 million cubic meters of water each year. Engineers have tried covering the lake's surface with plastic balls to cut down on the water-heating sunlight, but with little effect. The solution may be to refill the aquifers, where underground water is protected from evaporation, but its inventory still can recharge the river. The concept is to redirect waters during floods to replenish the underground ponds. The other major solution is to use less water. To that end, Nevada voters passed a law banning "non-functional turf", grass in areas not directly used by people, like lawns along median strips or in low-visibility business parks. California is studying similar laws, even as its residents replace grass yards with xeriscaped coverage or artificial turf. The US Department of the Interior is even paying farmers, ranchers, urban centers and Native American tribes to voluntarily reduce their water usage. Farmers may also be subsidized to raise higher-value, less thirsty crops. It's a new world in the drying west, and new solutions must be found.

Figure 9.9 The Colorado River runs through the desert southwestern region of the United States, supplying water for wilderness and major metropolitan areas. (photo by the author)

Escape Claws

Humans have tools at their disposal to provide options for dealing with climate change. Until global climate change is understood and—perhaps—brought under control, seaside villages can be relocated (although this is incredibly disruptive).[14] People can adopt lifestyle changes, enabling them to make wiser use of the planet's resources. Alternative energy can be brought into the mix of the world's populations. Sea walls and levies can be built to manage coastlines, and more advanced technologies can be applied. But what will happen to the vegetation and wildlife of the planet? Through its long and colorful history, the Earth has watched as microorganisms, plants, trees, and creatures with wings, fins and claws have responded to changes in their environment. In some cases, the creatures themselves have been at the heart of environmental change (like the cyanobacteria of the Great Oxygenation Event). But habitats have been shrinking at alarming rates as water drains from delicate wetlands, farmers and ranchers clear forests for agriculture, and

---

[14] One recent example of such relocation involves the residents of Panama's Gardi Sugdub island. 300 families of the indigenous Guna people are in the process of abandoning their island home to emigrate to the mainland, where their lifestyle will be quite different.

pollution damages healthy ecosystems. Wild populations that face shifting environs, whether plant or animal, have one of three responses. They can adapt. They can migrate to healthier environments. Their third option is the worst choice of all: extinction.

The Sixth Great Extinction?

Our planet appears to be in the midst of another mass extinction. By studying the fossil record, researchers have established a baseline of extinction rates. The number comes out to roughly one species per one million total species annually. Biologists estimate that roughly eight million species now inhabit the Earth. Of those, at least 15,000 are threatened or endangered, a number clearly above the baseline.[15] Extinctions rates are difficult to track, because many endangered species have either not yet been identified, or researchers have not yet had the opportunity to study them (some have not even been discovered yet). And while biologists and wildlife experts improve their methods of assessing the pace of extinction, researchers generally agree that extinction rates today are hundreds—or even thousands—of times higher than the natural baseline rate of extinctions.

We have good documentation for five major extinction events in Earth's history, and a few others along the way (see Figure 05.04). Recall that the great dying of the Late Ordovician befell life on Earth at 447 million years ago. This event was followed 69 million years later by the Late Devonian extinction, but the two events paled by comparison to the End Permian 252 million years ago, the largest extinction event known. The Late Triassic (199 million years ago) and the Cretaceous/Paleogene (66 million years ago) also saw vast dyings across the board. The late Permian is perhaps the greatest cautionary tale, as its cause appears to have been global warming, induced in part by massive volcanic eruptions (see Appendix F). All of the major extinctions have one thing in common: a change in global sea level. But other factors came to play in each case. The question on the minds of biologists and ecologists is this: Are we in the midst of a sixth great extinction? If so, our current crisis may have begun some 10,000 years ago, at the close of the last ice age. During the warmest parts of the Quaternary, camels wandered across North America while hippos bathed in swampy eastern England, but only centuries later, temperate rain forests of pine expanded down to the Mediterranean coasts. Rapid changes made adaptation difficult for some, and many animal

---

[15] A value 1875 times that of the estimated baseline.

lineages came to an end, including mastodons, mammoths, and the turtle-like mammals called Glyptodonts. But many of the megafauna, including the great rhinoceros-like Brontotherium and the Irish Elk, survived many of these climate shifts. Each near-extinction event seems to have been met by a repopulation, as survivors came back from areas where the changes in conditions were more muted. Some individual Irish Elk survived on the Isle of Man up until about 7700 years past, and on Wrangel Island in the Arctic Ocean, a small herd of mammoths managed to hold on until just 3700 years ago. But eventually, mortality caught up with the largest of the legendary quaternary fauna.[16]

Hunting for the cause of extinctions

While climate change was undoubtedly a major factor in the die-offs, another variable came to bear: the first large-scale human interactions. At the time, humans began to learn cooperative hunting, trapping large mammals in pits or driving them over cliffs. Paleontological evidence hints at large-scale hunting and butchering in early settlements and encampments. Colorado's Olsen-Chubbuck kill site is the poster child for ice age hunting. There, hunters forced entire herds of bison (over 200 have been found so far) into a natural amphitheater to be slaughtered.

The diaspora of humans from northern Africa took place over many millennia. People groups colonized Australia some 50,000 years ago, and spread the opposite direction, into Europe, 48,000 years ago. Humans came to the Americas some 15,000 years in the past, and populated New Zealand just 800 years ago. According to recent studies,[17] the introduction of humans into various areas seems to line up with massive extinctions of large animal populations. During the time period between roughly 12,000 and 10,000 years ago, at a time of climate warming and human arrival, North America saw 35 out of 45 genera disappear, while in South America 45 out of 58 genera became extinct. In Europe and Asia, the numbers are equally convincing: 21 genera vanished by 11,000 years ago, suspiciously close in time to the arrival of nomadic peoples to the areas.

A study by a Tel Aviv University team finds that over a 1.5 million year period, the size of animals hunted by humans gradually declined, from the

---

[16] Some ice age creatures are still with us today, including the Asian/Alaskan Musk Ox, North American pronghorn, Caribou, Grizzly Bear, the Asian Saiga Antelope, and the Tapir (South American and Asian).

[17] See "Fifty millennia of catastrophic extinctions after human contact" by David Burney and Timothy Flannery, *Trends in Ecology and Evolution*, July 2005

large ice-age fauna like Mammoths at the beginning of the study period down to gazelles some 10,000 years ago. The authors assert that humans hunted the largest game available, because they provided the most food for the effort. Their conclusion: humans repeatedly overhunted—to practical extinction— the largest animals. When the only targets left were smaller than deer, humans began to domesticate animals and grow crops. Hence, they say, human hunt- ing led to the agricultural revolution. The 400,000-year-old German site of Schoninger has unequivocal proof of early hunting in the form of wooden spears and animal remains, but older sites have now revealed evidence of active hunting and butchering. Anthropologists have uncovered ancient sites marking mass processing of butchered game, including the Olduvai Gorge in Tanzania and several sites in western Kenya. These anthropological treasures may mark the first incidence of humans causing the decimation of wild populations.

Sadly, specific historic examples abound. In New Zealand, where the birds have few natural predators, many species build nests on the ground. Their populations have taken a precipitous nose-dive since the arrival of people,[18] who brought with them their house-cats.[19] These pets are devastating many of New Zealand's native birds, as are the rats that hitched rides aboard ships. Humans have played their own hand in bird exterminations there, as well. Humans arrived in the islands in 1300, and by 1445, all of the giant flightless moa birds (whose feathers were prized by the Māori) had vanished. Genetic studies from Moa remains show that the bird population was stable for the 4000 years leading up to the introduction of humans. The creatures were driven to extinction in less than two centuries.

Of course, the moas were not alone. Today, thousands of species of plants and animals across the globe are threatened. The question experts are asking is, did these early human activities mark the beginning of the sixth mass extinction, one we are participating in currently?

Human populations shifted gradually to agrarian life as people learned farming techniques. In an echo of events today, these served to bring radical change to various habitats. Farmers drained wetlands and cleared forests to grow crops (the Amazon Rainforest is one drastic modern example). Hunters burned brush and forest to harvest animals for food and supplies like bone

---

[18] Although the very first humans to arrive, the Māori, didn't have cats.

[19] This is not a segue into the great cat vs. dog pet debate, which is far too complex and nuanced for this book.

tools and pelts for their stylish Cro-Magnon apparel. And humans got really efficient at it. Effects increased with the advent of the industrial revolution, notorious for its use of coal, steam and water power (water wheels and eventually hydroelectricity), textile and automobile manufacturing, and the added side effects of toxic air, water and land pollution.

The Anthropocene

So great is the human impact on the world that anthropologists and geologists have informally coined a term for our current geologic epoch: the Anthropocene.[20] The International Geological Congress is still debating whether to formally declare the current geological epoch as "The Anthropocene", but the term is gaining popularity among researchers. Some propose that the beginning of the age should be marked at the point when agriculture arose (~15,000 years ago). Others assert that the worldwide pulse of radiation, concomitant with the beginning of the first atomic blasts and use of nuclear power, is a better indicator. Either way, some anthropologists and astrobiologists have submitted that the advent of any intelligent species may automatically spell a mass extinction event on their planet.

How Venus saved the Earth…for now

In the decade of the seventies, manufacturers of various insecticides, spray paints, household cleaners and refrigeration systems came up with a great compound for powering all of their products. It's called chlorofluorocarbons, or CFCs. CFCs seemed like a fine solution to several challenges in manufacturing. They are relatively cheap, they are efficient in pressurizing canned products, and they are safe because they are inert (they don't interact with other chemicals to make nasty pollutants). At about the same time that industry was beginning to introduce CFCs into the market, two other significant developments came onto the stage. First, our weather satellites, which were just getting sophisticated enough to chart chemicals in the air, noticed a thinning in the Earth's protective ozone layer, particularly over the poles.

The second development had to do with Venus. Chemists studying the atmosphere of Venus realized that certain chemical reactions involving chlorine and fluorine, two constituents of CFCs, could break apart ozone when exposed to ultraviolet radiation. In other words, under the right circumstances, CFCs could wipe out an ozone layer. It made fascinating reading to armchair chemistry nerds and inspired quiet corner chats at obscure science conferences, but the simmering back-room discussions erupted into public awareness with the

(continued)

---

[20] Others have suggested the less-technical "Shopping Mall Culture".

Pioneer Venus orbiter mission. Pioneer Venus discovered holes in the Venusian ozone layer, and those holes were coming from naturally occurring chlorine and fluorine in Venus' air. Could this have anything to do with that troubling finding of a thinning ozone layer on Earth?

It turns out that CFCs are quite safe and stable when we use them in our refrigerators and insecticide cans. The problem comes later, when they escape and drift off into the higher atmosphere, where sunlight transforms them into something much more sinister. CFCs break down into chlorine and fluorine, the very villains of the Venus ozone story. Our studies of Venus led us to the awareness that industry was pumping something into our atmosphere that could spell disaster on a planetary scale. Once the ozone layer goes, so does our natural barrier against solar ultraviolet radiation. Things with skin will get deadly cancer, and things with leaves will stop producing stuff for those cancer-ridden life forms to eat. Discoveries like this are just one beneficial aspect of planetary studies. When we compare processes on Earth with similar processes on other worlds, we learn deep secrets about our own planet, secrets that just might save us from environmental cataclysm.

What we can do

As a review of the climate-change-science-fiction TV series *Extrapolations* put it, "Can this climate sci-fi make us care?" Perhaps the most remarkable aspect of the modern climate discussion is the apathy with which many meet it. But if our survey of other worlds has shown us anything, it is the uniqueness of our own world. It's a place to be cherished and preserved.

Scientists are coming to a deeper understanding of the workings of long-term climate on this planet, and with that understanding comes specific strategies to keep our ecosystems fit.

One such strategy concerns how we handle material resources. Plastic, a petroleum-based material, takes centuries or millennia to break down, and even microscopic bits of plastic can cause damage to soil, ocean organisms, and the general food chain.

Environmentalists proclaim recycling as a major solution, but we are just getting started. In the United States, less than 20% of material put into recycle bins actually gets recycled back into the industrial stream to become other useful forms. In 2021, 5% of US plastic waste was recycled. Europe has a better track record, with 32.5 % of plastics recycled. In 2020, the overall recycling rate for the EU was over 46%. The rest, typically, ends up in landfills. 8 million tons end up in the seas and oceans each year. An estimated six billion tons of untreated plastics infuse the planet's environment, from its forests and cities to its beaches and the deepest oceans. Microplastic fragments have been found in rainwater and breast milk. A scant 6% of all the plastic that has ever been manufactured has been recycled. Increased social awareness can help,[21] and the streamlining of our industries to incorporate more recycled material will also be of benefit. Many corners of the economy are taking the hint: MARS, PepsiCo, Coca-Cola, and Unilever are just a sampling of companies that have pledged to use 100% reusable and compostable packaging by the year 2025. But chemists are exploring additional creative approaches. One promising tactic incorporates seaweed as a biodegradable substitute for petroleum-based plastics. A startup British company called Notpla has come up with a seaweed substitute for plastic. Notpla produced 36,000 bottles filled with drinking water for competitors in the 2019 London Marathon. The group manufactured a million seaweed-plastic coated food take-out containers for a British online network of take-away restaurants, with plans to create one hundred times that many for the wider EU. Other companies are producing plastic wrap and other products from seaweed.

In the quest to lower human-generated greenhouse gases, the ultimate goal is to attain global net-zero emissions. Net-zero refers to a balance between the greenhouse gases generated by human activity and those that are being removed, naturally or by artificial means.

As the Knights Who Say Ni famously put it, "We want a shrubbery." Healthy wilderness areas require healthy vegetation, but wild deer and fluffy bunnies are not the only ones to benefit from foliage. A study of 93 European cities suggests that more than 2600 human heat-related deaths over just three months could have been prevented if these places increased their average tree coverage from 15 per cent to 30 per cent.

In fact, vegetation does more than cool the inhabitants of our planet; it cools the planet itself, researchers at the University of Virginia in Charlottesville say. Their studies demonstrate that tropical forests lower average global temperatures by about 1.5°C. Forests capture carbon dioxide from the air, and this forms a critical aspect of the temperature reduction. But a third of the cooling also comes from the release of water vapor from the leaves, a biological operation called transpiration. The rippling surfaces of irregular forest canopies also lower temperatures, breaking up heat fronts as they move across the jungle. Vegetation also produces aerosols that cool their surroundings by reflecting sunlight away. The aerosols also seed clouds above.

Designed landscaping, city planning including parks and greenbelts, and wildlife areas interspersed with inhabited areas all serve to reduce the heat that

---

[21] For example, print and broadcast reporting through outlets like Sky media's "Ocean Rescue Campaign"

emanates from urban areas. Cities viewed in the infrared[22] from orbit appear as "heat islands". Cities tend to average 4°C warmer than adjacent areas. Our structures often take the place of natural areas that kept the environment cool. But new city designs increasingly take this into consideration. A recent analysis of five boroughs in New York City[23] found that on summer days, trees and grasses absorbed the equivalent carbon dioxide put out by the cars, buses and trucks in the area. Lawns, trees, and even potted plants took up carbon at surprising rates. Earlier studies clocked metro carbon absorption by concentrating on contiguous tracts of forest and grassland, but these types of terrain only account for ten percent of urban uptake. The new data includes developed areas with tree-lined streets, individual gardens and overgrown vacant lots.

Often the solution for city architects is to establish gardens and lawns on the roofs of buildings. "Green" roofs lower energy consumption and act as insulation (Figure 9.10). And like forests, they sequester some carbon dioxide. Rooftop habitats also provide waypoints for migrating birds and insects like the migratory Monarch Butterfly. Some double as gardens, providing food and beauty for those city-dwellers living downstairs. And all of that rooftop vegetation helps to reduce carbon dioxide levels. The down sides of a green topside include increased maintenance and the need for reinforced structural support.

The runoff from storms or normal precipitation has become a major issue as well. Fertilizers from agricultural areas and detritus from animal farms pollute nearby streams and cause often-toxic algal growth in wetland environments. "Green infrastructure" has become a viable solution. In a 2019 US Congressional act define the ecological approach as, "…the range of measures that use plant or soil systems, permeable pavement or other permeable surfaces or substrates, stormwater harvest and reuse, or landscaping to store, infiltrate, or evapotranspirate stormwater and reduce flows to sewer systems or to surface waters."

Another passive cooling system has been used for three thousand years in Iran. Cooling towers called windcatchers use cross ventilation and other types of air flow to cool the interiors of buildings. The towers are cheaper to build than conventional heating and air conditioning (HVAC) systems, and require no electric power. Windcatcher technology is prevalent in western Asia and northern Africa. More recently, the passive cooling towers also find uses in Europe, Australia and the Americas.

---

[22] Infrared light reveals heat levels.
[23] Columbia Climate School's Lamont-Doherty Earth Observatory, Wei et al., *Environmental Research Letters*, 2022

Figure 9.10   A beautifully preserved Viking-era turf house in Iceland displays the use of grass as an insulator against the harsh Icelandic winters. (photo by the author)

Old School "Green" Dwellings

Living under a grassy roof is by no means a new idea. In ninth century Ireland, builders erected homes of stone and wood, with roofs blanketed in turf. The living grass provided better insulation than the stone or wood structures would. In Iceland, Viking settlers built long houses, with grass turf covering the roofs (and sometimes the walls as well). Turf buildings were common throughout Scandinavia, and were often built by and for people of lower income or social class. But in Iceland, turf houses served people of all social strata, along with public buildings. The fine quality of green insulation served the populace well, and will continue to as it is adapted to modern dwellings.

The United States itself produces roughly 5 billion metric tons of carbon dioxide every year, second only to China. India is third. But the US is also one of the leaders of pollution solutions. Every nation across the planet faces a different set of challenges ecologically, and each nation will come up with its own creative solutions. There are no universal fixes, although there are many solutions to similar problems, as we have seen. The first step toward a net-zero environment is widely seen as the creation of more clean energy. The proposition is economically within reach: in the last decade, the cost of solar power had fallen by almost 90 percent, and the cost of wind power technology has dropped by nearly 70 percent.

Industries are shifting to more efficient production techniques, incorporating hydrogen-based "clean" furnaces in steel production, for example, or shuttering the doors of the most inefficient coal-burning plants in favor of more modern ones.

Some industry, such as commercial air flight, cannot easily shift to electrical solutions yet, but must instead use other types of liquid fuels, changing fro fossil fuels to low- or zero-carbon options like hydrogen.

Finally, environmental researchers and engineers are exploring more ways of actually pulling greenhouse gases from the air or our factories. One promising solution—already in place at many sites, is the capturing of carbon dioxide and other greenhouse gases produced at industrial facilities (like fossil-fuel power plants), and sequestering it in underground reservoirs. Across the planet, some 35 such operations are removing 45 million tons of carbon dioxide from the air every year. More are planned. Designers hope to have 200 new facilities functioning by the end of the decade. In China's Henan Province, a new plant is capturing waste carbon dioxide from manufacturing facilities and converting it into methanol, a product usually made from coal. The plant processes 160,000 tons of carbon dioxide, equivalent to emissions of 60,000 cars. The greenhouse gas is then converted to 110,000 tons of liquid methanol, which is an important material in industry. This type of reuse is an important complement to sequestering of carbon dioxide, transforming it from greenhouse gas to useable chemical feedstock for manufacturing. The human race has dumped much pollution into our environment, but we are beginning to meet that damage with new technology and creative approaches to our travel and industry.

There is continued cause for hope. In 1987, a treaty called the Montreal Protocol was ratified. It banned the use of ozone-damaging CFCs. Since the agreement went into practical effect in 1989, the Earth's ozone layer has been recovering by roughly one percent per decade. If the global population can combine efforts to solve the CFC crisis, we can do the same for other greenhouse gases.

Back from the brink: success stories and fiascoes

Some of the most dramatic ecological disasters befell the Earth during the decades of the Cold War. One such calamity, the burning of the Cuyahoga River (yes, a river on fire!), was an event a century in the making. The Cuyahoga runs through northern Ohio, cutting through the midsection of Cleveland and ending in Lake Erie. The river has long been treated as a sewerage dump, and toxic waste buildup has caught fire in least 13 instances (the first recorded in 1868). Pollution levels rose so high that occasionally the entire stretch of the river from Akron to Cleveland was devoid of fish. Steel mills and other industries line the river. Gravel and asphalt processing centers added their own contributions to the oil, chemical waste and raw sewage flowing into the river.

The worst fire broke out in 1952, costing Ohio residents over a million dollars to watercraft, one bridge, and at least one office building along the riverfront. June 22, 1969 brought another fire. A spark from a passing train ignited an oil slick and floating industrial rubbish, ultimately causing $50,000 in losses. Although not as large as the 1952 fire, this time the river got the public's attention, due in large part to an article in *Time Magazine* about the incident. Late night TV hosts joked about the event. One of the most famous, Johnny Carson, quipped, "What's the difference between Cleveland and the Titanic? Cleveland has a better orchestra." Environmental groups rallied around the river fire, showcasing it as an example of water pollution across America. Their efforts, and those of some dedicated government officials (including Cleveland Mayor Carl Stokes, who testified before the US Congress), led to the formation of the Environmental Protection Agency, and spurred legislation like the Clean Water Act.

Since the 1969 fire, things have changed for the better. In 2019, the conservation group American Rivers named the Cuyahoga their "2019 River of the Year". More than 60 species of fish have made a comeback, and where once abandoned cars lined the river's banks to ward off erosion, the revitalized river now features parks and hiking trails. The river's historic fires have even been immortalized by Cleveland's Great Lakes Brewing Company, which named its Burning River Pale Ale after the river's more combustible past.

Another environmental success story has its beginnings in the darkest moments of the Cold War. The Rocky Flats Plant lies just ten miles south of Boulder, Colorado, and sixteen miles northwest of Denver. Rocky Flats was one of thirteen nuclear weapons production factories in the US. From 1952 through 1989, the facility secretly manufactured plutonium "triggers" (spheres of plutonium) for nuclear bombs. The plant constituted a major operation: some 800 buildings sprawled across 6,240 acres of prairie land. During the plant's clandestine operations, industrial fires and routine activities, chemical spills, and poor waste management practices led to multiple hazardous material leaks and contamination of soil, water and air. Several of the industrial accidents there resulted in the release of radioactivity into suburban areas (although the majority took place in a 385-acre industrial park near the center of the government facility lands). The worst incident was a 1957 fire that spontaneously ignited[24] in a hermetically sealed glove box used for handling

---

[24] Plutonium is flammable at room temperature.

radioactive materials. When plutonium shavings burst into flame, the conflagration quickly spread to adjacent glove boxes. As smoke spread through the air handling system, fire destroyed the HEPA filters designed to filter out radioactive particles. Radioactive plutonium hazes escaped through the smoke, contaminating some districts of Denver, but the public was told, officially, that there was only slight risk of low-level contamination. At the time, the true nature of Rocky Flats remained a secret, and the public was unaware of its actual purpose until a highly visible fire in 1969. This fire was smaller than the earlier blaze, but burned hundreds of pounds of plutonium. Outside investigators found plutonium poisoning at several urban and suburban locations. The US government made public the earlier fire and toxic waste spills that had befallen the facility. After years of public protests, the Federal Bureau of Investigation and the Environmental Protection Agency raided the facility in 1989. Rocky Flats halted its weapons production in 1992. In 1989, Rocky Flats was declared a Superfund site.

After a decade-long, seven billion dollar environmental reclamation and cleanup operation, the Department of Energy has transferred management of Rocky Flats to the US Fish and Wildlife Service. According to the Environmental Protection Agency, all 800 buildings, warehouses and laboratories in the weapons manufacturing area were "decontaminated and demolished," along with over "500,000 cubic meters of low-level radioactive waste; and remediation of more than 360 potentially contaminated environmental sites." Contamination levels outside of the central industrial area posed no health risk to wildlife or humans, showing contamination levels well below the requirement for any type of decontamination or cleanup. The EPA's Rocky Flats website[25] asserts that any remaining contamination, "includes low concentrations of radioactive materials, chemical solvents and heavy metal contaminants, generally below regulatory standards. Studies show that this contamination poses no threat to human health and the environment." Both the EPA and the Colorado Department of Public Health and Environment (CDPHE) declared the remediation of the site complete in 2006. Several environmental groups consider remediation efforts inadequate. The Department Of Energy still monitors air, soil and groundwater in the central portion of the acreage where the industrial operations took place (the "Central Operable Unit"), but the majority of the land has been evaluated as safe.

On September 20, 2018, the Rocky Flats National Wildlife Refuge opened to the public. Habitats in the wilderness area include prairie grassland,

---

[25] https://cumulis.epa.gov/supercpad/SiteProfiles/index.cfm?fuseaction=second.cleanup&id=0800360

woodlands and wetlands. Herds of deer and elk, coyotes, prairie falcons, and an assortment of songbirds inhabit the refuge, along with the rare Preble's meadow jumping mouse. Remnants of the xeric tallgrass prairie represent some of the most ancient grassland habitat, indigenous to the area since the last ice age. Amidst the refuge's ten miles of hiking trails, various government and independent groups continue to monitor levels of radioactivity on and near the site.

While Rocky Flats stands as a success story today, the highest radiation levels at the facility, along with the breakdown of techniques to contain radio-active material, created trauma for those Coloradoans living nearby in the 1980s and 90s. But the risks paled by comparison to those potential catastrophes arising at active nuclear power plants. Failures at nuclear reactors constitute existential nightmares of the highest order. Chernobyl, Three Mile Island, and Fukushima[26] are names seared into the consciousness of our society and culture, and rightly so. At the US nuclear plant Three Mile Island in Pennsylvania, the meltdown and resulting release of radioactive gas and iodine led to new regulations for nuclear facilities. The 1979 accident also brought development of new US plants to a grinding halt, as some 120 plans for reactors were shelved and eventually cancelled. The incident was the most serious crisis in nuclear power history up to that point. Despite the fact that the majority of the population lacked a foundational understanding of nuclear reactor operations, safety systems and track records,[27] the popularity of nuclear power has continued to drop dramatically in the United States since the Three Mile Island crisis. That drop has also been fuelled by problems in Europe and Asia.

Seven years after Three Mile Island, in northern Ukraine, the most serious nuclear disaster[28] in history transpired at the Chernobyl Nuclear Power Plant's number 4 reactor. During a safety test, engineers accidentally dropped the reactor power output to zero. While attempting to recover and bring the reactor back on line, too many control rods were removed, triggering a cascade of events that led to the melting of the reactor core and a series of steam explosions that spread deadly radioactive material into the surrounding environment.[29] Many firefighters sustained injuries, and several deaths occurred.

---

[26] A failure in 1957 at the Mayak secret plutonium production weapons plant sent radiation over 54,000 square km.

[27] And still do.

[28] In terms of deaths, injuries and economic cost.

[29] For a more detailed account of the chain of failures, see HBO's excellent five-part docu-drama *Chernobyl*.

During the initial conflagration, 237 emergency personnel were hospitalized, over half of them for radiation exposure. 49000 inhabitants from the nearby city of Pripyat—where most of the power plant's employees lived—were immediately evacuated, with another 68000 soon to follow. In all, 335,000 people had to be evacuated as a result of the radioactive danger. Radiation fell across the USSR and vast regions of Europe. A World Health Organization study predicted that 9000 people would eventually die from cancer related to the disaster.

Engineers erected an emergency concrete "sarcophagus" over the structure to contain the radiation, but in 2016 this building was superseded by the Chernobyl New Safe Confinement. A series of 13 arches hold the structure erect above the entire site—including the sarcophagus. The arched building was slid into place along tracks, and is designed to last a century, long after radioactive materials and ruins of the reactor have been safely removed and processed.

Chernobyl was not the last word in nuclear reactor crises. That honor falls to Japan's Fukushima Daiichi nuclear power plant. On March 11 of 2011, the strongest earthquake ever recorded in Japan hit just off the coast. The trembler triggered 14-meter-high tidal waves that crippled Fukushima's emergency backup diesel generators. The earthquake and tsunamis caused widespread power failures. These, in tandem with the damage to the backup generators, caused the cooling pumps to fail. Three reactor meltdowns ensued, along with three hydrogen explosions releasing radioactive particles. 110,000 residents fled from the path of radioactivity, fallout contaminated food supplies, and radioactive water flowed into the Pacific Ocean periodically for several years. Though many people lost homes and livelihoods, Japan's proactive evacuation and protection plans for its population kept health dangers to a minimum. A report by the International Atomic Energy Association stated that on a world-wide scale, the Fukushima disaster, "damaged confidence in nuclear power."

Today, tourists pay top dollar to wander the empty streets of Pripyat in the shadow of Chernobyl, to visit haunted Kindergarten playrooms and gaze across the abandoned amusement parks of a once-thriving city. Despite the best engineering attempts, Chernobyl is still a sleeping giant, critically dangerous. Nuclear apocalypse is big business, and radioactive nightmare has become a new kind of amusement park.

The crisis and meltdowns of Chernobyl and Fukushima Daiichi have become the great cautionary tales of the nuclear age. The Chernobyl "incident" was born out of acts of desperation in the context of cascading emergencies, and an overwhelming cascade of failures also visited Fukushima. The radiation from both will be a concern for generations to come. Cleanup continues at Japan's destroyed reactor. Chernobyl's high-tech tomb is a Band-Aid designed to outlast the most dangerous of the fallen reactor's radiation. But it is only a Band-Aid. The best resolution to nuclear dangers is the proactive one: to build reactors right in the first place, with many backup systems in place and highly trained staff always present. Other forms of reactors may offer safer nuclear solutions as well. Still other countries are opting to phase out the nuclear option in favor of renewable energy (energy that is not destroyed or transformed into another form when used, such as solar or wind power).

Life on the wild side

Rising temperatures, increased acid rain, loss of wilderness habitat and changing weather patterns have all contributed to the extinction or endangerment of wildlife on a grand scale. But amidst the bad news, the Arabian Oryx is the poster child for species rescue. The elegant Arabian Oryx stands out against the desert with its snow-white coat. Graceful curvilinear horns crown its black-and-white masked head. Those horns may be the origin of the unicorn legend. In 1962, dwindling wild populations of the Oryx prompted a network of five international zoos and six nations to mount an Oryx capture scheme that would lead to a captive breeding program. Participants included the San Diego Zoo/Wild Animal Park, the Flora Preservation Society of London, and the Phoenix Zoo. The program began in the nick of time: the Arabian Oryx went extinct in the wild in 1972. By then, several zoos had successfully bred them. In 1981, a wild herd was again established in Oman. Today, the herd holds a population of roughly 1220, with approximately 6500 of the oryxes held in "managed care" at various captive breeding sites including Syria, Bahrain, Qatar and the United Arab Emirates.

There have been other success stories. At the time of the European arrival in the Americas, over ten thousand Whooping Cranes—tallest cranes in the Americas—made up the migratory flock, which flies from southern Texas to its Canadian nesting grounds annually. By 1870, the flock had dwindled to just over 1300 individuals. The North American Whooping Crane is still critically endangered, but in 1941 its wild population was counted at 21, with only two captive birds. A major effort was mounted, involving captive breeding and introduction of Whooping Crane eggs into the nests of Sandhill Cranes. Today, with the reintroduction of the rare cranes into the wild, the migratory population—combined with the captive Whoopers—stands at 800. From near-extinction, the cranes are back, but remain in a precarious state.[30]

Still, with new legislation to prevent hunting and preserve habitat, and better species preservation techniques, progress continues. In the United States, since the enactment of the Endangered Species Act, 85% of bird populations increased or stabilized under the act's auspices. Average population increase of all bird types was 624%. Between 1978 and 2015, a dozen birds were delisted from the endangered list.

There are many battles yet to be won, and decisions to be made in terms of land management, resource and habitat preservation. The Sixth Mass Extinction is one that we, as a species, can suppress or even halt. But time is running out for many species, and we must act now.

Back from *beyond* the brink: undoing extermination

One of the most potent causes of extinction on Earth today is loss of habitat. Wildlife refuges, game preserves and National Parks play increasing roles in preserving habitats for endangered and other species. One of the most extensive undersea conservation preserves[31] is the Papahanaumokuakea Marine National Monument. Larger than all the US national parks combined, the site protects 7000 species in Hawaii's shallows, coral reefs, and deep-sea environments. Papahanaumokuakea joins a growing number of marine sanctuaries. Out on dry land, the 200,000 square mile Kavango-Zambezi Transfrontier Conservation Area centers at the junction of Angola, Botswana, Namibia, Zambia and Zimbabwe. Within the park, wildlife can hear the thunder of Victoria Falls and wander across the Okavango Delta. Other regions, including Yellowstone (US) and Australia's Featherdale Wildlife Park and Great Barrier Reef, pale in light of the 375,000 square mile "biosphere reserve", which spreads across half of the Greenland sub-continent.

These preserved wilderness areas are critical for the health of our planet's biome, and their expansion could only be good news. But the underlying problem remains: how will even protected creatures and plants survive if the climate within their protected areas changes drastically? One aid is the establishment of preserved corridors to promote migration. But migrating animals need a livable climate to migrate to.

---

[30] Members of the wild flock are susceptible to hurricanes and illegal hunting.

[31] 938,250 square kilometers

But what of the species who came before? Humans have witnessed the last surviving individuals of species that finally vanished from this planet. The Passenger Pigeon, the Thylacine, Mexico's Imperial Woodpecker and the Columbian Mammoths are just a few. Genetic engineers have made progress in cloning various life forms from single cells. The first mammal reproduced from a single cell was Dolly the sheep, cloned in 1996. DNA from an assortment of extinct creatures lies hiding away in dusty drawers, storage bins, and even in skins hanging from the walls or in taxidermied reconstructions of past living things. Could extinct species be resurrected from these samples, as Hollywood scientists did in *Jurassic Park*? And what are the implications? How would the reintroduction of a Wooly Mammoth to the steppes of Asia affect the indigenous populations of wildlife and vegetation already living in the area?

Mapping the DNA of an extinct creature is the first step in any cloning project, and the genetic mapping itself is an endeavor that yields much insight into past species' natures. Each April 1st, reports surface on the Internet that claim the successful cloning of an ice-age mammoth. These good-natured pranks are met with giggles and skepticism, but serious thought has been given to sequencing and cloning the DNA of extinct species. The Mammoth is the most high profile and popular subject to the general public, and its genome has been mapped by a team at Penn State University. According to the genetic team's website: "The Mammoth Genome Project...is the first to decipher the genome of an extinct animal. Our data allow a view back in time as far as 60,000 years and describe the genetic changes that occurred in mammoths. Our analyses show that the rate of evolution within the three lineages of elephants (mammoth, Indian elephant and African elephant) since they separated about 6 million years ago is only half of that between humans and chimpanzees." In other words, the genetic makeup of mammoths may be more similar to modern elephants than initially thought. The Penn State work demonstrates that studying the genetic details of extinct animals is possible. The researchers assert that lessons learned from their sequencing can shed light on the processes that are impelling today's endangered wildlife toward extinction. After their initial mapping phase, the geneticists have embarked on Phase two of their study, which seeks to map the mammoth genome at far greater resolution. They have their work cut out for them: the mammoth's DNA contains some 4 billion base pairs. And cloning such a beast is a long way off. Even with a complete genetic map in hand, advances in DNA editing and reproductive technologies must be accessed in new ways to resurrect these rulers of the ice-age plains.

Figure 9.11 Australia's now-extinct "Tasmanian Tiger", the thylacine, was a marsupial with tiger-like stripes and a pouch for its young (both males and females had pouches). These animals were photographed in the Beaumaris Zoo in Hobart in 1910. The last known survivor died in a zoo in 1936. (public domain, via https://commons.wikimedia. org/wiki/File:Thylacines.jpg)

Other extinct species are under study as well. Andrew Pask at the University of Melbourne leads a team currently researching the genome of the recently extinct marsupial called the thylacine, or Tasmanian Tiger (Figure 9.11). It was the size of a German Shepherd, with a wolf-like build and striped camouflage similar to that of a tiger. The University of Melbourne has established the TIGRR lab (for Thylacine Integrated Genomic Restoration Research lab) to develop technologies designed to conserve and restore rare and extinct marsupials like the Thylacine. To date, their genome map represents the highest-quality DNA map of any extinct species (including mammoths).[32] The team is sequencing DNA from close relatives of the thylacine, including the Dunmart mouse. Species with similar DNA can provide the living cells and genetic templates that would then be edited on a genetic scale, altering it into a thylacine genome ready for cloning.

European settlers hunted the thylacine—world's largest marsupial apex predator—to extinction, with the help from destruction of thylacine habitats and the introduction of wild dogs by Asian sailors much earlier. A bounty had been put on all thylacines, as ranchers considered it a threat to the Tasmanian sheep industry. Just 58 days after the government of Australia had granted the

---

[32] Pioneering genome work on thylacines was earlier carried out by Sydney scientists Don Colgon, Karen Firestone and Michael Archer.

rare animal protected status, the last confirmed thylacine died of exposure in a zoo in September of 1936, after a zookeeper accidentally locked it out of its protective enclosure overnight. Its body was gifted to the Tasmanian Museum and Art Gallery, but staff soon misplaced it, due to improper documentation. For decades, museum curators and zoologists searched for the missing thylacine, presumably the best-preserved specimen in existence. Finally, the search was narrowed down in light of contemporary documents, and the specimen was discovered tucked away in a cupboard at the Tasmanian Museum. The last thylacine contributed some of its DNA to the research being done in Melbourne.

While there have been reported sightings in the wild since, no sightings have been confirmed, nor any related evidence (such as fresh tracks). But thanks to ongoing genetic research, we may not have seen the last Tasmanian Tiger in the wilds of Australia.

A host of complications will meet wildlife experts who attempt to reintroduce resurrected species into the world. Depending on how far removed the original species was from the current environment, new animals must be taught to survive in an environment not designed for them. Dietary concerns are also at the forefront: what foods are the reborn animals designed to eat? Can they adapt to what—to them—is a quite alien menu? How will they interact with other wildlife, and how will their immune systems respond to new strains of viruses and bacteria? In the long run, it is far more advantageous and prudent to assure the survival of species still here than it is to bring them back from the dead (although that does make for great movies). And if astrobiologists like Brownlee and Ward are correct—that animal life is rare in the cosmos—this makes the current high rates of extinction on our world even more significant, and may serve as inspiration to do something about it, through preservation, resurrection, or both.

Weaning off of fossil fuels: a mixed solution?

Climate change is often tied to the use of fossil fuels. We have seen how much influence human activity has had on climate over time. But surprisingly, the most potent greenhouse gas is not carbon dioxide. Nor is it methane. The factor that affects our atmospheric heating the most is simple water vapor. Nearly half of all greenhouse gas in our air is water vapor. A scant .04% is carbon dioxide. Methane checks in at only 0.00017% (although it is much more potent). This means that our ability to control greenhouse gases is somewhat limited. But what is important to note is that water vapor disappears quickly.

Carbon dioxide and methane build over time, because they take long periods to break down.

Still, the burning of fossil fuels is one arena in which we humans can have a positive influence. We know of concrete ways to mitigate an increase in global greenhouse gases. At the forefront of the discussion is an ongoing debate between advocates of "clean" gas and coal vs. promoters of alternative energy (i.e. bioreactors, solar, wind, hydroelectric). Somewhere off to the side of that conversation lie other options like new forms of nuclear power, including liquid thorium or fusion.

Technical observers as well as the popular press often raise the question, *Can we replace fossil fuel with something else, so that we are free of our dependence on non-renewable energies?* "Green" energy has a lot of potential, both figuratively and literally. The current power usage of the world's civilizations comes out to roughly 15 terawatts (a terawatt is 1 trillion watts). The average incandescent light bulb uses between 25 to 100 watts. The world's current power use is equal to burning 435,000 liters of gasoline *every second*. In contrast, energy reaching the Earth's surface from the Sun is greater than 20,000 times the world's power consumption. Global winds generate more than ten times this amount of energy. Researchers estimate that accessible wind power (close to the ground) offers 20 terawatts of power. But in addition to the cost of entirely revamping our infrastructure without completely disrupting civilization, we face another obstacle: most renewable energy sources are intermittent. Solar works well when the Sun shines, but enough storms are bad news for the grid. Wind faces the same issues.

Renewable energy sources are increasing in use and efficiency, but many are still expensive to build and operate. Integrating these power alternatives into existing systems takes time and investment in infrastructure. But one partial solution may be found in a graduated approach to phasing out fossil fuels. On Justin Brierley's popular podcast "Unbelievable?"[33], astrophysicist/author Hugh Ross points out that, "CO_2 is only part of the problem. Canada is warming five times faster than other parts of the world." Carbon soot from China and India is darkening the surface ice, Ross says, causing it to melt and damaging delicate arctic environments as well as agricultural areas, where acid rain damages crops. "Coal is the problem. What we can do immediately, without waiting for a high-tech solution that may be years in the making, is to use natural gas instead of oil and coal. It releases only half

---

[33] https://www.premierunbelievable.com/

the greenhouse gases that coal does, and this can be applied today, as an intermediate step."

The automotive industry is a bellwether of the fossil fuel/alternative energy dialogue. Factories are retrofitting corporate fleets with hydrogen-fueled vehicles. New hybrid technologies cut down on emissions and fuel costs (although batteries are expensive to replace). All-electric cars are hitting the market in record numbers. Charging stations are on the increase, and new battery technology has shortened charging times significantly. In the Americas, Detroit is under pressure to increase mileage efficiency on conventional cars as well. Other nations have passed legislation to encourage factories to turn out more efficient cars, trucks and vans.

Electricity has always been a viable option on the rails, and it's beginning to make an even greater impact on long-distance and commuter trains. Japan's famous bullet trains provide fast-moving electric alternatives to other forms of transportation. Even the aircraft industry is making inroads into more fuel-efficient, cleaner-burning jet engines.

Powering a vast grid—say, a city—takes the challenge up to the next level. Passive solar energy such as trees to cool sun-facing sides of buildings, control of interior air flow, and shunting of sunlight at various times, have been strategies used for thousands of years. At cave dwellings in sites like Mesa Verde, clever indigenous architects arranged structures to take advantage of rock overhangs in summer, which cast shadows across the settlement. In winter, lower sun angles allowed sunlight to fall directly onto buildings to heat them.

But modern technology has brought new options. Among these are solar panels that convert sunlight directly into electricity (photovoltaics). Panels are dropping in price as they are manufactured in bulk, and more roofs in suburbia and urban areas are sprouting the blue and gray wings of solar arrays, lowering their personal bills and contributing to the grid.

Large-scale solar reactors are coming on line as well. One of the largest photovoltaic facilities sprawls across the rolling green plains of Spain's Usagre region. Supplying energy to 250,000 homes, the plant is designed to reduce carbon emissions by 215,000 tons each year. Other mega-plants include Benban in Egypt (covering 37 square kilometers), India's Shakti Sthala (50 square km), and China's Hainan Solar Park. India's Bhadla Solar Park is now the largest solar voltaic plant in the world, generating 2000 Mw over 57 square kilometers. The site could power the equivalent of 700,000 suburban dwellings. Italy, Germany, the United Kingdom and France are also leaders in solar power plants.

Another approach to solar power is called solar thermal (see Moroccan Megawatts). This system uses mirrors to direct sunlight to a boiler, converting the Sun's energy to steam. One of the world's largest plants of this type rises from the Mojave Desert in California, USA. Covering an impressive 3500 acres, Ivanpah utilizes three 140-meter towers surrounded by 173,500 heliostats, sun-tracking mirrors that focus sunlight onto the heating towers. At the top of the stacks, each tower houses a boiler. Sunlight focuses on the boilers, heating water that drives turbines for electricity. Although it has not yet generated as much energy as it was designed to, and has been plagued by technical issues, Ivanpah has proven new technologies on a large scale.

Engineers continue to pursue new, more efficient ways of harvesting energy from the Sun. One area of study involves space-based solar power. The concept involves building a huge orbital solar collector that would beam energy down to surface stations. The concept was studied in the 1970s at NASA's Jet Propulsion Laboratory. The baseline called for use of the space shuttle as the delivery system, with many launches required and an ambitious zero-G construction project. But at the time, the technologies involved had not matured, and the cost came in at a trillion dollars. The proposal was shelved indefinitely.

Things have changed. While the complexities of constructing a large-scale solar collector in orbit were theoretical in the 70s, astronauts have honed the necessary skills today with assembly and maintenance of various space stations, including MIR and the ISS. On-orbit construction is no longer an unknown quantity. Additionally, solar cells are 25% more efficient than they were back then, and costs have taken a nosedive. The UK, China, Japan, the US and South Korea all have modest ongoing research programs. Europe is the leader in research these days, with a recent Airbus team carrying out a test in a warehouse in Munich. Their hardware collected energy using solar panels, and then transmitted that energy via microwaves to a receiver 36 meters away on the far side of an aircraft hangar. The received electricity lit up a model of a city. One game changer is the cost of launches. While the space shuttle lofted a 1 kg payload to orbit at a price of roughly $58,000, the SpaceX Falcon 9 brings cargo to orbit at the rate of $2600/ kg, about 5% the cost of the space shuttle. The Munich test yielded about 5% efficiency, but designers project that if they can reach 20% efficiency, space-based solar power could compete with other types of energy now on line. And the approach has one other major advantage: in orbit, there are no clouds, winds or storms to interfere with the collection of solar energy. High above the Earth, the Sun's light can be harvested 24/7.

Morocco Megawatts

Unlike most of the continent's real estate, the northern African nation of Morocco has no oil (it imports over 90% of its energy from outside). But it does have 305 days of sunshine each year. In an effort to make his country energy-independent, Morocco's King Mohammed VI set a national goal for Morocco to generate 52% of its energy from renewable sources—both solar and wind power—by the year 2030. The focus of this ambitious goal is the Noor Ouarzazate Concentrated Solar Power Station. Spread over the equivalent of 3500 football fields, the plant uses three types of solar energy to generate power. The systems include a field of cylindrical parabolic mirrors that heat fluid for steam power (Noor 1 and 2), a photovoltaic station, and a solar tower heated by a vast field of 7400 mirrors, also for steam generation (Noor IV). The tower uses salt to carry heat to generators. The molten salt is then stored so that power can be generated up to 7 hours after sunset. Combined, the plant generates enough power for 120,000 individual homes. The plant generates renewable energy equivalent to 2.5 billion tons of imported oil.[34]

But a problem haunts projects like Ouarzazate: power transmission. Power is lost as it is forced through a network to the customers who would use it. Morocco's long-term plan was to supply Europe with energy, but cables across the Mediterranean are expensive. They have been done with success, but many European nations have opted to use domestic solar power, primarily with photovoltaics that can supply energy where they are needed. Morocco's high-tech power plant has had technical issues and hardware failures, but its concept is a solid one. Ouarzazate serves as a model solution to Africa's growing energy needs, an example that promises to be an inspiration for many nations across the great continent.

Solar energy is a high-profile renewable energy, but there are others. Hydroelectric power is one of the most ancient, dating back 2000 years to the Greeks, who used water wheels to grind grain. Hydroelectric stations use similar technology, forcing water through turbines to generate electricity. Fully 70% of Iceland's electricity comes from hydropower. The picturesque rivers and waterfalls are some of Iceland's most famous natural wonders. Torrential rivers flow from its glaciers, and are a perfect source of hydroelectricity. (See the box "Running on Empty: The Starving of the Colorado".)

In the wilderness of Iceland's Landmannalauger, a "hut" rises from the mossy lava hills. It's a way station for hikers with 64 beds for sleeping bags. In the kitchen area, a sign above the sink reads, "This tap is for cold potable water. The hot water is in the stream on the north side of the hut." Iceland utilizes geothermal energy, power issuing from its ubiquitous underground

---

[34] The Noor 1 and 2 sections do use some fossil fuels to power various systems, and as a backup in case primary systems fail.

reservoirs of heated water. It's completely clean, and aside from building and maintaining the equipment to harness and transmit it, it is free for the taking. Heat from the Earth's interior warms underground water reservoirs. In some locations the geothermally-heated water is so close to the surface that it takes very little construction to access it. Wells also provide access to deeper sources.

Iceland is a world leader in geothermal technology (Figure 9.12). A geothermal plant in Myvatn has been a pioneering facility for this type of energy. Europe leads Asia, the Americas, Oceania and Africa in its use. Heated water from geothermal sources is directly used in 32 European countries.[35]

Europe's main geothermal fields lie in the Larderello region of Italy, France's Paris Basin, the Pannonian Basin (which overlaps Hungary, Serbia, Slovakia, Slovenia, and Romania)

several provinces within the European Lowland of Germany, and Poland), the Palaeogene systems beneath the Carpathian Mountains (Slovakia and Poland), and other Alpine and older geological units in Bulgaria, Romania, and Turkey. Geothermal power plants have also been erected in Japan, Hawaii, New Zealand and other geologically active areas. Already, geothermal sources supply some 60% of the power needs along the northern California coast.[36] While some locales have heated water or steam near the surface, at other sites wells must be drilled in order to access the heat. The natural geothermal heat—in the form of steam or hot water— is then transferred to turbines, which spin to generate electricity. Hot water is also used for direct heating of dwellings.

Thousands of brilliant white turbines rotate in the winds China and Mongolia, propelling the graceful windmills to generate 20 GW of power. Wind energy is another favorite of alternate energy technologies, and China's Jiuquan Wind Power Base is the largest to date. Also called the Gansu Wind Farm, the site sprawls across the Hebei, Xinjian, Jiangsu, and Shadong provinces of China, and includes a stand in Inner Mongolia. India, the United States and the United Kingdom also rank among the ten biggest wind farms in the world. Some sites are offshore. The largest of these, the Hornsea 2, became fully operational in 2022. Its forest of turbines stand 90 kilometers off the coast of Yorkshire, England.

As alternate energy sources go, wind power is among the most cost-effective, assuming adequate winds at the site. Wind is an abundant energy source, and turbines emit no polluting waste. Wind turbines are adaptable to different

---

[35] World Geothermal Congress 2005 (Lund et al., 2005)
[36] From San Francisco to the California/Oregon border.

Figure 9.12 Top: All of the power on the grid in Iceland is renewable. The bulk is hydroelectric, but some 30% is geothermal, as seen at this plant in the northern region of Myvatn. (photo courtesy Marilynn Flynn) Bottom: a major source of energy in the southwestern US is the Hoover Dam, which harnesses the power of the Colorado River. Dropping water levels from recent droughts threaten its operation. Note the white "bathtub ring" above the water line beyond the dam, indicating where the original level of the lake stood. (Photo courtesy Bill Gerrish)

types of terrain. Unlike solar energy, wind tends to blow at all hours of the day and night (although wind speeds typically drop between sunset and sunrise) According to the US Office of Energy Efficiency and Renewable Energy, "When comparing the cost of energy associated with new power plants, wind and solar projects are now more economically competitive that gas, geothermal, coal or nuclear facilities." However, sites with the strongest and steadiest winds often lie in remote settings, where the cost of transmitting the power increases. While interconnection technology—connecting the wind-generated electricity to the grid—are improving, costs can still be prohibitive in some areas. Another drawback is aesthetic: turbines are often loud, and their towers impact the landscape. The massive blades on the turbines impact the behavior on wildlife and their bird-killing ability is legendary. Turbine collisions account for between 140,000 and 500,000 birds killed annually in the US alone. These occur primarily during night migrations. But this number is substantially less than bird deaths caused by domestic cats and collisions with buildings. Planners are prioritizing wind farm construction that lies outside of major migratory routes. Engineers are also coming up with new technologies that will help to reduce damage to birds and other wildlife.

Biofuels bring another energy source to the table. Engineers create biofuels from microbes or other living organisms. The most well known is ethanol, a corn-based additive for gasoline. Although ethanol reduces emissions and cost in gasoline, raising corn is an energy-intensive type of agriculture. Additionally, raising corn for ethanol uses farmland that could be more efficiently used to grow crops for food. But bio-engineered microbes or biofuels based on algae may provide opportunity for major new energy sources. Although biofuel releases carbon dioxide when burned, the microbes used to produce it actually absorb carbon dioxide, metabolizing it as they grow. If biofuel manufacturing is carried out carefully, it can be a carbon-neutral production. Both the US Navy and commercial airline manufacturers have successfully used algae-based biofuels in tests.

Nuclear energy is another aspect of alternate energy, and it's a powerful one. Among the European Union's 27 member states, 13 use nuclear power, contributing nearly a forth of all power to Europe. (46% of Europe's energy comes from fossil fuels and biomass, and renewables bring in 29%). France produces half of Europe's nuclear energy. In addition to supplying a substantial amount of energy to the world, nuclear power produces no greenhouse gases.

Nuclear power does, however, have some obvious drawbacks (as we saw with Chernobyl, Three Mile Island and Fukushima). In addition to the danger of radiation release during a malfunction, spent nuclear fuel (uranium or plutonium) retains deadly levels of radioactivity over thousands or tens of thousands of years. Improvements to nuclear reactor engineering have seen great strides in recent years, with redundant safeguards and advanced processing techniques for nuclear fuel. Some new reactors reprocess their spent fuel, converting it into more stable (and less dangerous) types. Another consideration is that nuclear waste, while dangerous, is threatening only on a local scale, while other types of energy have a global impact, endangering the entire biome of the Earth. This latter point is certainly not ideal. If only we had a nuclear power plant that didn't give off so much radiation, a facility that was more stable and exuded less toxic waste.

We do. They're called liquid thorium reactors. As the name implies, these power plants do not use dangerous uranium or plutonium. Liquid thorium reactors don't use water to cool them. Instead, liquefied salts circulate through their systems. This is advantageous in several ways, but perhaps the most important is that the nuclear fuel is not exposed, Thorium yields three hundred times the energy, it doesn't "melt down" because of the way it is cooled, and its fuel cannot be used for weapons.

The Holy Grail of power is, of course, nuclear fusion. Fission, the process used by nuclear reactors today, produces radiation because it gains energy by splitting apart atoms of uranium, plutonium or thorium.[37] But fusion is the sort of energy that powers the Sun. Energy comes from the fusing together of hydrogen atoms into helium. Sound familiar? This is exactly the process by which stars shine. We get hydrogen from water, and since the "waste" is non-toxic helium, nuclear fusion is probably the most efficient, clean, natural energy source in the universe.

The efficiency of fusion is truly awesome. If we could fuse only the hydrogen contained within the water of a kitchen sink faucet, we could power the entire United States with that flow.[38] It sounds like a dream, but so far, making it work has been a nightmare. Laboratories across the world are experimenting with various systems to trigger nuclear fusion. So far, results have not yielded a solution for commercial-scale fusion power plants, but work continues, and several avenues show promise. In December of 2022, engineers at Lawrence Livermore National Laboratory focused 192 high-powered lasers onto a small cylinder of diamond-encased hydrogen. This triggered a chain reaction that fused a pellet of deuterium and tritium. For the first time in history, the reaction created more energy than it to to start it. It didn't last long.

---

[37] Nuclear fission is also what powers nuclear weapons.
[38] For details, see *A Global Warming Primer* by Jeffrey Bennett, (Big Kid Science, 2016) pp 83-4

The reaction carried on for a few trillionths of a second, but it released 1.5 times the energy that engineers put into it. Analysts hope the process will lead to sustained fusion reactions.

All forms of alternative energy have inherent challenges, whether with hardware or economics. With technological advances, care of site selection, and appropriate use in reasonable contexts, none of the challenges facing various renewable industries are insurmountable. Many show great potential for near-future solutions of carbon emissions on a grand scale.

In speaking about parenting his son, author Mark O'Connell writes, "I want to defer the knowledge that he has been born into a dying world."[39] I absolutely understand his sense of urgency, especially considering the artificially induced climate damage we have seen in this very chapter. And yet, in light of the history of the planet, the sentiment may be disconnected from the reality of our natural world. Our planet is continuing to transform itself. It has always been changing, it is presently changing, and, in fact, it will continue to change. As unfortunate as it is, some of its alterations—or at least their rate—are clearly destructive ones brought on by humanity. True, through planet Earth's history we have seen changes in temperature by a few degrees before. Many times. But the point is this: as far as we can tell, we have never seen such *abrupt* global temperature swings unless they were due to mass extinction events (the kind associated with an impact, for example). This time around, we have the capacity—as a self-aware race—to do something to slow or subdue them (although asking ourselves if we have the will to make those course corrections is a different question entirely). But the world is not dying any more than it was at the end of the Hadean or Permian or Cretaceous. It is transforming into something new, with or without our influence.

And why not involve humanity in the calming of the climate and the health of the ecosphere? We have seen, through the previous pages here, reason for hope. We've learned lessons from the clouds of Venus, and applied those lessons to the preservation of our planet's ozone layer. Venus has taught us something about atmospheric carbon dioxide, and about chlorofluorocarbons. We've successfully removed many CFC's from industry, and continue to find ways to reduce greenhouse gases, although the latter constitutes our most urgent and biggest challenge ahead, politically as well as pragmatically. Clean energies, in their various forms, promise to provide a key to greenhouse gas reduction. We can look to success stories within the realm of wilderness preservation and species rescue, such as the Arabian Oryx. While the

---

[39] *Notes from an Apocalypse* by Mark O'Connell (Doubleday, 2020)

environmental crises we face are formidable and make for good press (just ask activist Greta Thunberg), our small successes should also give us encouragement.

It can be argued that our triumphs are too few, and if we would rather not have a return to the global temperatures of the Jurassic, it's time to bring our A-game. We must take ownership for the damage we have done, and bring to bear our technology, self-discipline and creativity to repair and restore what we can. We know we have the capability to reverse the trends that human activity has wrought. The question remains: do we have the determination? The husbanding of our planet's resources and the care of its environment will require a reevaluation of lifestyle for many of us. For example, the American appetite for driving, the western appetite for beef, some eastern industrial practices, and so many overlooked cultural traditions that influence our way of life must undergo major change. Are we ready for *that* discussion? We can focus on doom and despair, and we do have some reason to do so. But for all our flaws, we are a creative, resourceful species, and thus we can choose to step up and meet our challenges head-on. We can approach the world with renewed hope, born not of denial or naiveté, but rather from a scientifically informed, proactive stance. In doing so, we will be celebrating the wonder and joy that comes from living on this unique, resilient, amazing world.

It is of the utmost importance to humans as to how the climate crisis will unfold and resolve. What we do will affect the quality of life on this planet, for us, for our wildlife, our forests and jungles, and for all other living things from microbes to manatees. Because we have self-awareness, and because only we have the capacity to husband the Earth's resources and, when needed, to change trends we're seeing, it is our responsibility to do so, especially if we want our cosmic home to be an enjoyable, livable setting for the foreseeable future.

The distant future of Earth's siblings

But in the grand scheme of planetary history, 4.6 billion years long so far, our actions will have little long-term effect on our planet. Terra firma will keep spinning about its axis, revolving around its star, its continents drifting, its water, carbon and rock cycles cycling, its oceans coming and going. We've been studying our planet's past, and a few trends, but what about its future? For insights, let's look again to the planets around us, and what's in store for them.

Whether we humans cause it or not, the entire solar system is in for climate change sooner or later. The Sun's temperature will rise to its highest when it reaches 8 billion years of age. During its youth to middle age, the Sun will burn its primary fuel—hydrogen—converting it slowly to helium. As the helium increases, the star will begin to cool, but as it does it will also expand, becoming a red giant at about 11 billion years old (Figure 9.13). As it expands, the Sun will engulf Mercury and almost surely Venus. Experts debate whether the Earth will remain intact, outside of the outer fringes of our swelling star. It is a difficult situation to project, because even as the Sun expands toward the Earth, it will be losing some of its mass, allowing Earth and other planets to drift farther out. But long before our star's swan song, the other planets will feel the heat, even as they lose their own internal warmth.

Although we've called Venus "planet volcano", some studies estimate its modern volcanism to be roughly equal to what it currently is on Earth. If Venus retains all of its atmosphere without losses to the vacuum of space (which is likely because of carbon dioxide's weight), the pressure at the surface will double in a billion years or so. But if volcanism has tailed off and the environment is stable, the carbon dioxide is leaking away at a rate that will leave the planet nearly airless.[40] A billion years from now, many of the volcanoes on Venus will fall silent. As with all worlds, the radiogenic material of the core will fade, and the fires of volcanism will die. We have seen how much sulfur drifts in the Venusian atmosphere. Without the replenishing of volcanic eruptions, most of the sulfur compounds will dissipate. The sulfuric acid clouds will not last long; sulfur dioxide combines with surface carbonates as it has on Earth, so the skies will clear. The lack of clouds will bring lower

Figure 9.13 Observations from the Hubble Space Telescope revealed an explosion and mass ejection from the red giant star Betelgeuse (left two images). The star appeared to dim, but astronomers now think that dimming was due to material belched out from the star (second view from right), which left the star's surface unstable and undulating. (art courtesy NASA/ESA/Elisabeth Wheatley (STScI)

---

[40] This takes a long time: the loss of carbon from the Venusian atmosphere to space follows at a slow pace because carbon likes to combine with oxygen, turning into carbon dioxide (CO_2) or carbon monoxide (CO). Both gases are too weighty to escape rapidly.

temperatures at the surface, as their greenhouse effect weakens. But the former clouds had another effect: their high albedo, or brightness, reflected sunlight away. Without them, temperatures will tend to rise. Which effect will win out? Temperatures on the ground may rise above even where they are today, but dropping carbon dioxide in the air will eventually mitigate the heat. In time, the climate of our sister world will morph into a more temperate environment. One model describes a surface pressure of between 2 and 3 bars, with temperatures at about 158°F (70°C). It's still a toasty prospect, but far more comfortable than today's blast-furnace environment. In another billion years, as pressures continue to fall, Venus may actually host conditions tolerable for humans, as long as they bring their water with them.

If we have not terraformed Mars, but have rather left it to travel on its natural path, its climate will also see changes in the future, but less than those on Venus.

Climate change on Titan will look substantially different. As we saw in Chapter 5, Titan's axis precesses over long periods, prompting long-term cycles in its climate. But the changes Titan has seen are nothing compared to what's coming as the Sun swells into a red giant. If temperature is the main roadblock to life on Saturn's behemoth moon, then the future will open all sorts of possibilities. Titan's long-term cycles will continue. In a warming environment, one of two things will happen. First, wind patterns will change on timescales of centuries or millennia. A million years from now, Titan's methane may vanish as the moon's inventory runs out. This will trim the greenhouse effect and lower temperatures for a while. But another possibility is that increased cryovolcanism, brought on by changing surface conditions (like the entire landscape melting) will actually increase greenhouse gases, raising temperatures to levels approaching that on Earth today. Five billion years from now, the Sun will have brightened considerably. It will have cooled a bit and become redder in color due to its decreased temperature, but our star will also expand, ballooning beyond the orbit of Venus. Its closer proximity to Titan will turn Saturn's moon into an ocean world. No rock will rise out of the global seas, as it is all cloistered hundreds of kilometers down. But the waters above will be seething with organic material, the very kind that astrobiologists suggest may have given rise to the living world that is our Earth.

The many ocean worlds of the outer solar system house internal seas today, but as the Sun swells to a red giant, their oceans will come to the surface. Atmospheres will blossom as surfaces become global seas. The extent and longevity of these new worlds is energetically debated, and the predictive models feature wide error bars.

Capricious star: what Betelgeuse can tell us about Earth's past

The red supergiant Betelgeuse has been a famous star from ancient times. Second century astrologer Claudius Ptolemy described its color ("tawny-orange"). Its regular dimming and brightening over a 400-day cycle was legendary, as was its prominent place as the left shoulder of Orion, the Hunter constellation. Astronomer John Herschel[41] described the star's regular variability in 1836. But something recently disturbed the star's regularity.

In the winter of 2019, Betelgeuse departed from its comfortable 400-day pulsating. The star suddenly darkened, and this so-called Great Dimming lasted for several months. At first, the event was a mystery; nothing like it had been seen before in stars whose luminosity regularly oscillated. But after months of observation and study, scientists think they understand what happened: Betelgeuse vomited out a huge flare or wave of plasma equivalent to several Moon masses. The stellar material cooled and condensed into dark dust. That dust blocked our view of the star's glowing face. The expanding stardust cloud should become visible to observers here on Earth by 2024.

The dust wasn't all that contributed to the Great Dimming. Betelgeuse also cooled as a result of the event, and began to put out less light. The ejection of so much material at once may have left a darkened spot on the surface that is now fading. The infamous ebb and flow of Betelgeuse's light is no longer so predictable. Its slow motion flickering may indicate that the star is still recovering from its violent expulsion of material.

Scholars have made another discovery recently: the color of Betelgeuse may have become dramatically redder in recent centuries. As a star like our Sun transitions from the "main sequence", a stable hydrogen burning stage, it transforms hydrogen into helium, becoming less stable. When the star starves of it primary hydrogen fuel, it begins to burn the less stable helium. It begins to swell and cool, becoming a red giant or supergiant. It appears that Betelgeuse may have made the transition to red supergiant only recently on the cosmic clock. While Ptolemy described the star as orange, more ancient Roman documents label Betelgeuse as yellow, comparing it to the golden hue of Saturn. A 100 BC Han Dynasty narrative colors Betelgeuse yellow as well. Other ancient accounts that list red stars skip Betelgeuse altogether. It seems to have reddened dramatically within the last two millennia, a symptom of the star's entry into senior citizen status. Because the Earth's own Sun is burning through its hydrogen fuel en route to becoming a red giant like Betelgeuse, the star's activity may well be a preview of what our planet's star is in for.

What of the Earth?

But under the light of our angry red sun, what of the Earth? The Sun's luminosity is increasing by approximately 1% every 110 million years. As the Sun expands and increases in intensity, the Earth's water will be heated to mist.

---

[41] Son of Sir William Herschel, discoverer of Uranus.

The planet's atmosphere will increase dramatically in pressure as the oceans vaporize. A canopy of fog and cloud will cocoon our planet, holding in even more heat with that famous greenhouse effect. But the process won't stop there. As temperatures continue to surge, solar energy will tear apart the water molecules, forcing the hydrogen into space and leaving behind the oxygen. But even the oxygen won't hold on forever. The Sun's fierce energy will slowly strip the atmosphere from our world. The living biome will be long gone. The desiccated surface stones will radiate a dull red glow at 1500°C. Volcanoes will continue to expel gases into the dwindling atmosphere as the radiogenic heat within Earth's core languishes. If plate tectonics are dependent on water for lubrication, the continents will eventually seize up, paralyzed by planetary dehydration. The volcanoes—and the atmosphere with them—will quietly fade away.

Across that singed and desolate landscape, no trace will linger of the cool green oceans of the Great Oxygenation Event, of the forested alpine mountains with their grinding glaciers and splashing waterfalls, or the sparkling ices of Antarctica and Greenland. Perhaps, by then, we humans will have learned to get along, learned to travel among the stars, and learned just how precious life among the planets truly is. If so, we will have made our home in other distant star systems, far from the planet that gave us birth (Figure 9.1).

Does a planet hold five billion years of memories? It does, in the minerals and chemistries it keeps in its dark, ancient places. It recalls the songs sung by the oldest asteroids and comets, shrouds them within the sediments of primordial shores and the basalts belched from volcanoes long before Vesuvius. Even the echo of living things rests within its rocky layers, memories of the first cyanobacteria, the wriggling trilobites and sea-serpent Mosasaurs, the bouncing gazelles and wheeling pterosaurs. It's all there, history inscribed on stone below and cloistered among the gases above, upon atom and molecule. It is there for us to see and learn from, with reminiscences of what was and cautionary tales of what things might come. Through it all, the Earth has stood, and stands today, not unchanged, but fluid, a great monument to the planets and moons that arose in our solar system, worlds that remind us of our own Planet Earth.

Appendices

Appendix A: Important Terms

Accretion: The planetary building process wherein solid particles and larger fragments growing as they collide and stick together.

Amino acids: molecules that combine to form proteins. As such, they are critical building blocks of living systems.

Astronomical Unit: the distance between the Earth and Sun, 150 million kilometers.

Banded iron deposits: deposits in 2.5 billion-year-old rock which record a period in Earth's history before the advent of free oxygen in the atmosphere.

Biogenic: relating to biological material or processes

Biogenesis: the origin of life from non-living material

Carbonaceous chondrites: the most primitive meteorites made up of material from before the Sun's formation

Chirality: the twisting of a biological structure (found in amino acids, DNA, etc) toward the left or right. As the vast majority of terrestrial life-related structures have left-handed chirality, the search for right-handed chirality may be a good test for extraterrestrial life.

CHNOPS: the six main chemical elements of terrestrial life: carbon, hydrogen, nitrogen, oxygen, phosphorus and sulfur

CHON: the four primary elements of biology: carbon, hydrogen, oxygen and nitrogen

M. Carroll, *Planet Earth, Past and Present*, Springer Praxis Books, https://doi.org/10.1007/978-3-031-41360-5

Climate: the environmental conditions—in particular, weather—dominant over long periods

Comet: a nucleus of ice that develops a cloud of vapor that becomes a long dust and gas tail as it approaches the inner solar system.

> Long period: comets that enter the inner solar system from the Oort cloud with orbital periods of hundreds or thousands of years.
>
> Short Period: comets with short (years-long) orbits that remain within the planetary region of the solar system.

Continental crust: silica-based granitic rock, lighter than oceanic crust, that builds the continents.

Core: The center of a planet. In the case of terrestrial worlds, the core consists of molten and/or solid metals.

Crater saturation: a condition on the surface of a planet or moon in which craters are so dense that any new craters obliterate a like number of existing ones, so that the total number of craters cannot increase with new impacts.

Cratons: massive rock assemblages that represent remnants of the Earth's primordial crust

Croll-Milankovich cycles: long-term climate change resulting from the precession of a planet's axis or irregularities in its orbit.

Crustal dichotomy: difference in nature or appearance between two hemispheres of a world

Dark energy: unseen repulsive force driving objects away from each other, causing the universe to expand more quickly than normal gravitational acceleration would.

Dark matter: the majority of matter in the universe, invisible to normal instruments.

Deuterium: a heavy isotope of hydrogen.

Differentiation: The internal heat of a planet, moon or large asteroid frees material to sink, while lighter elements in molten rock rise. Differentiation sorts a planet's interior into a core, mantle and crust.

Doppler effect: shift in light or sound waves due to movement.

Feeding zones

> of planets: region in the solar nebula in which a planetary embryo "feeds upon" material from the surrounding cloud of dust and gas to develop into a full-grown planet.
>
> of giraffes: typically the tops of trees.

Fossil fuels: hydrocarbon-based matter, originating from biological substances, which can be burned for energy production. Fossil fuels include natural gas, coal and oil.

GUT force: combined force of the strong, weak, and electromagnetic forces.

Grand Tack model: a variation of the Nice model, proposing that Jupiter originally formed at 3.5 AU, then migrated inward to 1.5 AU, before reversing course and "tacking" out—like a sailboat against the wind—to its current orbit at 5.2 AU.

Great Oxygenation Event (GOE): a transformational event 2.4 billion years ago in which microbial life pumped oxygen into the atmosphere, changing Earth's environment dramatically.

Greenhouse effect: certain gases allow visible light through, but shift it to infrared light, trapping it as heat in a planet's lower atmosphere

Habitable Zone

> galactic: region in a galaxy close enough to the hub to form rocky planets, but far enough out to be sheltered from radiation of the galactic core
> microbial: region within a solar system able to support microbial life
> migrating: the drift of a habitable zone as a star advances through different stages of development, changing in its energy output.
> Permanent: subset of habitable zone, the region that remains habitable throughout the entire life of a star.
> Planetary: zone in which water can exist as liquid, vapor and solid, a condition thought to be critical to life.

Hermian: relating to Mercury

Hill Radius: the area influenced by the gravity surrounding an object, usually a planet, moon or star

Hollows: bright, sunken areas on Mercury likely related to volcanic outgassing.

Kuiper Belt: a donut-shaped band of icy objects orbiting beyond Neptune from roughly 30 to 55 AU distance from the Sun

Kuiper Belt objects: planetoids orbiting within the Kuiper Belt

Late Heavy Bombardment: a hypothesized period of increased asteroid and comet impacts that may have crested 4.1 billion years ago, tailing out roughly 3.8 billion years ago.

Lava: molten rock erupted onto a surface, or cooled surface rock that originated as magma

Low Reflectance Material: dark, carbon-rich patches of surface material within and adjacent to craters on Mercury.

Magma: molten rock beneath the surface of a planet or moon. In the case of ice planets and moons, magma is typically liquid water

Magnetosphere: the region around a planet, moon, or star dominated by magnetic fields emanating from within

Mantle: usually the largest layer of a planetary body, the mantle is the region of a planet sandwiched between the central core and the outer crust.

Mantle plume: a heated mass of partially molten material which rises up through the mantle and impinges on the crust above. A mantle plume is responsible for the Hawaiian island chain, for example.

Meteor: a meteoroid entering an atmosphere

Meteorite: a meteor on the surface of a planet, moon or other celestial body.

Meteoroid: a rocky or metallic fragment smaller than an asteroid or comet

Nicemodel: theory—named after the French town in which it was first presented—that details planetary migration, in which the planets' orbits drift toward or away from the Sun, and the planets actually pass into and out of resonance with each other

Obliquity: the tilt of a planet's spin axis

Oceanic crust: the crustal portions of Earth making up the sea floors, broadly basaltic in content.

Oort Cloud: spherical shell of comets surrounding the solar system, extending from 2000 to 10,000 AU (or farther). Long period comets come from the Oort cloud.

Organic: containing carbon compounds, often related to living matter

Photodissociation: the breaking apart of bonds in a chemical compound by sunlight.

Pioneer Venus 12.5-kilometer anomaly: a Venus atmospheric phenomenon that seems to affect landers and probes at an altitude of ~12.5 kilometers.

Planetary embryo: a solid body that has dominance within its feeding zone in a planetary nebula.

Planetesimal: a small body in the process of developing into a planet

Plates: a rigid slab of crust, usually a combination of oceanic and continental rock, forming the Earth's outer layers.

Plate tectonics: the movement of Earth's crustal plates, involving subduction and uplift.

Planetary migration: Expanding or contracting orbits of planets within a protoplanetary disk or a young planetary system, causing them to migrate to obits other than the one in which they formed.

Plutinos: Kuiper Belt objects in a 3:2 orbital resonance with Neptune.

Precession: the long-term wobble of a planet's axis

Proplyds: protoplanetary disks or still-forming planetary systems

Protoplanet: a large planetary embryo in the process of developing into a planet.

Radiogenic: radioactive, energized or heated by radioactive material such as thorium or uranium.

Regolith: pulverized rock that makes up the sterile "soil" covering the bedrock of an airless or lifeless world.

Renewable energy: energy that is not destroyed or transformed into another form when used, such as solar or wind power.

Seafloor spreading: an important part of plate tectonics, basaltic crustal stone of the seafloor spreads out from a central ridge, creating new ocean floor material.

Snowball Earth: hypothesis that the Earth passed through an ice age severe enough to completely encase the planet in ice. Scientists estimate the period to have occurred before 650 million years ago.

Sublimation: transforming directly from a solid (like ice) to a vapor without melting into liquid.

Subduction: The process of one plate sliding beneath another.

Super Continent: The result of the merging of two or more continents and subcontinents by collision.

Super Earth: a planet larger than Earth but smaller than Neptune, but sharing some characteristics of Earth.

Tectonics

- (see plate tectonics)
- columnar: theoretical vertical movement of crust on Venus, which serves a similar role to the Earth's plate tectonics.
- block: tectonic movement of blocks of crust, with features resembling rafting of terrestrial sea ice.

Terraforming: engineering an entire planet's environment to resemble the Earth's.

Terrestrial: Earth-like; referring to the terrestrial planets of Mercury, Venus, Earth, Mars, and sometimes the Moon. Terrestrial can also mean relating specifically to the Earth.

Theia: theoretical planetoid that collided with the proto-Earth, resulting in the Moon

Tholins: brownish, complex carbon chains in reduced atmospheres like methane or ethane, generated when ultraviolet light or electrical discharge (i.e. lightning) interacts with carbon-rich molecules.

Unknown ultraviolet absorber: appears as dark patches in the Venusian clouds that absorb light in the ultraviolet.

Uplift: The raising of vast regions of a plate, either by subsurface forces like magma plumes, or by collision of two plates to establish structures like mountain chains.

Zircons: minerals (technically zirconium silicate) that crystallize within silica-based rock. Zircons are important because they fold uranium into their structure, making them easy to date radiometrically.

Appendix B: Further reading

Note: while some of the books below are slightly dated scientifically, older titles provide valuable insights into the ongoing history and development of research in the areas outlined.

GENERAL PLANETARY GEOLOGY

Atlas of the Galilean Satellites by Paul Schenk (Cambridge University Press, 2010)

A Brief History of Earth by Andrew H. Knoll (Custom House, 2021)

The Cosmic Perspective, 9th Edition edited by Bennett, et al (Pearson, 2019)

Evolution of the Earth by Donald Prothero and Robert Dott Jr, (McGraw Hill, Seventh edition 2004)

History of the Earth by William K. Hartmann and Ron Miller (Workman publishers, 1991)

Ice Worlds of the Solar System by Michael Carroll (Springer, 2019)

ASTROBIOLOGY

Astrobiology for a General Reader by Vera Kolb and Benton Clark III (Cambridge Scholars Publishing, 2020)

Introduction to Astrobiology edited by Rothery, et al. (Cambridge University Press, 2018)

Life in the Universe by Bennett and Shostak (Pearson, 2016)

Rare Earth by Brownlee and Ward (Copernicus, 2003)

SPECIFIC PLANETARY BODIES

Mercury: the view after MESSENGER edited by Solomon, et al (Cambridge University Press, 2018)

Venus Revealed by David Harry Grinspoon (Addison Wesley, 1998)

The Scientific Exploration of Venus by Fredric W. Taylor (Cambridge University Press 2014)

Mars by Giles Sparrow (Quercus 2015)

Planet Mars: story of another world by Francois Forget, Francois Costard, and Philippe Lognonne (Springer Praxis 2007)

Enceladus and the Icy Moons of Saturn, edited by Paul Schenk, et al (Arizona University Press, 2018)

Saturn's Moon Titan (Owner's Workshop Manual series) by Ralph Lorenz (Haynes Publishing, 2020)

Titan from Cassini-Huygens edited by Robert H. Brown, et al (Springer, 2010)

EXOPLANETS

Earths of Distant Suns: how we find them, communicate with them, and maybe even travel there by Michael Carroll (Copernicus books, 2017)

Envisioning Exoplanets: Searching for Life in the Galaxy by Michael Carroll and Elisa Quintana (Smithsonian books, 2020)

Exoplanets edited by Sara Seager (University of Arizona Press, 2011)

Exoplanets: Diamond Worlds, Super Earths, Pulsar Planets and the New Search for Life Beyond our Solar System by Michael Summers and James Trefil (Smithsonian Books, 2018)

CLIMATE

A Global Warming Primer: Answering your questions about the science, the consequences, and the solutions by Jeffrey Bennett (Big Kid Science, 2016)

LIVING ON OTHER WORLDS

Living Among Giants: Exploring and Settling the Outer Solar System by Michael Carroll (Springer, 2015)

On the subject of living on Titan, see the "science behind the story" section at the back of the novel *On the Shores of Titan's Farthest Sea* by Michael Carroll

On the subject of living in the clouds of Venus, see the "science behind the story" section of *Lords of the Ice Moons* by Michael Carroll

Appendix C: Data for analogous planets/moons

Venus	Earth	Mars	Titan
Major gases:			
Carbon Dioxide 96.5%	nitrogen 78%	carbon dioxide 95.3%	nitrogen 97%
Nitrogen 3.5%	oxygen 20%	nitrogen 2.7%	methane 2.7%
	Water vapor 1%	argon 1.6%	hydrogen 0.1%
	Argon 1%		
Mass (Earth =1)[a]			
8/10th	1	1/10th	2/100th
Surface pressure (Earth =1)			
92	1	0.006	1.5
Surface gravity			
0.88	1	0.38	0.14*
Surface temperature (average)			
864°F	59°F	-67°F	-290°F
Diameter (kilometers)			
12105	12759	6788	5103
Rotation period (day)			
117 Earth days	24 hours, 4 minutes	24 hours, 36 minutes	15.9 Earth days

[a]Despite its high density, tiny Mercury is just 1/20 Earth's mass
*Gravity on the Earth's Moon is 0.165 that of Earth, slightly stronger than that on Titan.

Appendix D: Major successful missions to the terrestrial planets (including Titan), comets and asteroids

Launch Date	Mission	Origin	Comments
Mercury			
10/03/73	Mariner 10	NASA[a]	Three Mercury flybys from 03/29/74, 09/21/74 and 03/16/75
08/03/04	MESSENGER	NASA	Orbital entry 03/18/11; active until 04/30/15
10/20/18	BepiColumbo	ESA[b]/JAXA[c]	Dual orbiters studying Mercury and environs; arrival scheduled 2025

(continued)

Launch Date	Mission	Origin	Comments
Venus			
08/27/62	Mariner 2	NASA	first successful planetary flyby
06/12/67	Venera 4[d]	IKI[e]	atmospheric entry
06/14/67	Mariner 5	NASA	flyby
01/05/69	Venera 5	IKI	atmosphere science (did not survive long enough to land)
01/10/69	Venera 6	IKI	atmosphere science (did not survive long enough to land)
08/17/70	Venera 7	IKI	first successful landing
03/27/72	Venera 8	IKI	55 min descent data, 63 minutes data from surface
06/08/75	Venera 9	IKI	orbiter/lander; first surface photos returned (bw)
06/14/75	Venera 10	IKI	orbiter/lander; surface photos returned (bw)
05/20/78	Pioneer Orbiter	NASA	first radar mapping from orbit
08/08/78	Pioneer multiprobes	NASA	entry bus and four atmospheric probes
09/09/78	Venera 11	IKI	flyby, lander descent science, surface science for 95 minutes
09/14/78	Venera 12	IKI	flyby, lander descent science, surface science for 110 minutes[f]
10/30/81	Venera 13	IKI	flyby, lander; first color panoramas of surface
11/04/81	Venera 14	IKI	flyby, lander; color panoramas of surface
06/02/83	Venera 15	IKI	orbiter; high resolution radar mapping from north pole to ~30°N
06/07/83	Venera 16	IKI	orbiter; high resolution radar mapping from north pole to ~30°N
12/15/84	VeGa 1	IKI	lander and balloon deployed on way to Comet Halley; premature instrument deployment resulted in failed surface science
12/21/84	VeGa 2	IKI	lander and balloon deployed on way to Comet Halley; lander sampled soil
05/04/89	Magellan	NASA	Orbiter/highest resolution radar mapping of entire globe
10/18/89	Galileo	NASA	Venus gravity assist flyby on way to Jupiter (02/10/90)
10/15/97	Cassini/Huygens	NASA/ESA	Multiple Venus gravity assists on way to Saturn
11/09/05	Venus Express	ESA	Orbiter returned atmospheric science until 2014

(continued)

Launch Date	Mission	Origin	Comments
05/20/10	Akatsuki	JAXA	"Venus Climate Orbiter" studying atmosphere, cloud decks
Mars			
10/28/64	Mariner 2	NASA/JPL	First successful flyby 11/28/64, returned 21 full photos
02/24/69	Mariner 6	NASA/JPL	Equatorial flyby 07/31/69
03/27/69	Mariner 7	NASA/JPL	Polar flyby 08/05/69
05/19/71	Mars 2 (orbiter)	IKI	second spacecraft to orbit another planet (science for 362 orbits)
	Mars 2 (lander)	IKI	failed; first lander to impact Mars
	Prop-M	IKI	first Mars rover; may have automatically deployed after Mars 2 crash; no contact made
05/28/71	Mars 3	IKI	Successful orbiter (science for 20 orbits)
	Mars 3 lander	IKI	survived landing, returned data for 23 seconds
	Prop M rover	IKI	survived landing aboard Mars 3; may have automatically deployed, no data received
05/30/71	Mariner 9	IKI	first successful planetary orbiter (arrived 2 weeks before Mars 2) first orbital global map of Mars
07/21/73	Mars 4	IKI	failed to achieve orbit, returned flyby images
07/25/73	Mars 5	IKI	orbiter returned 180 images, contact lost after nine days
08/05/73	Mars 6 (bus)	IKI	bus returned science during flyby
	Mars 6 (lander)		failure upon landing; some atmospheric data returned
08/09/73	Mars 7 (bus)	IKI	bus returned science during flyby
	Mars 7 (lander)		missed planet due to premature separation
08/20/75	Viking 1 (orbiter)	NASA	orbit 06/19/76, mapping science for 1385 orbits
	Viking 1 lander		first fully successful lander, first life science lab on another world Lander operational for 2245 sols[g]
09/09/75	Viking 2 (orbiter)	NASA	planetary mapping and orbital science for 700 orbits
	Viking 2 (lander)		Operated for 1281 sols
07/12/88	Phobos 2	IKI	Orbital science successful, communications failure before deployment of Phobos lander and rover
11/07/96	Mars Global Surveyor	NASA	High resolution orbital imaging and science for a decade

(continued)

Launch Date	Mission	Origin	Comments
12/04/96	Pathfinder	NASA/JPL	Returned surface science for three months; deployed rover
	Sojourner		first rover on another planet; operated for 84 days
04/07/01	Mars Odyssey	NASA	currently operational in orbit
06/02/03	Mars Express	ESA	currently operational in orbit
	Beagle 2 Lander	UK	orbital images show successful landing, but craft failed to unfold properly
06/10/03	Spirit Rover	NASA	rover, operated for 2208 sols
07/08/03	Opportunity	NASA	rover, operated for 5351 sols
03/02/04	Rosetta	ESA	Feb. 2007 flyby science en route to Comet 67P
08/12/05	Mars Reconnnaissance Orbiter	NASA	Operational since March 2006[h]
08/04/07	Phoenix	NASA	landed in polar north, transmitted for 5 months
09/27/07	Dawn	NASA	Mars flyby en route to asteroids Vesta and Ceres
11/26/11	Curiosity	NASA	Rover, continues to operate
11/05/13	Mangalyaan (Mars Orbiter)	ISRO[i]	Continues to return orbital science
11/18/13	MAVEN	NASA	Atmospheric orbital science, continues to operate
03/14/15	ExoMars	ESA/ Roscosmos	Trace Gas Orbiter
	Schiaparelli	ESA	Lander impacted, but entry science successfully returned
05/05/18	InSight	NASA	Lander with seismometer; currently operational
	MarCO A		Cubesat relay with lander during descent/landing
	MarCO B		Cubesat relay with lander during descent/landing
07/19/20	Hope	MBRSC[j]	UAE orbiter; currently operational
07/23/20	Tianwen-1	CNSA[k]	China's Mars orbiter, currently operational
	lander		successfully touched down 05/14/21, deployed rover and 2 remote cameras
	Zhurong		rover, currently operational
07/30/20	Perseverance	NASA	rover similar to Curiosity, currently operational
	Ingenuity		helicopter; multiple flights, first flight on another planet Currently operational

(continued)

Launch Date	Mission	Origin	Comments
Comets			
08/12/78	Internat'l Comet	NASA/ESA	21P/Giacobini–Zinner flyby Explorer
12/15/84	VeGa 1	Soviet Union	Comet Halley flyby after Venus flyby
12/21/84	VeGa 2	Soviet Union	Comet Halley flyby after Venus flyby
01/07/85	Sakigake	ISAS	Comet Halley flyby
07/02/85	Giotto	ESA	Comet Halley close flyby/extended mission to comet Grigg Skjellerup
08/19/85	Suisei	ISAS	Comet Halley flyby
10/24/98	Deep Space 1	NASA	Failed at primary, successful flyby mission to Comet Borrelly
02/07/99	Stardust	NASA	Sample return from Comet Wild 2, flyby of Comet Tempel
03/02/04	Rosetta	ESA	orbit of Churyumov-Gerasimenko (67P) 2014-16
03/02/04	Philae lander	ESA	Landed on 67P; surface operations 11/12/14-11/15/14
01/12/05	Deep Impact	NASA	Impactor/flyby of Temple 1; flyby of Hartley 2
Asteroids			
03/02/72	Pioneer 10	NASA	distant encounters of unnamed asteroid and Nike no data returned
10/18/89	Galileo	NASA	Gaspra and Ida flybys en route to Jupiter/discovery of Dactyl
02/17/96	NEAR Shoemaker	NASA/APL	Mathilde flyby; Eros orbit 02/2000 to 03/2001; first soft landing on an asteroid, 02/12/01
10/15/97	Cassini/Huygens	NASA/ESA	distant flyby en route to Saturn
10/24/98	Deep Space 1	NASA	partial failure flyby of Braille
02/07/99	Stardust	NASA	flyby of Annefrank en route to Comet Wild 2
05/09/03	Hayabusa	JAXA	sample return from asteroid Itokawa
03/02/04	Rosetta	ESA	encountered asteroids Steins and Lutetia en route to 67P
01/19/06	New Horizons	NASA/APL	distant recon of 132524APL; flyby of KBO Arrokoth after encounter of Pluto system
09/27/07	Dawn	NASA	07/11 to 09/12 orbit operations at Vesta 03/15 to 10/18 orbit operations at Ceres (still orbiting)
10/01/10	Chang'e-2	CNSA	12/13/12 flyby of Toutatis after Lunar orbit mission

(continued)

Launch Date	Mission	Origin	Comments
12/03/14	Hayabusa 2	JAXA	Ryugu orbiter/sample returned 12/05/2020. Successfully deployed series of MINERVA rovers
09/08/16	OSIRIS-Rex	NASA	Entered Bennu orbit 12/2018, sample return in progress
10/21	Lucy	NASA	planned eight asteroid flybys beginning 04/2025
10/21	DART/LICIA cube	NASA	asteroid impact/diversion mission to Dimorphos/Didymos successful change in asteroid orbit
10/23 (planned)	Psyche	NASA	first planned mission to a metallic asteroid, projected to orbit the asteroid 16 Psyche in August 2029
Titan			
04/05/73	Pioneer 11	NASA	First direct measurements and flyby imaging
09/05/77	Voyager 1	NASA	passed within 4000 kilometers, successful close encounter
08/20/77	Voyager 2	NASA	distant encounter to enable Uranus and Neptune flybys
10/15/97	Cassini	NASA/ESA	orbital tours from 2004-2017, frequent radar mapping of Titan
Landing 01/14/05	Huygens	ESA	first successful landing in outer solar system; 2 ½ hours of atmospheric and surface science

[a]National Aeronautics and Space Administration (USA)

[b]European Space Agency

[c]Japan Aerospace Exploration Agency

[d]On March 1, 1963, Venera 4's predecessor—Venera 3—became the first object to impact the surface of another planet. Although it returned no scientific data, it was an historic first.

[e] Russian Space Research Institute (Soviet Union)

[f]Although both Venera 11 and 12 carried cameras, the lens caps did not release.

[g]A sol is a Martian day, which lasts 24 hours, 39 ½ minutes

[h]Mars Express, Mars Odyssey and Mars Reconnaissance Orbiter are all used as orbital communications relays for surface operations.

[i]India Space Research Organization

[j]United Arab Emirates' Mohammed Bin Rashid Space Center

[k]China National Space Administration

Appendix E: Supercontinents throughout Earth's History

Formation	Age (approximate)	Comments
Vaalbara	3.6 to 2.8 billion years ago	supercontinent during the archean, consisting of two cratons: the Kaapvaal (with remains in South Africa) and the Pilbara (remnant in West Australia)
Ur	2.8 to 2.4 bya	At its inception, probably sole continent on the planet.
Kenorland	2.7 to 2.1 bya	Material from Kenorland later became the central cores of Greenland and N. America, Scandinavia, and southern Africa and western Australia
Nuna (Columbia)[a]	1.8 to 1.3 bya	an assemblage of India, western North America, southern Australia and western Canada
Rodinia	1.2 billion to 700 mya	The breakup of this Precambrian supercontinent is thought to have triggered the Earth's Snowball periods. Rodinia's arrangement and location are not well understood.
Pannotia[b]	633 to 573 mya	this supercontinent arose at the end of the Precambrian and dissembled in the early Cambrian.
Gondwana	550 to 175 mya	At its height, Gondwana covered a fifth of the Earth's area (equal to Africa, Asia and Europa combined). It combined with the subcontinent Laurasia to form Pangaea.
Pangaea	336 to 175 mya	Centered on the equator, Pangaea was surrounded by the Panthalassa, Paleo-Tethys and Tethys oceans. Its breakup took place between the Triassic and Jurassic periods.

[a]Between the formation of the Kenorland and Nuna supercontinents, the subcontinents Arctica and Atlantica arose.
[b]Also called the Vendian or PanAfrican supercontinent. Recently, its supercontinent status has been questioned.

Appendix F: Probable Causes of Major Extinction Events

Event	% loss of species	time (millions of years ago)	causes
Bombardment	-	4600-3800	impacts, pre- and post-accretion
Great Oxygenation Event	-	2500-2200	microbes infused atmosphere with oxygen
Snowball periods	-	750-600	glaciations changed land and atmosphere

(continued)

Event	% loss of species	time (millions of years ago)	causes
Cambrian	40?	~499	oxygen depletion? glacial cooling?
Close of Ordovician	86%	444	Glaciation caused large swings in sea level; uplift of Appalachian Mountains absorbed CO_2, changed sea chemistry and climate.
Late Devonian	75%	360	Severe global cooling due to rapid expansion of land plants
Close of Permian (Permo-Triassic)	85-95%	250	Siberian volcanoes caused global warming from high CO2 and sulphur levels; acid rain and other chemical changes to land and water
Close of Triassic	80%	202	Sea floor volcanoes in the Atlantic triggered global warming and changed ocean chemistry
Close of Cretaceous	76%	65	surge of global volcanism and continental uplift, capped by an asteroid impact in the Yucatan (Mexico) caused widespread fires and global winter for an extended period.

Appendix G: Early Life Timeline

Date (millions of years ago)	geological or life event
4,600-4,550	Formation of the Earth, differentiation, solidification of surface
4,500	Impact of Theia forms Moon, leaves Earth's surface molten again
4,400	most ancient mineral grains show evidence of the establishment of water cycle, ponds and seas begin to form
3,900	Late Heavy Bombardment tails off
3,850	first carbon isotopic evidence of life
3,500	oldest microfossils; oldest stromatolites
2,350	earliest trace of oxygen in atmosphere/beginning of the Great Oxygenation Event
520	first evidence of pond scum on land
470	non-vascular plants (i.e. liverworts, mosses, spore fossils)
450	soft-bodied creatures? (not preserved)
430	first evidence of vascular plants on land
425	Kampecaris obansensis, first fossilized land creature found (so far)

Appendix H: Cold epochs in Earth's History

Glaciation events involve 3-4 % ice coverage of Earth's surface; Snowball events involve at least 66% coverage.

Date (millions of years ago)	cooling event	geological period
2900-2780	Pongola Glaciation	Mesoarchean
2400-2375	Huron Glaciation I	Siderian
2370-2345	Huron Glaciation II	Siderian
2340-2310	Huron Glaciation III	Siderian
2300-2200	Huron Snowball	Rhyacian
715-680	Sturtian Snowball[a]	Cryogenian
650-635	Marinoan Snowball	Cryogenian
582-580	Gaskiers Glaciation	Cryogenian
547-546	Baykonurian Glaciation	Ediacaran
450-440	Late Ordovician Glaciation	Late Ordovician
440-420	Silurian Glaciation	Silurian
359-310	Karoo Glaciation I	Mississippian
305-299	Karoo Glaciation II	Pennsylvanian
295-280	Permo-Carboniferous Glaciation	Permo-Carboniferous
15 to present	Antarctic Glaciation	Neogene/Quaternary
2.58 to present	Quaternary Ice Age	Quaternary

[a] Breakup of Rodina supercontinent begins roughly at this time.

Appendix I: Exoplanet ESI index

The Earth Similarity Index (ESI) tracks multiple characteristics in planetary geology, atmosphere, mass and location (in relation to the planet's star) to determine how similar an exoplanet is to the Earth. The ESI scale ranges from zero to one, with one equaling the conditions on Earth.

Planet	location	diameter(km)	distance from Earth	ESI
Earth	right here	12,742	-0-	1.0
Mars	solar system	6,779	54.6 million km minimum	0.73
Venus	solar system	12,104	61 million km minimum	0.44
Gliese 667Cc	Scorpius	~19,000	22 light years (ly)	0.82
Kepler 22b	Cygnus	~15,300	587 ly	0.71
Kepler 62E	Lyra	~17,800	1,200 ly	0.82
Kepler 186F	Cygnus	~18.000	500 ly	0.84
Kepler 438B	Lyra	~14,271	640 ly	0.88
Kepler 442B	Lyra	~16,950	1,194 ly	0.84
Kepler 452B	Cygnus	~20,400	1,400 ly	0.84
Kepler 1649C	Cygnus	~13,500	300 ly	TBD*
Luyten B (Gliese 273b)	Canis Major	~17,200	12.2 ly	0.91

(continued)

Planet	location	diameter(km)	distance from Earth	ESI
Proxima Centauri B	Alpha Centauri	~16,000	4.2 ly	0.87
Teegarden B	Aries	~12,997	12.5 ly	0.97
TRAPPIST-1D	Aquarius	~9,939	39.5 ly	0.89
TRAPPIST-1E	Aquarius	~11,468	39.5 ly	0.95

*recent estimates put the planet near the top of the ESI list

Appendix J: Astronomical distances/scales

Earth diameter: 12,756 km (7917.5 miles)
Sun diameter: 1,392,700 km (865384 miles)
Astronomical unit: 150,000,000 km (93,000,000 miles)
Light year (the distance light travels in a year): 9.46 trillion km (5.88 trillion miles)
Parsec: 3.26 light years (206,265 AU, 30.9 trillion km, 19.2 trillion miles)

Appendix K: Star types

Stars similar to the Sun range from cooler K types to G types like our own star.

M stars (red dwarfs) are the most common star type, making up 90% of all stars in the Milky Way. Small and cool, they hold habitable zones close to their faint surfaces.

Giant red, blue and yellow stars burn more quickly, and may have fleetingly stable environments, perhaps not ideal surroundings in which life might arise.

name	type	mass	distance from Earth
Assorted Sun-similar stars			
Sun	G (yellow dwarf)*	1.99 x 1030 kg (332,950xEarth)	150 million km/8 light minutes
Alpha Centauri A/	G*	1.079 x Sun mass	4.2 light years (Rigil Kentaurus)
Tau Ceti	G*	0.78 x Sun mass	11.9 light years
Sigma Draconis	G	0.85 x Sun mass	18.8 light years
Gliese 34A	G	0.97 x Sun mass	19.3 light years
82 G. Eridani	G*	0.7 x Sun mass	19.7 light years
Delta Pavonis	G*	1.05 x Sun mass	19.9 light years
Assorted M stars (red dwarfs)			

(continued)

name	type	mass	distance from Earth
Note: because of red dwarfs' often-irregular luminosity, determining the presence of a planet orbiting an M star is more difficult than detecting planets orbiting steadily shining stars.			
Proxima Centauri	M*	0.12 x Sun mass	4.2 light years
Barnard's Star	M*	0.14 x Sun mass	5.96 light years
Wolf 359	M*	0.09 x Sun mass	7.85 light years
Lalande 21185 (Gliese 411)	M*	0.39 x Sun mass	8.3 light years
Gliese 65 A	M	0.1 x Sun mass	8.72 light years
Gliese 65 B	M	0.1 x Sun mass	8.72 light years
Ross 154	M	0.17 x Sun mass	9.7 light years
Ross 248	M	0.136 x Sun mass	10.3 light years
Gliese 887	M*	0.49 x Sun mass	10.7 light years
Ross 128	M*	0.168 x Sun mass	11.0 light years
Gliese 866 A, B & C (3 M-class stars)	M	0.1, 0.1, 0.11 x Sun mass	11.1 light years
Gliese 229 A	M	0.57 x Sun mass	18.8 light years
Gliese 213	M	0.35 x Sun mass	18.89 light years
TRAPPIST-1	M*	0.9 x Sun mass	40.7 light years
Assorted giant stars			
Red giants/supergiants			
Pollux	KO*	1.91 x Sun mass	33.8 light years
Arcturus	K1	1.08 x Sun mass	36.7 light years
Aldebaran	K5	1.16 x Sun mass	65 light years
Betelgeuse	M1-M2	~17.5 x Sun mass	548 light years
Antares	M1.5	~13 x Sun mass	~550 light years
Mu Cephei	M2-1a	19.2 x Sun mass	3060 light years
Blue giants/supergiants			
Rigel	B8	~20 x Sun mass	863 light years
Deneb	A2	19 x Sun mass	2615 light years
Yellow giants/supergiants			
Rho Cassiopeiae	G2Iae (variable)	40 x Sun mass	3400 light years
Delta Cephei	F5	4.5 x Sun mass	887 light years

*indicates one or more possible or confirmed planets

Appendix L: Potency of greenhouse gases

Greenhouse gases act to maintain the Earth's current climate. Without its natural greenhouse gases, Earth's temperature would be 33°C cooler. Ironically, the most powerful greenhouse gas is simple water vapor. But carbon dioxide has more long-term effects: it sets the Earth's thermostat. Water vapor cycles out of the atmosphere easily, precipitating out as rain and snow. Carbon dioxide remains aloft for centuries or millennia. The amount of water vapor changes in response to the temperature of ocean and air, so if increased carbon dioxide levels raise temperatures, water vapor will follow. Water vapor is also important in that it acts as a reinforcing feedback, bolstering the warming effects of other greenhouse gases. Below are the major greenhouse gases in Earth's atmosphere, in order of warming potential.

Gas	Radiative forcing (Wm^{-2}) (influence on environment)
Water vapor	-variable-
Carbon dioxide	1.46
Methane	0.48
Nitrous oxide	0.15
Chlorofluorocarbons (CFCs)	0.007
Hydrofluorocarbons (HFCs)*	0.003
Ozone (lower atmosphere)**	0.000

*includes HFC-23 (a fire suppressant and refrigerant), HFC-134a (a refrigerant) and HFC-152a (a propellant for aerosol sprays)
**refers to ozone introduced into the lower atmosphere. Naturally-occurring high-altitude ozone has little effect on planetary temperatures, but acts as a barrier to solar radiation.

Index

© The Editor(s) (if applicable) and The Author(s), under exclusive license to Springer Nature Switzerland AG 2023
M. Carroll, *Planet Earth, Past and Present*, Springer Praxis Books,
https://doi.org/10.1007/978-3-031-41360-5